electron-diffraction analysis of clay mineral structures

Monographs in Geoscience

General Editor: Rhodes W. Fairbridge

Department of Geology, Columbia University, New York City

B. B. Zvyagin
Electron-Diffraction Analysis of Clay Mineral Structures–1967

E. I. Parkhomenko
Electrical Properties of Rocks–1967

In preparation

A. I. Perel'man
The Geochemistry of Epigenesis

electron-diffraction
analysis of
clay mineral structures

Revised Edition

Boris Borisovich Zvyagin
*Institute for Geology of Ore Deposits, Petrography,
Mineralogy, and Geochemistry
Academy of Sciences of the USSR, Moscow*

Translated from Russian by
  Simon Lyse

PLENUM PRESS · NEW YORK · 1967

Library of Congress Catalog Card Number 65-17783

The original Russian text, first published for the Institute for Geology of Ore Deposits, Petrography, Mineralogy, and Geochemistry of the Academy of Sciences of the USSR by Nauka Press in Moscow in 1964, has been extensively corrected and updated by the author for the English edition.

Борис Борисович Звягин
Электронография и структурная кристаллография глинистых минералов
ELEKTRONOGRAFIYA I STRUKTURNAYA KRISTALLOGRAFIYA GLINISTYKH MINERALOV
ELECTRON-DIFFRACTION ANALYSIS OF CLAY MINERAL STRUCTURES

© *1967 Plenum Press*
A Division of Plenum Publishing Corporation
227 West 17 Street, New York, N. Y. 10011
All rights reserved

Printed in the United States of America

PREFACE TO THE AMERICAN EDITION

As a method of structure analysis, electron diffraction has its own special possibilities and advantages in comparison to the X-ray method for the study of finely dispersed minerals with layer or pseudolayer structures. However, possibly because of the prior existence of the X-ray method, which found universal application in different fields and attracted the main efforts of specialists, electron diffraction has been unevenly disseminated and developed in different countries. In particular, the oblique texture method, which gives very complete and detailed structural information, has been mainly used in the Soviet Union, where electron-diffraction cameras specially suited to the method have been constructed. In other countries, studies have been made of micro-single crystals, because these studies could be carried out with existing electron microscopes. It should be recognized that the scale of distribution and use attained by electron-diffraction methods, at present limited by existing experimental conditions, is more than justified by the value of the results which may be obtained by their aid. The author hopes that the present book will give the reader a fuller idea of the valuable advantages of the method, and of the structural crystallography picture which has been built up for clay minerals, and layer silicates in general, from electron-diffraction data.

The time between the appearance of this book and that of the Russian edition has been comparatively short. During this period, however, new data and results have been obtained in the fields of electron diffraction and structural crystallography, forming a significant addition to the original content of the book. Unfortunately, within the time at his disposal, the author has been able to include only that new material produced by himself or his colleagues.

The additional material has allowed certain gaps to be filled and some confused topics to be clarified. It includes an explanation of the structural diversity and specific features of the serpentine-type minerals, a consideration of the significance and role in structural crystallography of semi-random layer silicate structures, a clarification of the position of halloysite as a special member of the kaolinite group, a detailed examination of the features of diffraction patterns from chrysotile tubes and their structural interpretation, a note on the diagnostic features of semi-random varieties of micas of different chemical composition and chlorites differing in the structures of their packets, etc. It will be appreciated that there are many unresolved problems still remaining. There is a great need to make use of the possibilities offered by structure analysis in order to lead to further progress in our knowledge and understanding of layer silicate structures.

PREFACE TO THE RUSSIAN EDITION

Clay minerals are used in many branches of science and technology and figure significantly in the economy. Clays have found well-known uses as catalysts and absorbents, fillers and pigments, and binding and cleansing agents. They play an important part as raw materials in cement and ceramic production, they are the most significant factor in determining the fertility and stability of soils, and they serve to indicate the formation conditions of sedimentary rocks and the presence of certain types of minerals in prospecting (Grim, 1960).

The study of the crystal structures of clay minerals is of the highest practical importance in every respect. Clays and clay minerals are deserving subjects for structure analysis, which is necessary for them because of their diversity and structural complexity.

Original and valuable information on the structures of clay minerals and important practical results may be obtained by methods based on the diffraction of electrons, which have special possibilities in this field. Electron-diffraction analysis of clay minerals forms a separate branch of electron-diffraction methods, and occupies an independent position in the group of methods used in the study of clays.

The application of electron diffraction to the investigation of clay minerals developed under favorable circumstances in the Soviet Union.

It was assisted, above all, by the very high level of development of electron diffraction methods in this country. The books written by Z. G. Pinsker (1949) and B. K. Vainshtein (1956) played an important part, laying the foundations of structure analysis by electron diffraction. Under their influence, the unique image of Soviet electron-diffraction work was developed, characterized in particular by the discovery and utilization of the rich store of information obtainable from electron-diffraction texture patterns through the application of Fourier analysis methods. Electron-diffraction cameras of great practical utility were devised and constructed in the USSR. Electron diffraction

vii

became an essential tool in the study of the structures of materials, used for the most diverse substances and applications. In its whole range of applications, this remarkable physical phenomenon on one hand led to the discovery of special laws of its own, and on the other provided objective results on the properties of the objects scattering the electrons; it also allowed evaluation of the stage reached by changes which might occur during the course of an investigation.

In the study of clay minerals which, because of their special properties, were natural candidates for application of the method, it was inevitable that electron diffraction would mark the advent of a new, fruitful approach.

Clay minerals show great variability and diversity in their chemical composition, geometry, and degree of structural perfection, and thus are by no means always suitable for immediate use in detailed structural investigations. In this connection another important factor in the development of electron-diffraction clay mineral analysis, leading to an increase in its effectiveness, was the work of N. V. Belov on structural mineralogy and structural crystallography (1947, 1949, 1950, 1951a,b). This laid the foundations for the derivation of general laws for the formation and diffraction properties of structures, with the guidance of which it was possible to evaluate the atomic structures of clay minerals even with only a meager amount of diffraction data.

Thus, the intimate partnership of a structural analysis method with a structural synthesis theory, electron diffraction with structural crystallography, was dictated by the specific characteristics of clay minerals. It also made possible a concrete, unrestricted approach to the treatment of diffraction patterns and to the analysis of their structural meaning, which also was of fundamental importance in clay mineral work.

It should be noted that electron diffraction of clay minerals began its development during the period when the basic concepts of silicate atomic structures, in particular those of layer silicates, had already been worked out by X-ray methods of structure analysis, where they resulted in the general theories of Bragg (1937), Pauling (1930a,b), and Belov (1947, 1949). In more recent times as well, the X-ray method has continued to be used in innumerable and widely varying investigations, a large number of which have been specially devoted to clay minerals (Brindley, 1955; Frank-Kamenetskii,1958; Brown, 1961); these have continued to build up valuable factual material, which has apparently had a beneficial effect on the development of electron-diffraction mineral studies also.

Together with X-ray structure analysis, electron diffraction occupies a special position among investigational methods. These two methods reveal the primary characteristics of a substance, which directly and unambiguously determine all its properties, its special features, and its individuality. The scope, function, and applications of the two methods in the study of minerals may be compared as follows.

The X-ray method gives the general structural characteristics of the rock being investigated. The X-ray photograph gives indications of all the mineral components present in the specimen, over a wide range of particle sizes. Electron diffraction, on the other hand, shows up only the finest fraction, so that electron-diffraction patterns principally reveal the minerals with platy or needle-shaped particles.

There is no doubt as to the importance of obtaining a complete picture of the mineral composition of a test material. At the same time, it should be remembered that the components forming the coarse fractions are usually known from optical studies. In diffraction patterns these coarse particles usually appear in the role of annoying obstacles to the straightforward determination of the structural characteristics of the finely dispersed minerals. To resolve many problems it is necessary to determine the structural characteristics of the fine particles in question, which in most cases are clay minerals. In addition, the possibility cannot be excluded that different size fractions of the same mineral may differ somewhat in structure or ordering, and to clear this up it is necessary to use both methods.

The advantage of the X-ray method is that the specimen may be investigated in its natural state or over a whole range of conveniently chosen states. The electron-diffraction method places the specimen under rigidly circumscribed and, at the same time, abnormal conditions—those of a high vacuum. In the majority of minerals the vacuum does not cause any structural changes. Only specimens containing weakly bound water molecules are likely to suffer any effects. It is in fact found that the so-called swelling layer silicates (montmorillonites and vermiculites) may be deprived of their interlayer water, so that the whole of the dynamics of structure transformation with change in moistness of the surrounding medium, and the characteristics of mixed-layer products, become inaccessible to electron-diffraction investigation. In all probability this limitation on the use of electron diffraction is not absolutely exclusive. Thus, some vermiculite specimens, kindly presented by G. F. Walker, and studied in detail by him previously (Mathieson, Walker, 1954), did not show loss of interlayer water molecules under electron-diffraction conditions. It may also be assumed that as experimental methods are improved, this difficulty may also be obviated for other swelling minerals as well.

However, in spite of the foregoing, there is some positive value in the fact that in electron diffraction the swelling minerals are studied in a strictly fixed state which is itself also worthy of attention, particularly in connection with the fact that in this state the structure is most highly ordered, thus allowing concealed features to be brought out. It should also be taken into account that under vacuum conditions, electron diffraction may be used to investigate structures which are unstable under natural conditions, as, for example, in the study of the monohydrate $BaCl_2 \cdot H_2O$ (Vainshtein, Pinsker, 1949).

Another factor concomitant to electron-diffraction studies, i.e., bombardment with electrons, does not have any marked effect on the specimens because the energy absorbed by them is comparatively small. If the electrons interact more intensely with the substance (as, for example with microdiffraction methods) the time of exposure of the specimen must be reduced, while structural changes observed to occur may serve as an additional characteristic of the crystalline substance.

With finely dispersed minerals, the electron-diffraction method will give a special kind of diffraction pattern, the texture pattern, which contains a two-dimensional distribution of a regularly arranged set of reflections; this is found to be highly productive for the derivation of the structural features of the specimen under study. However, these patterns, when obtained by the transmission method from a platy texture, lack the $00l$ basal reflections from the planes parallel to the silicate layers. It is true that these reflections can be recorded on texture patterns obtained by the reflection method, but there are considerable experimental difficulties with this at present.

In X-ray analysis of clay minerals the opposite situation prevails. The general hkl reflections are arranged in a one-dimensional sequence in order of decreasing value of their interplanar distances d_{hkl}, and there are difficulties in indexing and measuring the intensities of these reflections. On the other hand, the $00l$ basal reflections are easily recognized on ordinary X-ray powder photographs and are mostly separated out with special types of photographs in studies of oriented aggregates.

The $00l$ basal reflections express the features of the structure which are contained in its projection along a direction perpendicular to the silicate layers. This projection gives a great deal of information on the layer silicate structure in a very small space. It reveals the sequence of atomic planes within in a single layer and in the interlayer spaces, the orientation of organic inter- layer molecules, the strict constancy of certain interlayer distances, and the ordered or randomly disordered sequences followed by some interlayer distances.

For their part, the general hk$l$ reflections, which appear saliently on electron-diffraction texture patterns, express the spatial characteristics of the structure, particularly those related to directions parallel to the silicate layers, which do not show up in the 00l reflections. These general reflections char- acterize the finer individual differences between minerals, in particular those due to individual polymorphic modifications, and the order or disorder in the relative displacements of the networks within layers or of the layers within a complete structure.

Thus, the two methods successfully complement each other in all re- spects. The weak or vulnerable points of one are the strong points of the other. The two methods neither compete nor clash. Neither has replaced or made obsolete an application of the other.

From what has been said, it is apparent that electron diffraction is a structure analysis method which is best applied to finely dispersed minerals having crystals penetrable by electrons. The main uses of the method are in recognition and in detailed structural studies of minerals, in showing their fine differences from other related minerals, in rigorous determinations of poly- morphic modifications, in characterizing loss of order, and in determining the content in a specimen of those mineral components which are amenable to electron diffraction.

Whereas X-ray analysis of clays and clay minerals has been the subject of a whole range of specialized monographs and systematic handbooks (Grim, 1956; Brindley, 1955; Brown, 1961), electron-diffraction studies of clay miner- als have been described only in separate, disconnected articles. The aim of the present book is in some measure to fill this gap. It is based on results and practical experience built up by the author during many years of study of clay minerals by electron-diffraction methods, in the All-Union Geological Re- search Institute. These studies were a continuation of work undertaken in the 1940's on the initiative of Academician V. I. Vernadskii and carried out under the supervision of Professor Z. G. Pinsker, with the aim of developing clay mineral electron-diffraction work as an independent branch of science and at the same time as an independent method of investigation (Pinsker et al.,1948; Zvyagin, Pinsker, 1949a,b; Zvyagin et al., 1949; Pinsker, 1950, 1954).

The distinctive features of electron-diffraction work in the particular field of application considered are intimately related to the structural features of the clay minerals, which belong to the layer and pseudolayer silicates. The structural crystallography of these minerals is dealt with in Chapter 1. The chapters which follow describe experimental methods (Chapter 2), geometrical analysis of electron-diffraction patterns (Chapter 3), the distribution and intensity analysis of diffraction-pattern reflections (Chapter 4), and results of experimental electron-diffraction investigations (Chapter 5).

The texture patterns reproduced in the book were obtained with an EM-4 electron-diffraction camera, while part of the single-crystal patterns were prepared with N. M. Popov's 400-kV apparatus, and part with a JEM-6 electron microscope.

Besides the author, those who took part in the electron-diffraction studies described were K. S. Mishchenko, R. A. Shakova, and V. A. Shitov, the author's colleagues at the Geological Research Institute, and N. M. Popov. Valuable observations on the content of the book were made by Academician N. V. Belov, Corresponding Member B. K. Vainshtein, Professor Z. G. Pinsker, Professor V. A. Frank-Kamenetskii, and V. A. Drits. To all these the author extends his sincere thanks.

BIOGRAPHICAL NOTE

Born in 1921, Boris Borisovich Zvyagin studied at the Moscow State University, graduating with distinction from the Physics Faculty in 1944. He carried out his postgraduate work, specializing in the electron-diffraction study of clay minerals, at the Soil Institute of the Academy of Sciences. In 1949, his thesis on "An Electron-Diffraction Study of the Structure of Montmorillonite" was accepted for the Candidate degree in Geology and Mineralogy. For the next 13 years he worked at the All-Union Geological Research Institute (VSEGEI) in Leningrad, where he created an electron-diffraction and electron-microscopy laboratory and carried out many experimental and theoretical studies on the application of electron diffraction to the structural crystallography of finely divided minerals, principally layer silicates. Since 1963, he has continued these investigations in Moscow, in the Institute for Geology of Ore Deposits, Petrography, Mineralogy, and Geochemistry (IGEM) of the Academy of Sciences of the USSR. In 1963, his thesis on "Electron Diffraction and Structural Crystallography of Clay Minerals" was accepted for the degree of Doctor of Physical and Mathematical Science. After a number of modifications and additions, this thesis has been rewritten and serves as the basis for this monograph.

During the course of his scientific activities, Dr. Zvyagin has revised the methods used in interpretation and calculation of texture patterns from layer silicates with nonorthogonal unit cells. The results of this revision have appeared in "Systematic Handbook on Petrographic-Mineralogical Study of Clays," published by Gosgeoltekhizdat in 1957. This work formed the basis of structural studies of clay minerals and layer silicates which, in the case of kaolinite, celadonite, muscovite, and phlogopite, proceeded to full structure determinations. Studies in less detail which, however, explained the basic structural features, were made of dehydrated halloysite, nacrite, sepiolite, and palygorskite. The author has also made a theoretical examination of polymorphism—polytypism and order—disorder in layer silicates, using an original analytical method of describing the structures of these minerals. The results of this work have been described in individual articles, listed in the bibliography, and appear in a generalized form in the present book.

CONTENTS

CHAPTER 1

STRUCTURES AND STRUCTURE MODIFICATIONS OF LAYER SILICATES

1. General Introduction to the Structures of Layer and Pseudolayer Silicates

The basic concepts in the field of layer silicate structures were introduced and developed in the work of Pauling (1930a,b), Bragg (1937), and Belov (1947, 1949, 1950, 1951) and were based on crystallochemical analyses of experimental results. The simplest and clearest way of representing these structures is in terms of their structural polyhedra, i.e., the tetrahedra and octahedra defined by the large O, OH, and F anions lying at their corners, with the small cations lying inside. The tetrahedra in most cases contain Si atoms, but in some cases these may be partly replaced by Al atoms or, more rarely, by Fe atoms. The octahedra may contain Al, Fe, Cr, Mg, Zn, Li, or other atoms. In the structures of layer silicates, the tetrahedra are connected together through the corners of their faces, and the octahedra through their side edges, to form two-dimensional networks (Fig. 1) in which the centers of the polyhedra are arranged in a hexagonal pattern (Figs. 2-4). The tetrahedral and octahedral networks, which are of similar dimensions, are connected together into layers, and in any given mineral group all the layers are formed from a given combination of these networks.

In the layer silicate structures the axes a and b are chosen to lie in a plane parallel to the layers, so that the c-axis is the direction along which the layers alternate. The a axis lies parallel to the median of a face of an octahedron (Fig. 5), the b axis lying perpendicular to a, with the unit cell constants a and b related by the approximate formula $b \simeq a\sqrt{3}$.

One-storied layers, in the form of individual octahedral networks, are found in the chlorites. The kaolinite and serpentinite minerals consist of two-storied layers, made up of one octahedral and one tetrahedral network. Three-storied layers, made up of two tetrahedral networks attached above and below

1

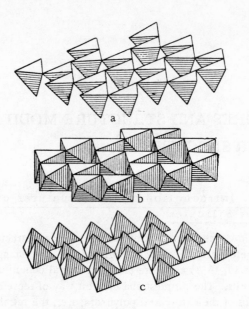

Fig. 1. Method of connecting networks of
tetrahedra (a,c) and a network of octahedra
(b) to form a silicate layer.

a central octahedral layer, are found in the structures of the micas, hydromicas, montmorillonites, and vermiculites, and they alternate with one-storied
layers in the chlorites (Fig. 6).

When a tetrahedral network is attached to an octahedral network, the
tetrahedron corners can match up with only two-thirds of the corners of the
faces of the octahedra. From Pauling's rule concerning the balance of val-
ences of the cations belonging to an anion and the valence of the anion itself, these matched corners must be occupied by O atoms, with monovalent
OH and F anions in the other third of the corners. It will be obvious that the
corners of octahedra in one-storied layers or on the free side of two-storied
layers will be occupied by monovalent anions only.

In the octahedral networks of cations, either all, or only two-thirds of
the octahedra may be occupied. In the latter case, with two out of every
three octahedra filled, these alternate regularly with the empty octahedra
(Figs. 3 and 4). The layer silicates are divided up into dioctahedral and tri-

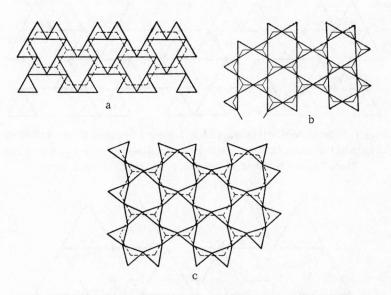

Fig. 2. Arrangement of tetrahedral networks (represented by the connected tetrahedron bases). (a) For close-packing of the anions; (b) for the case where the anions take up $\frac{3}{4}$ of the close-packing positions; (c) an intermediate case.

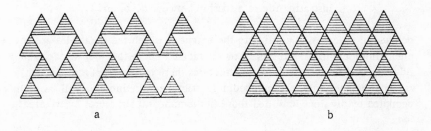

Fig. 3. Types of octahedral networks. (a) Dioctahedral network; (b) trioctahedral network (only the upper faces of the octahedra are shown).

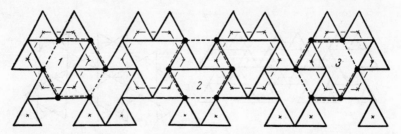

Fig. 4. The three possible ways of attaching a tetrahedral network to an octahedral network (the dots denote the octahedron corners to which one "loop" of the tetrahedral network is attached, in each case).

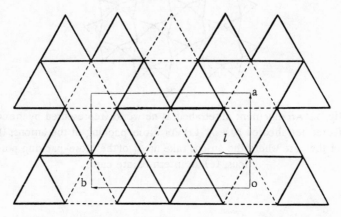

Fig. 5. The dioctahedral pattern of the coplanar faces of an octahedral network (dashed lines show empty octahedra), illustrating the relationship between a, b, and l.

octahedral silicates, according to the extent to which the octahedra are occupied. In dioctahedral structures, two octahedra come together at each shared side edge, while in trioctahedral structures there are three. From Pauling's valence-balance rule, the octahedra in triotahedral structures will be mainly occupied by divalent ions, and those in dioctahedral structures by trivalent ions.

Minerals built up of layers of the same type may nonetheless differ in chemical composition and in degree of replacement of cations of high charge by cations of lower charge; the latter leads to formation of negatively charged layers.

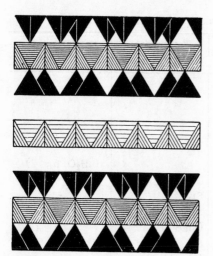

Fig. 6. The sequence of one-storied and three-storied layers in chlorite structures (after N.V. Belov, 1951a).

These charges are compensated by other cations, normally of larger dimensions, which lie in the spaces between the layers. These spaces may also contain water molecules and molecules of certain organic compounds. The nature and extent of the interaction between layers is closely bound up with the magnitude and distribution of the negative charge between the octahedral and tetrahedral networks and with the particular features of the interlayer spaces. An ionic type of bonding between layers is characteristic of mica-type minerals, which are made up of three-storied silicate layers. In micas and hydromicas, the charges on the layers are due to replacement of Si atoms in tetrahedra by Al (or Fe) atoms, and so the charge sources lie close to the surfaces of the layers. In consequence, the intensity of their interaction with the cations and hence the attractive forces between the layers are greater in these minerals than in the montmorillonites, in which negative charges on the layers are due mainly to replacement of cations in the octahedra.

The two-storied layers in the kaolinite-type minerals are neutral. They are arranged relative to one another in such a way that a hydroxyl group, OH, on the free surface of one layer matches up with a tetrahedron O atom of the next layer, with a hydrogen bond being formed between them. In the chlorites, there are both ionic and hydrogen bonds between the layers. The negative charges on the three-storied layers are compensated by positive charges on the one-storied layers, and at the same time the hydroxyls of the one-storied layers and oxygen atoms of adjacent three-storied layers are paired off to form O−OH groups like those in the kaolinites.

The distinguishing features of different layer silicates having a similar makeup can be described in terms of the distribution of their constituent atoms between the octahedral and tetrahedral networks and the interlayer spaces. Here, we can be guided by the relative numbers of atoms lying in the planes

Table 1. The Chemical Makeup of Layer Silicates

Mineral	Octahedral network	Tetrahedral network	Interlayer space
Two-storied layer minerals			
a) dioctahedral:			
kaolinite,dickite			—
nacrite	Al_2	Si_2	—
halloysite			$(nH_2O, n \leq 2)$
b) trioctahedral:			
serpentines	Mg_3	Si_2	—
amesite	Mg_2Al	$SiAl$	—
cronstedtite	$(Fe^{2+}Fe^{3+})$	$(SiFe^{3+})$	—
greenalite	$(Fe_{2.2}^{2+}Fe_{0.5}^{3+})$	Si_2	—
chamosite	$(Fe^{2+},Mg)_{2.2}$ $\times (Fe^{3+},Al)_{0.7}$	$(Si_{1.4}Al_{0.6})$	—
Three-storied layer minerals			
a) dioctahedral:			
muscovite			K
paragonite	Al_2	$AlSi_3$	Na
celadonite,			
glauconite	$(Fe,Mg)_2$	Al_xSi_{4-x}, $(x < 1)$	$[K(H_3O)]_x$
hydromica	(Al, Fe, Mg)		$(K, Na, Ca)_x$
Al-montmoril-lonite	$(Al_{1.67}Mg_{0.33})$	Si_4	$(Na, Ca, ...) +$ nH_2O or organic
nontronite	Fe_2	$(Al_{0.33}Si_{3.67})$	molecules
beidellite	Al_2	$(Al_{0.5}Si_{3.5})$	(ethylene glycol, glycerol, etc.)
b) trioctahedral:			
phlogopite	Mg_3		
biotite	$(Mg, Fe)_3$	$AlSi_3$	K
zinnwaldite	$LiFeAl$		
polylithionite	Li_2Al	Si_4	
vermiculite	$(MgFe)_3$	(Al_xSi_{4-x}) $(x \geq 1)$	$(Mg,Ca)_{0.35} \cdot 4.5H_2O$
saponite	Mg_3	$(Al_{0.33}Si_{3.67})$	$(Na,Ca), nH_2O$
hectorite	$(Mg,Li)_3$	Si_4	
Trioctahedral chlorites	$Mg_{6-x}Al_x (x < 1)$	$Si_{4-x}Al_x$	

parallel to the layers of the structure. In the general case, for every 3 atoms lying in the plane of the tetrahedron faces there are 2 cations lying at tetrahedron centers, 2O and 1(OH) (or F) in the plane of octahedron faces to which tetrahedra are attached, or 3 (OH) if no tetrahedra are attached, and 2 or 3 cations lying at octahedron centers, depending on the proportion of occupied octahedra. Thus, the anionic parts of two-storied or three-storied layers or chlorite packets can be represented by the radicals $O_5(OH)_4$, $O_{10}(OH)_2$, or $O_{10}(OH)_8$, respectively.

The distribution of cations in the structures of the main representatives of these minerals is shown in Table 1 which, while giving a general picture, does not pretend to cover the layer silicates in a full, detailed manner. The numbers of cations shown in Table 1 are those corresponding to the anion radicals given above.

Within the limitations described above for these silicate layer structures, there is room for considerable individual variation. The possible variations may be due: (1) to the method of joining each pair of networks into a layer [they may be joined in three different ways, which differ by relative displacements which are multiples of b/3 (see Fig. 4)]; or (2) to the relative positioning of adjacent networks (each network can be oriented with its triangular faces pointing in either of two directions, opposite to one another); or (3) to the configuration of the tetrahedral networks, determined by the relative orientation of their tetrahedra; or (4) to distortion of the regular shapes of the polyhedra, and displacement of their cations from central positions; or (5) to the structure of the interlayer spaces; or (6) to the particular way in which the layers are arranged relative to one another. Differences which are due to variations in the method of joining or arranging the networks within layers are, in practice, equivalent to rotations of the layers as a whole by angles which are multiples of 60°. In the majority of cases these angles are multiples of 120°. Whereas the structures of micas are determined uniquely by the orientations of their layers, since any two adjacent layers, cemented together by interlayer cations, can be arranged relative to one another in only one way, differences can arise in kaolinites and chlorites because of layers being displaced relative to one another, so long as the necessary conditions are maintained for the formation of hydrogen bonds. As in the case of displacements of networks within a layer, these displacements are multiples of b/3. These displacements of networks and layers by multiples of b/3 are tied up with the loss of ordering typical of layer silicates. The individual structural features found in particular varieties of the layer silicates are due to variations in chemical composition, in interaction between atoms, and in conditions of formation of the minerals. In any given case these structural features can be revealed by struc-

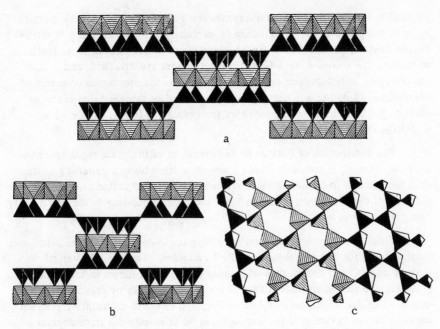

Fig. 7. Diagrammatic structures of (a) sepiolite; (b) palygorskite; (c) the corrugated oxygen—silicon network in sepiolite; reproduced from article by N.V. Belov, 1958.

ture analysis of the minerals, in greater or lesser detail depending on the significance of the diffraction data.

The clay minerals include certain varieties which cannot strictly be considered as layer silicates. These include sepiolite and palygorskite, which have pseudolayer structures.

The structures of sepiolite and palygorskite contain, like the layer silicates, continuous two-dimensional networks of Si tetrahedra connected together via their faces, but in their case the plane of tetrahedron faces is split up into bands of tetrahedra pointing in opposite directions; the bands face in opposite.directions in adjacent networks and are arranged so that the Si tetrahedra pointing toward each other in adjacent bands touch the oxygen—atom corners of bands of Mg octahedra on both sides (Fig. 7). In sepiolite the bands are formed by condensation of three "pyroxene" chains, and in palygorskite from two such chains, and for this reason the b constants of these minerals are three times or two times greater, respectively, than the b constants of the true layer silicates (Belov, 1958). Hence, if the relationship which holds between the unit cell constants of the layer silicates is b = $\sqrt{3}a$, for sepiolite it is b = $3\sqrt{3}a$, and for palygorskite it is b = $2\sqrt{3}a$. For convenience here, and in contravention of the usual practice, the fiber axes of sepiolite and palygorskite are denoted by a instead of c, so that they have the same letter symbol as the correspondingly positioned axis in the silicate layer structures (i.e., the axis with a constant of $a \sim 5.2$-5.3 A).

2. Some Features of the Structures of Individual Layers

The particular type of layer present in a structure determines the thickness of the layers and consequently the magnitude of the constant c, along the direction of stacking of the layers. Two-storied and three-storied layers and chlorite packets made up of a one-storied and a three-storied layer represent distances along the c axis of about 7, 10, and 14 A, respectively. Because the octahedra of a silicate layer are connected together by shared side edges, the construction formed by them is more rigid than that formed by the tetrahedra, and the linear dimensions of the octahedra basically determine the constants a and b. As a first approximation, it can be assumed that the octahedron bases parallel to the ab plane have the shapes of equilateral, similarly oriented triangles. From Fig. 5, if l is the length of side of a triangle, then

$$a = l\sqrt{3}, \ b = 3l. \tag{1}$$

On the other hand, because the tetrahedra are joined together only via the corners of faces, then, as Belov pointed out (1949), the tetrahedral network formed by them can adjust itself to various values of a and b, without significantly altering the linear dimensions of the tetrahedra; these variations in configuration are accomplished by rotating the tetrahedra about the normals to their faces which pass through their free corners (Fig. 2). Conversely, and for the same reasons, it is possible to have several networks of tetrahedra, each made of tetrahedra of different edge size l', where $l' \leq l$, which will match up with the same octahedral network, with fixed values of l, a, and b.

It should be appreciated, however, that due to mutual repulsion of Si atoms, neighboring tetrahedra have a tendency to rotate relative to one another in such a way that the Si$-$O$-$Si angle comes closer to π. Another point is that the rotation of tetrahedra may be limited by the interlayer cations. For a given value of l', this favors increased values of a and b.

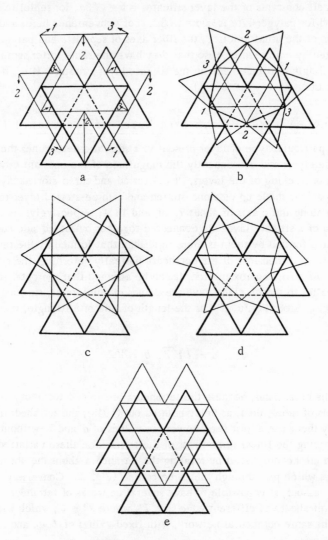

Fig. 8. Various configurations of one "loop" of a tetrahedral network (thin lines) attached to a loop of an octahedral network (thick lines).

If the octahedra were regular, as in anion close packing (Belov, 1947), the octahedral network would have a thickness of

$$h = \frac{l \sqrt{6}}{3} = \frac{a \sqrt{2}}{3} = \frac{b \sqrt{6}}{9}.$$ (2)

This value of h for the cation octahedra may be too large. According to Yamzin (1954), h is determined by the interatomic distance r from the octahedral cation to an oxygen atom, according to the formula

$$h = 2 \sqrt{r^2 - \frac{l^2}{3}}.$$ (3)

If r is very small, the limiting case will then be when the two O atoms forming the side of an octahedron are in contact. In this case,

$$h = \sqrt{(2R_{O^{-2}})^2 - \frac{l}{3}},$$ (4)

where $R_{O^{-2}}$ is the radius of O^{-2}.

In reality, the octahedra of layer silicates are usually very much compressed. If not all the octahedra are occupied, or if a proportion of the octahedral cations are of different size, then the faces of the octahedra may also be rotated, like those of the tetrahedra, through certain angles which are,however, smaller than those in the case of the tetrahedra.

From a comparison with the dimensions of the octahedra (at least in the plane of their bases), the tetrahedron edge length l' may be estimated from purely qualitative considerations. The values of l and l' will be closer together, the smaller the values of a and b, and the greater the degree of replacement of Si by Al in the tetrahedra. Conversely, l and l' will differ to a greater extent, the larger are the values of a and b and the smaller the degree of replacement. As a first approximation, l' can be taken as equal to the normal O–O distance in an Si tetrahedron, i.e., $l' \sim 2.65$ A.

Figures 8 and 9 represent projections on the ab plane of both the set of octahedron faces lying in a plane and the plane of interconnected faces of the tetrahedra attached to these octahedra. If we take the pattern formed by the tetrahedron faces (Fig. 8a) as the starting point, then any other pattern of regular tetrahedron faces can, as represented in Fig. 9, be considered as the result of a uniform displacement of the corners of the tetrahedron faces, by distances of t', along the three angular bisectors of the octahedral pattern lying at 120° to one another. In the original arrangement, where $l' = l$, the anions are

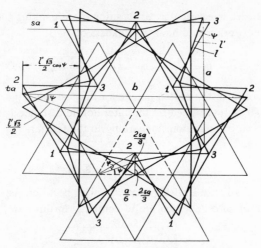

Fig. 9. Diagram demonstrating the relationship
between the quantities l, l', φ, ψ, a, and b.

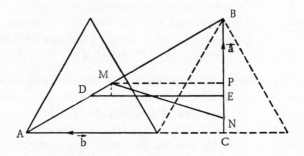

Fig. 10. Diagram showing the movement of a tetrahedron edge during the pattern transformation shown
in Figs. 8 and 9.

close-packed. When $t' = (l\sqrt{3})/6 = a/6$, the other limiting case is reached,
with the minimum value of $l' = l\sqrt{3}/2$, at which the anions in the tetrahedron
bases occupy $^3/_4$ of the close-packing positions (Fig. 8c).

In any intermediate case the corners must be shifted in the same directions, but by lesser amounts, which leads to an arrangement of the type shown
in Fig. 8b.

If the displacement of the corners is continued beyond the position shown in Fig. 8c, then, on passing through intermediate stages of the type shown in Fig. 8d, the limiting case shown in Fig. 8e will be reached, in which the tetrahedron edges will be equivalent to octahedron edges, as in the Fig. 8a arrangement; this again indicates that the anion corners are close-packed. However, in Fig. 8a the tetrahedral and octahedral patterns have opposite orientations, corresponding to cubic close-packing, whereas in Fig. 8e they have the same orientation, corresponding to hexagonal close-packing (Belov, 1947).

In Fig. 10, DE represents the edge of a tetrahedron in the original motif, MN is the edge in some arbitrary arrangement, and EN = DM = t'. The assumption is made that the bases of the tetrahedra are equilateral triangles.

As the other quantity required to fix any arbitrary tetrahedral arrangement, we can choose S', the projection of the edge MN on the a axis, so that the segment PN = s' = $\frac{3}{4}$t' (since $\angle$ MDE = 30°, PE = t'/2).

The quantities t' and s' are best expressed in terms of the unit cell constant a, i.e., we choose the quantities $t = t'/a$ and $s = s'/a$.

Thus, the arrangements shown in Fig. 8 are defined by the following values of t and s:

Arrangement in Fig. 8a $t = 0,\ s = 0.$
      "   8b $0 < t < 1/6,\ 0 < s < 1/4.$
      "   8c $t = 1/6,\ s = 1/4.$
      "   8d $1/6 < t < 1/3,\ 1/4 < s < 1/2.$
      "   8e $t = 1/3,\ s = 1/2.$

It is a very simple matter to express the changes in the coordinates of the points 1, 2, and 3 in Fig. 8a, in terms of t and s, for the change to any other arrangement (compare Figs. 8a and 8b):

$$\Delta x_1 = t/2 = s/3,\ \Delta y_1 = -t/2 = -s/3;$$
$$\Delta x_2 = -t = -2s/3,\ \Delta y_2 = 0;$$
$$\Delta x_3 = t/2 = s/3,\ \Delta y_3 = t/2 = s/3. \tag{5}$$

From Fig. 10 it follows that the length of edge of the tetrahedron is

$$MN = l' = \sqrt{\left(l - \frac{s'\sqrt{3}}{3}\right)^2 + s'^2}. \tag{6}$$

Let

$$l = kl' \; (k \geqslant 1),$$ (7)

then, from equation (6),

$$s' = \frac{l\sqrt{3}}{4}\left(1 \pm \frac{\sqrt{4 - 3k^2}}{k}\right).$$ (8)

Since $l\sqrt{3} = a$, then

$$s = s'/a = \frac{1}{4}\left(1 \pm \frac{\sqrt{4 - 3k^2}}{k}\right).$$ (9)

On the other hand, according to (6),

$$k = \frac{1}{\sqrt{1 - 2s + 4s^2}}.$$

From equation (9) it follows that the ratio of the edge lengths satisfies the condition

$$1 \leqslant k \leqslant 2/\sqrt{3}.$$

The maximum value of $2/\sqrt{3}$ is achieved when $s = \frac{1}{4}$, i.e., in the arrangement shown in Fig. 8c, when the tetrahedron edge has its smallest length, equal to the height of an octahedron face, i.e., exactly $l\sqrt{3}/2$. This value defines the minimum length of l, below which it would be impossible to continuously cover a plane with equilateral triangles, the centers of which would form a noncentered hexagonal pattern with constants a and b. This can be expressed in another way by saying that when $k > 2/\sqrt{3}$, the quantity s becomes a complex quantity. All other values of k correspond to two values of s, which is to be expected, since there are two patterns with identical tetrahedron edge lengths; one of the type shown in Fig. 8b, and the other of the type represented in Fig. 8d. The first of these corresponds to the minus sign within the brackets in equation (9), and the second to the plus sign. In the particular case where $k = 1$, then $s_1 = 0$, $s_2 = \frac{1}{2}$, in full agreement with the characteristics noted above for the arrangements shown in Figs. 8a and 8e.

The parameter s, equal to the projection of a tetrahedron edge on the a axis expressed in terms of a, can also be expressed in terms of the angles of rotation of the tetrahedron faces. If φ is the angle measured from the positions corresponding to close-packing ($s = 0, 1$), and ψ is measured from the other limiting case with hexagonal loops ($s = \frac{1}{4}$), so that $\varphi = 30° - \psi$, then, from Fig. 9,

$$\tan \varphi = \frac{s\sqrt{3}}{1-s}, \tan \psi = \frac{1-4s}{\sqrt{3}}.\tag{10}$$

It is easy to see (Fig. 9) that $2l'\cos\psi = a$, $2\sqrt{3}l'\cos\psi = b$, and so if we are given a definite value of l', then from the values of a and b we can find the angles φ and ψ.

A similar relationship holds for rotation of the octahedron faces, when this occurs. In this case the face edges parallel to the plane ab are reduced in length below their "ideal" value of l. The shared side edges of adjacent octahedra are even more compressed (see results below on the kaolinite and muscovite structures, Figs. 39 and 45), which is in accordance with Pauling's rules governing the structures of ionic crystals (Pauling, 1947).

There is another feature of the tetrahedral arrangement which should be noted; the hollow formed by the ring of tetrahedron faces may have the shape of a triangle of side $2l$ (Figs. 8a, 8e), a nonregular hexagon with threefold (ditrigonal) symmetry (Figs. 8b, 8d), or a regular hexagon of side l' (Fig. 8c). Interlayer cations may be attached at these hollows. The six corners of a hollow define two oppositely oriented equilateral triangles (Fig. 8b), the smallest of which forms the bottom of the octahedron occupied by the interlayer cation in the case of the mica minerals (Belov, 1949). The sides of these triangles have lengths given by the relationships

$$l_1'' = l + 2\Delta y_3 \cdot b = l + \frac{2sb}{3} = l(2s+1),$$

$$l_2'' = 2l - 2\Delta y_3 b = 2l - \frac{2sb}{3} = 2l(1-s).\tag{11}$$

For the arrangement in Fig. 8b, $l_1'' < l_2''$, and for that in Fig. 8d, $l_1'' > l_2''$. As would be expected, $l_1'' = l_2''$ when $s = \frac{1}{4}$, i.e., for the motif shown in Fig. 8c. By substituting the value $s = \frac{1}{4}$ in equation (11), we find that in this case $l'' = 3l/2$.

Table 2 is a summary of various characteristics of the different tetrahedral arrangements.

The maximum rotations of tetrahedra, to form hexagonal loops, are both to be expected and actually found in practice in layer silicates with large values of the constants a and b, such as the trioctahedral micas phlogopite and biotite, the chlorites, talc, etc., or in the minerals with the smallest tetrahedra,

Table 2. Characteristics of the Tetrahedral Patterns Shown in Fig. 8, Formed from the Faces of Polyhedra

Motif	t	s	l'	k	$l''_{1,2}$
8 a	0	0	l	1	$l;\ 2l$
8 b	$0 < t < 1/6$	$0 < s < 1/4$	$l > l' > \dfrac{l\sqrt{3}}{2}$	$1 < k < \dfrac{2}{\sqrt{3}}$	$l(2s+1);\ 2l(1-s)$
8 c	$1/6$	$1/4$	$l\sqrt{3}/2$	$2/\sqrt{3}$	$3l/2$
8 d	$1/6 < t < 1/3$	$1/4 < s < 1/2$	$\dfrac{l\sqrt{3}}{2} < l' < l$	$2/\sqrt{3} > k > 1$	$l(2s+1);\ 2l(1-s)$
8 e	$1/3$	$1/2$	l	1	$2l;\ l$

i.e., those with tetrahedra predominantly occupied by Si atoms, and with average size octahedra, for example, lepidolite.

In contrast, small rotations, with the oxygen atoms of the tetrahedron faces approximately close-packed, are most likely for minerals with minimal values of a and b and no replacement of Si by Al, such as pyrophyllite, or for those in which considerable replacement of Si by Al has led to tetrahedra with a large average size, coupled with quite small interlayer cations. The latter case is in agreement with results obtained for xanthophyllite (Takeuchi and Sadanaga, 1959), which had $\tfrac{3}{4}$ of the Si replaced by Al. It is true that a similar small rotation was not found in a mineral related to xanthophyllite, i.e., seybertite (Akhundov, Mamedov, and Belov, 1961), which had somewhat larger values of a and b. An intermediate case is that of muscovite (see Chapter 5, below), which has fairly small, though not minimum, values of a and b, with a quarter of the Si replaced by Al, and a tetrahedron rotation angle of $\varphi \approx 20°$. Approximately the same tetrahedron rotation angle is shown by kaolinite, which has values of a and b smaller than those in muscovite, but to compensate, its tetrahedra are wholly occupied by Si atoms.

A controlling factor limiting the rotation of the tetrahedra is the size of the interlayer cations. If the cations hinder complete matching between the tetrahedral and octahedral networks, then, as pointed out by Radoslovich (1961), the degree of replacement in the tetrahedra will also affect the a and b constants of the lattice. This point is apparently the reason why the a and b constants in Na-allevardite or pyrophyllite are less than those in muscovite.

Rotation of octahedra occurs mostly in the dioctahedral minerals, which have vacant octahedra, and in trioctahedral minerals which have cations of different radii in the octahedra.

Other deviations of real structures from the ideal models include the compression of octahedra along their heights, tilting of the tetrahedra, displacement of anions from their arrangement in planes parallel to the ab unit-cell face, and displacement of cations from the centers of their polyhedra. The particular nature of these deviations depends on the dimensions of the anions and cations making up the structure and the forces acting between them, and this information can be used to some extent to forecast certain structural details, or to explain these if they have been established experimentally.

Examples of such studies of the real structures of layer silicates will be found in the series of articles by Radoslovich (1963) devoted to analysis of the factors affecting the size of the unit cell in layer silicates. Based on a critical examination of the available structural evidence, Radoslovich has formulated a set of rules which must be obeyed by real layer silicate structures, and which state the limits within which their structural features may vary. The basis of these rules is as follows.

1. The least variable characteristic of a structure is the distance between a cation and an anion. The shapes of the polyhedra of a structure may be deformed as a result of changes in the distances between anions, and in the angles between their bonds to cations. Here the possible positions of anions are distributed over a sphere of radius equal to the cation–anion distance.

2. The angles between the bonds to an anion can be altered even more easily than the angles between the bonds to a cation; this corresponds to changes in the relative orientations of the structural polyhedra.

3. The cations lying at the centers of tetrahedra or octahedra of the same network are arranged in a hexagonal pattern and lie approximately in the same plane.

3. Use of Structural Polyhedra in the Description and Analysis of Layer Silicate Structures

All the different structures which are theoretically possible for layer silicates can be defined in terms of the structures of their individual layers and variations in their relative arrangement.

It has already been pointed out above that the concept of a silicate layer of a given type is ambiguous in view of the different possible ways of making it up from networks of tetrahedra and octahedra, since each of these networks may have two opposite orientations and may be attached to an adjacent network in three ways.

The relative arrangement of the layer depends itself on the bonding forces acting between the layers, and on the filling of the interlayer spaces.

By varying these features it is possible to formally derive an enormous number of different silicate layer structures. It is only by the application of crystallochemical methods of analysis and comparison that we can select those which are theoretically more probable and which possibly exist in real structures.

In carrying out this task, a great deal depends on the method used to describe the structures, in particular its ability to give a clear, graphic representation.

The widely used structural descriptions in terms of atomic coordinates or of arrangements of atoms represented by conventional symbols are, generally speaking, understood only by the authors preparing these descriptions, and their significance is not easily comprehensible to an outsider. They are difficult to understand from the viewpoint of distinguishing both individual variations and the overall pattern.

In contrast, the method of examining crystal structures originated by Pauling and developed by Belov (Pauling, 1928, 1929, 1930a,b; Belov, 1947, 1949, 1950, 1951), which extended the sphere of consideration from individual atoms located at isolated points to whole fragments made up of polyhedra and patterns of polyhedra, has turned out to be particularly appropriate and fruitful. It will be shown below how the method can be used, not only for the simple, clear description of any sequence of networks and layers making up a structure, but also in a unified scheme of calculation of the structure amplitudes which express the diffraction properties of these structure components.

We have already noted that the layers found in layer silicates are mainly of three types, the kaolinite, the mica, or the chlorite type, these being frequently denoted, from the relative numbers of tetrahedral or octahedral networks they contain, as the 1:1, 2:1, or 2:2 type, respectively. The discussion which follows therefore only embraces structures made up of layers of these three types. Structures containing corrugated, deformed layers such as antigorite, or pseudolayer structures of the sepiolite or palygorskite type, are special cases and therefore are excluded from the present discussion.

By using these polyhedra, the structures of layer silicates can be represented in perspective or as different projections of regularly arranged two-dimensional tetrahedral and octahedral networks. The diagrammatic representations which are simplest to draw, but which are nonetheless the most expressive in depicting the finer features of a structure, are those suggested by Belov (1949, 1950, 1951), in the form of combined patterns of the parallel faces of tetrahedra and octahedra, projected along their normals onto the ab plane.

Every individual network made up of regular polyhedron faces gives a normal projection on the ab plane of symmetry p31m, provided that $s \neq \frac{1}{4}$ ($\psi \neq 0$). When $s = \frac{1}{4}$, the tetrahedral network has the symmetry p6mm. When connected together, an octahedral and a tetrahedral network form a combined two-dimensional arrangement of symmetry c1m1 as long as all the octahedra in the network are not uniformly occupied, so that its characteristic repeat distance $b' \neq b/3$. In the special case where $b' = b/3$, the combined symmetry of both networks is p31m (International Tables for X-Ray Crystallography, 1952).

To start with, we can assume that the tetrahedral network has the hexagonal symmetry p6mm ($s = \frac{1}{4}$, $\psi = 0$), which it is unchanged on rotating through 180°, and that the octahedral network is dioctahedral, i.e., two occupied octahedra and one vacant octahedron alternate regularly within it. We can examine the consequences later of changing to a ditrigonal pattern of tetrahedra and to a trioctahedral network of octahedra.

Under the conditions we have selected, there are six possible arrangements of both networks, due to the two possible orientations of the octahedral network and the three methods of combining it with the tetrahedral network. It is not difficult to demonstrate that all six of the resulting configurations are rotationally identical and may be transformed into one another by rotations through multiples of 60°. In order to derive them in a plain form, we need only to depict the combined pattern of both networks in any given position, and, choosing a conventional origin of coordinates at the center of an empty octahedron, rotate the system of coordinates by angles which are multiples of 60°.

All six versions of the relative arrangement of the two networks are depicted in Fig. 11 through the configuration of their combined pattern with respect to each of the six axes a_i ($i = 1, 2, \ldots, 6$). These versions differ among themselves in their characteristic values of the displacement σ_i of the tetrahedral network relative to the octahedral network.

If we measure these displacements from the centers of the vacant octahedra of the octahedral network and the centers of the hexagonal loops of the tetrahedral network, then in a rectilinear system of coordinates with a cell ab ($b = a\sqrt{3}$) they are defined by the components along the axes a and b which are given in Table 3.

Table 3. Components of the Displacements σ and τ Along the a and b Axes

Displacement symbol	Components *	Displacement symbol	Components *	Displacement symbol	Components *
σ_1	1,1	$\tau_1,\ 2\sigma_1$	$-1,-1$	τ_0	0,0
σ_2	$-1,1$	$\tau_2,\ 2\sigma_2$	$1,-1$	τ_+	0,1
σ_3	1,0	$\tau_3,\ 2\sigma_3$	$-1,\ 0$		
σ_4	$-1,-1$	$\tau_4,\ 2\sigma_4$	1,1	τ_-	$0,-1$
σ_5	1,-1	$\tau_5,\ 2\sigma_5$	$-1,1$		
σ_6	$-1,0$	$\tau_6,\ 2\sigma_6$	1,0		

* The displacement components are expressed in multiples of $a/3$ and $b/3$.

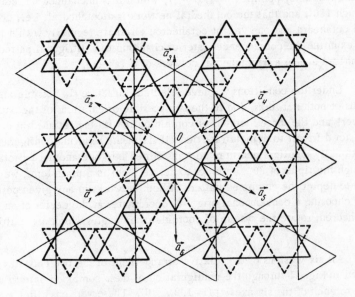

Fig. 11. Configurations of a two-storied layer relative to the a axis for rotations of the coordinate system about the $c*$ axis by angles which are multiples of $\triangle\varphi = 2\pi/6$.

Since the combination of two networks which we have described is identical with a two-storied layer of the kaolinite type, the displacements σ_i of
the networks within a layer, expressed as their components along the axes a
and b, can at the same time be considered as defining or "labeling" the layer
with respect to its azimuthal orientation.

In a three-storied layer of the mica type, the same displacements σ_i
are characteristic of any pair of adjacent layers. To avoid any ambiguity we
will take it that they relate to the displacement of an upper layer relative to
the one lying beneath it. Then tetrahedral networks of three-storied layers
may be displaced relative to one another, in the general case, by $\sigma_g + \sigma_i =$
σ_{gi}, and these quantities, which uniquely define the individual differences between the layers, may also be used as conventional "labels" for these layers.
If g = i, or if the layers are trioctahedral, when this equation is obeyed automatically, the three-storied layers (Fig. 12) are assigned the value of σ_i of
one pair of networks, and this can be used to designate the layers.

The same three-storied layer is also found in the chlorite structures. In
addition, however, they contain one-storied layers of the brucite type, forming
2:2 packets in combination with the three-storied layers. The chlorite packets may therefore show further variations in addition to those existing between
the mica-type layers, depending on the orientation and position of the brucite-
type layers.

The majority of the chlorites are trioctahedral, and variations in the
relative positions of the layers in the packets, i.e., differences in the rotation
of the single-storied layer about its normal by a multiple of 120°, or in its
displacement along the b axis by a multiple of b/3, all amount to the same
thing. The features which retain their significance are the two opposite orientations of the one-storied layer, either parallel or antiparallel to the octahedral network of the three-storied layer and, with a fixed orientation, the two
possible positions, differing by $a/3$, which satisfy the condition that the OH
groups of the brucite-type layer and the O atoms of the adjacent three-storied
layer are grouped in pairs to form hydrogen bonds (Fig. 13).

Thus, for a given three-storied layer, there are four possible ways of arranging an adjacent one-storied layer relative to it. To avoid confusion, we
will stipulate that, here and in what follows, any sequence of networks and
layers will be examined from the bottom upwards. It is convenient to base
our notation for the different possible arrangements on the displacements of
the networks inside the three-storied layers, with certain additional symbols,
depending on the particular version.

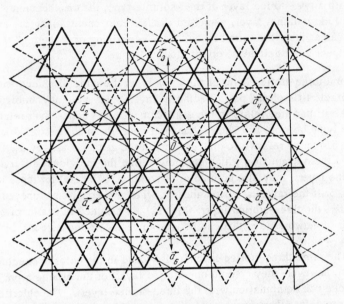

Fig. 12. Configurations of three-storied layer relative to the
a axis for rotations of the coordinate system about the c*
axis by angles which are multiples of $2\pi / 6$.

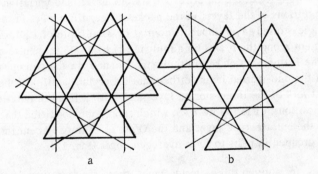

Fig. 13. Two versions (a and b) of the relative arrangement of the lower faces of a single-storied layer (thick
lines) relative to the tetrahedral network of a three-
storied layer (thin lines).

1. Layers antiparallel. Centers of upper-layer octahedra project onto the same positions as the lower corners of lower-layer octahedra, and upper corners of upper-layer octahedra project onto upper corners of lower octahedra. Symbol, σ_{gi}.

2. Layers antiparallel. Centers of upper octahedra project onto centers of upper tetrahedra, and upper corners of upper octahedra onto centers of lower octahedra. Symbol, $\sigma_{gi}|$.

3. Layers parallel. Centers of upper octahedra project onto centers of lower octahedra, and upper corners of upper octahedra onto upper corners of lower octahedra. Symbol, σ'_{gi}.

4. Layers parallel. Centers of upper octahedra project onto centers of upper tetrahedra, and upper corners of upper octahedra onto lower corners of octahedra of the lower three-storied layer. Symbol, $\sigma'_{gi}|$.

In the notation used here, a prime indicates that the octahedral networks are parallel, and its absence implies that they are antiparallel. When upper cations are superimposed on upper tetrahedral cations in a projection on the ab plane, this is denoted by a single vertical line. Because of electrostatic repulsion between the cations, arrangements 2 and 4 are the least stable.

The same interpretation and notation are applicable to the relative arrangement of a three-storied layer and the one-storied layer preceding it. The above notes 1-4 need only be modified by interchanging the words "upper" and "lower," and by placing the vertical line in the symbol to the left of the packet symbol.

If the layers of the chlorite packet are not trioctahedral, then their possible arrangements are not limited to the above four cases, since displacement of the layers parallel to the b axis by a multiple of $b/3$ creates additional versions. The displacement of the $0:1$ layer relative to the $2:1$ layer has to be indicated by an independent parameter which we will call τ. If, in addition, the $2:1$ layer is noncentrosymmetric, and $\sigma_g \neq \sigma_i$ in it, then the sequence will be made up of links of the type $(\sigma_g + \sigma_i)(\tau_j + \tau_k) = \sigma_{gi}\tau_{jk}$.

The use of the symbols σ_{gi} or σ_i for $1:1$, $1:2$, and $2:2$ layers need not lead to confusion so long as the types of layers to which the symbols are being applied are stipulated in the text.

For a complete description of a layer structure it is necessary, in addition to giving symbols for the alternation of layers which completely define the structure and orientation of these layers relative to the chosen system of

coordinates, to somehow define their relative arrangement. This cannot be
defined in terms of relative displacements of layers, since the very concept of
such a displacement is imprecise when the layers are not translationally identical, particularly so when they are not of the same type. Since the types and
orientations of the layers are known, then their relative arrangement in space
can be quite satisfactorily stated in terms of the value of the displacements τ_j,
measured between the closest networks of adjacent layers. For chlorites it is
convenient to use the displacement τ between the closest tetrahedral networks
of different layers. In the particular case of the micas, neighboring layers are
arranged so that the hexagonal loops of the tetrahedra of the upper surface of
the lower layer and those of the lower surface of the upper layer coincide in
a normal projection on the ab plane. In the planes formed in this way lie the
cations, these being most frequently K atoms. We therefore have it that $\tau = 0$
and these parameters can therefore be omitted from the analytical symbol for
a mica structure.

In kaolinites and chlorites the values of τ are determined by the condition that O atoms and OH groups from adjacent networks of different layers
should match up to enable hydrogen bonding to take place. This condition is
satisfied by the nine displacements τ_1, τ_2, τ_3, τ_4, τ_5, τ_6, τ_+, τ_-, τ_0, the
components of which, along the axes a and b, are given in Table 3. The nu-
merical values of these displacements are the same as those for the σ displacements, expressed as multiples of $a/3$ and $b/3$.

Any sequence of layers can be represented by a sequence of σ,τ symbols, which will fully define the features of the structure formed, when the
characteristics of the internal structures of the layers and the limitations on
their relative arrangement noted above are taken into consideration. The
quantities σ,τ can therefore be considered as the minimum number of independent parameters necessary to define uniquely any arbitrary sequence of
layers, including the repeat unit, symmetry, atomic coordinates, and diffraction properties of the sequence.

The analytical notation used here is as useful and expressive as the analytical methods of describing various structures containing close-packed atoms
which were devised by Belov (1947) and by Zhdanov (1945). A fundamental
difference here is that Belov's sequence of the letters h and c and Zhdanov's
numerical symbols define structures which differ in the methods of close-
packing of their atoms, while differences in the values of σ and τ relate first
and foremost to differences in the manner of packing cations within what is
basically the same arrangement of anions. In the structural modifications

which we are considering, the nature of the packing can only vary for an even number of layers and therefore does not embrace all structural varieties. In the vicinity of layer silicate interlayer spaces, moreover, the anions do not have a close-packed arrangement, and an analytical close-packing notation cannot be extended to these structures without modification.

It should be noted that when the system of coordinates is rotated, so that the a_i axes are interchanged as shown in Fig. 11, the quantities σ_j, τ_k ($i,j,k = 1,....,6$) undergo corresponding changes, as if they were arranged in a circle like the ends of the $\mathbf{a}_i$ vectors. They transform one into the other according to a cyclic permutation with $\Delta i = \Delta j = \Delta k$. The quantities τ_+, τ_- reverse their signs if $\Delta i = (2n+1)[\Delta\varphi = (2\pi/6)(2n+1)]$, and leave them unchanged if $\Delta i = 2n[\Delta\varphi = (2\pi/3)m]$ (m,n integral). The quantity τ_0 does not alter for rotation of the coordinate system about a normal to the layer.

If the structure does not have a definite value of the constant c, the sequence of the symbols σ, τ will not repeat regularly. Where there is a definite value of c, the layer sequence will be represented by a repeating group of σ and τ symbols. The sum of the numerical values of these quantities gives the displacement of the origin of coordinates, conveniently taken at the center of a vacant octahedron, on passing through one repeat unit of the structure. If both components of this displacement differ from zero, then a rotation of the system of coordinates by a multiple of $2\pi/6$ will always be possible, such that these components become $-\frac{1}{3}$, 0. If the displacement has the components 0, $\frac{1}{3}$, it will be similarly possible to transform these to 0, $-\frac{1}{3}$. In such an event the subscripts of all the quantities are changed according to the above transformation rules, and in its new orientation the structure will have a unit cell which is either monoclinic, with the angle β or $\alpha > \pi/2$, or orthorhombic (both components equal to zero).

If the sequence repeats every n layers, then, depending on which of the two components differs from zero, one of the two following relationships will be satisfied:

$$c_n \cos \beta_n = -a/3, \quad c_n \cos \alpha_n = -b/3. \qquad (12)$$

It is not difficult to show that, since $c_n' = nc_i' = c_n \sin \beta_n$, where c_n' is the projection of c_n on the normal to the layers, the nonrectangular angles α and β, corresponding to different layer repeat units, are related by the equations

$$\tan(\alpha, \beta)_n = \frac{n}{m}\tan(\alpha, \beta)_m, \qquad (13)$$

and, in a particular case,

$$\tan \beta_n = n \tan \beta_1.$$

From equations (12) and (13), a triple-height orthorhombic cell can be chosen from the lattice, with a cell constant perpendicular to the ab plane of $3c_n'$.

4. Polymorphic and Polytypal Modifications and Repeat Sequence Methods in Layer Silicates

The characteristic and individual features of layer silicate structures are governed by the following three basic properties: types of layer present, structural variations within a given single layer type, and relative disposition of the layers. Layers of a single type should be understood as meaning layers consisting of a given combination of elementary networks of polyhedra (octahedra or tetrahedra), arranged in a given sequence along the c-axis direction. These layers may, as already noted, differ in the relative orientations of their networks and in their positions relative to the axes of coordinates. Some diagrammatic examples of different types of layers are given in Fig. 14.

If the layers were constructed out of regular tetrahedra and octahedra, and these polyhedra lay in a trigonal or hexagonal arrangement, then many of the possible combinations of networks into layers and layers into structures would, because of their identical interaction distances, be energetically equivalent and equally probable. Within the bounds of these idealized layer silicate structures, there is no way of determining the factors which decide whether one or another particular σ, τ type, or one or another structure defined by a sequence of σ, τ values, is more favorable. If all were equally favorable a disordered structure would be formed in which there would be a random sequence of energetically equally favorable values of the displacements σ, τ.

However, layer silicates do in fact have, to a greater or lesser extent, ordered structures. This is a reflection of the fact that the regular structures possessed by the ideal models do not exist in reality, and real structures deviate from the ideal models in many respects.

The distortions are due to the force field acting within the structure, a complex phenomenon resulting from the combination of forces of different types (ionic and covalent) and forces of attraction and repulsion. Under its influence, the octahedra are compressed, shared edges are reduced in length,

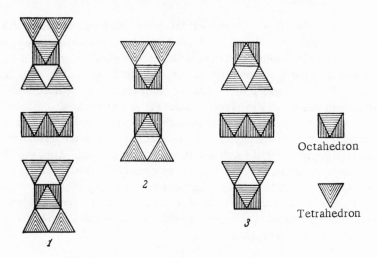

Fig. 14. Examples of layer sequences of various types. (1) Layers differ in their component networks (one-storied and three-storied layers); (2) layers differ in order of arranging networks within them; (3) layers differ in number and order of arrangement of networks.

tetrahedra are rotated into a ditrigonal arrangement (edges forming a hexagon of threefold symmetry), and so on. An important part is played by the ordered distribution of isomorphous cations among the structure positions available to them, and disordered fluctuations in chemical composition are also important.

Because of these structure distortions, various possible combinations of structural units as described by a set of σ, τ values, which, in the idealized model, were equally probable, here assume different probabilities. One of the combinations may unarguably be more favorable than the others, but the appearance of several forms may occur due to particular conditions of formation. In such a case different regular layer sequences may be possible for a fixed chemical composition, giving rise to different structure modifications. Here, the real, nonmodel properties of the layers will be the controlling factors which will fix the particular layer sequence which, for a given chemical composition and given formation conditions, will be thermodynamically most favorable, even though the energy difference may not be large. It should be noted that trioctahedral layers are normally less distorted than dioctahedral layers, so that there is not such a large energy difference between the various trioctahedral possibilities. Because of this, structures based on trioctahedral layers may include some which are "forbidden" for dioctahedral layers, if an energetically unfavorable disposition of the cations does not prevent this. In connection with the fact that layer silicate modifications are made up of large, closely similar structural units, which can be examined as separate layers and packets, their energy characteristics cannot differ very greatly, in contrast to the case where transition from one structure to another would involve a major reorganization; and so the physical and chemical limits of the formation conditions of different ordered structures may partially overlap. Also, therefore, the regions of preferential crystallization of each of the modifications on their phase diagram cannot be clearly differentiated with sharp boundaries between them. In these circumstances, layer silicates are typically formed which are made up of a combination of several modifications, which alternate as separate zones within an individual crystal.

The modifications making up such a combination have identical values of the unit cell constants a and b and must therefore have identical chemical compositions, differing only in the relative disposition of the layers in their structures. The formation of such combinations of modifications can be ascribed to their formation conditions, which would be intermediate between the conditions most favorable for each particular modification on its own.

This peculiarity of the structural modifications of layer silicates is the reason for the dual basis of their appraisal, as polymorphic modifications on the one hand, and as polytypal modifications on the other. It is noted that, in the layer silicates, pure polytypal structure modifications are possible and actually exist. They are formed when a chance combination of the structure characteristics σ, τ is carried on and regularly repeated by the spiral growth mechanism (Amelinckx, Dekeyser, 1955) regardless of how closely it corresponds with the conditions of formation. Structure modifications which can, with high or low probability, be related to certain formation conditions and, therefore, which can be considered as polymorphic modifications, differ from pure polytypes in that they must satisfy a condition of uniformity. This condition requires that the position of any layer with respect to the others must be identical for all layers; a transition from one layer to the next is equivalent to a transition from the next layer to the one after that.

The uniformity condition in a regular layer structure (and also in a nonregular one) means that, in particular, within a three-storied layer σ_{gi}, $\sigma_g = \sigma_i$. Otherwise, the lower and upper or left-hand and right-hand parts of a layer would not be equivalent. In the chlorites the one-storied layer must have a single definite orientation and form a rigidly defined packet with a three-storied layer, equivalent to a single layer. Under these conditions, layers which are of the same kind (1:1, 2:1, or 2:2) are rotationally identical as regards rotations about a normal to the layer by multiples of 60°.

In addition to its consequence of regular layer repetition, the uniformity condition also allows us to derive the theoretically possible regular polymorphic modifications of layer silicates. These structures, with all their properties, can be uniquely defined notationally in terms of the values of σ, τ in one link of the chain corresponding to their regularly repeating sequence.

The structural modifications which can be derived theoretically for layers or packets of a given type may include some which cannot be differentiated by diffraction methods (see Chapter 4), enantiomorphic structures for example. However, the number of independent theoretical structures exceeds, as a rule, the number established experimentally as existing in nature, since the deviations of real structures from their idealized models, not being equally favorable for different combinations of layers, greatly reduce the limits within which the uniformity condition will be satisfied.

Nonuniform sequences can exist in nature and have actually been detected. Their formation may be due to peculiarities in the formation conditions, with a nonuniform distribution of isomorphous cations giving rise to nearest-neighbor conditions more favorable than those applying if the uniformity condition were observed.

It should be noted that layer silicate structure modifications may show several forms of layer repetition. In addition to the strict c repeat unit along the natural axis, expressed in terms of a repeating group of σ, τ symbols in an analytical description of the structure, a partial repeat unit is possible which will relate to only part of the properties of the structure and which will embrace a smaller number of layers than the strict repeat unit.

Usually, in layer silicate structures, a definite order of alternation of layers or packets, corresponding to a particular modification, is not observed strictly. To a greater or lesser extent, it is subject to alterations involving replacement of the expected values of σ, τ by others which are equally probable for the ideal structure. The most probable deviation, and in practice the only one observed, is equivalent to a displacement of networks within layers or of layers relative to one another (where they permit this) by values of $\pm b/3$ along the b-axis direction.

Because of the small distortion of their layers, and the resulting weak structural control, trioctahedral structures are particularly prone to breakdown of the stacking sequence of their layers. Thus, for the chlorites, an ordered structure is a rare exception, and they usually have semirandom structures (see Chapter 4). For the same reason, if ordered structures are observed in serpentine-type minerals they are not alone, but in combination with other modifications. Semirandom modifications can also be met with as combinations. Here we should distinguish between the following concepts.

a. Strict c Repeat Unit. The sequence of layers forming a structure has a strict c repeat unit if, after a certain number of layers, each layer is accurately repeated. Generally speaking, different features and properties of a structure may repeat after different numbers of layers, but the strict repeat unit, if it exists, characterizes the repetition of all features and properties. For example, the property of cleavage has a repeat unit of one layer, while the particular situation of any given atom may repeat itself after a greater number of layers, if the layer sequence is repetitive, or may not repeat at all, if the layer sequence does not repeat regularly. In the latter case the structure will not possess a strict c repeat unit.

b. Minimum c Repeat Unit. In structures made up of layers of one type only, whatever the particular structure, both the actual layer repetition and the properties of a layer will have a repeat unit of one layer. This unit can be taken as the minimum repeat unit. It will be equal to the fraction of the c repeat distance corresponding to one layer.

c. Irregular c Repeat Unit. If the strict c repeat unit corresponds to one layer, the minimum repeat unit will simultaneously be the strict unit. In the cases where the strict unit relates to a large number of layers or is absent altogether, the minimum unit will relate to an irregular and incomplete repetition of the properties of the crystal lattice, and should be considered as an irregular c repeat unit, relating incomplete, only partly identical different unit cells in the structure.

d. Almost Strict c Repeat Distance. It can happen that the strict repetition of layers is destroyed by a certain amount of loss of ordering in the atomic distribution which is, however, not enough to completely prevent the ordering showing up on diffraction patterns, although these will also register the loss of ordering. Whatever the number of layers which it includes, the corresponding value of c can be considered as an almost strict unit, characterizing the approximate fulfillment of the rule describing the repetition of properties within the crystal lattice.

On similar qualitative criteria it is possible to have any intermediate form between a strictly ordered structure and a structure without any trace of a strict c repeat unit, possessing only an irregular, minimum c repeat unit of one layer.

Even among these latter structures, however, various degrees of perfection are possible, since even those properties of the crystal lattice which are repeated by the minimum unit may be maintained over the whole structure with different degrees of consistency in different cases. In particular, if the layers are displaced relative to each other in a completely disordered manner by arbitrary amounts, but with the distances between the layers constant, then the set of layers can be examined as a one-dimensional lattice of independent two-dimensional lattices. In addition to the geometrical factors already noted which lead to loss of strict repetition of the layers, important factors may include disordering in the distribution of isomorphous atoms, removal of certain atoms under the influence of some external force, and so on.

To what has been said we can add that, for any structure which has suffered a loss of ordering, the concepts of the lattice and the unit cell are, strictly speaking, not applicable. In using these terms for such structures it should be understood that what is really meant is a pseudolattice or a pseudo-unit cell.

If a layer sequence does not have a minimum repeat unit, and layers of different types alternate completely randomly, then it can only be examined as a set of two-dimensional lattices, each of which represents a single separate layer. In a space of three dimensions such a sequence does not possess either a lattice or a unit cell in any sense whatever.

5. Regular Structures of Two-Storied Kaolinite-Type Layers

A diagram of a two-storied layer is shown in Fig. 11. On this diagram the six possible directions of the a axis relative to the layer are indicated, for the different orientations of the layer within the structure.

It was noted by Hendricks (1938) that the relative arrangement of the layers is typically such that the OH groups of the lower faces of the octahedra of the upper layer match up with O atoms of lower-layer tetrahedron faces, forming hydrogen bonds.

Brindley (1961) has described the six different relative displacements of adjacent layers for which this condition is satisfied. A detailed examination

Table 4. Combinations of the Dioctahedral Layers σ_i with the Layer σ_1

No.	Layer combination	Δ_{Al-Si}	Δ_{Al-Al}	Δ_{Si-Si}	Δ_{Si-Al}
1	$\sigma_1\tau_0\sigma_1$	0,0	1,1	1,1	−1, −1
2	$\sigma_1\tau_0\sigma_2$	0,0	1,1	−1,1	0, −1
3	$\sigma_1\tau_0\sigma_3$	0,0	1,1	1,0	−1,1
4	$\sigma_1\tau_0\sigma_4$	0,0	1,1	−1, −1	0,0
5	$\sigma_1\tau_0\sigma_5$	0,0	1,1	1,1	−1,0
6	$\sigma_1\tau_0\sigma_6$	0,0	1,1	−1,0	0,1
7	$\sigma_1\tau_+\sigma_1$	0,1	1, −1	1, −1	−1,0
8	$\sigma_1\tau_+\sigma_2$	0,1	1, −1	−1, −1	0,0
9	$\sigma_1\tau_+\sigma_3$	0,1	1, −1	1,1	−1, −1
10	$\sigma_1\tau_+\sigma_4$	0,1	1, −1	−1,0	0,1
11	$\sigma_1\tau_+\sigma_5$	0,1	1, −1	1,0	−1,1
12	$\sigma_1\tau_+\sigma_6$	0,1	1, −1	−1,1	0, −1
13	$\sigma_1\tau_-\sigma_1$	0, −1	1,0	1,0	−1,1
14	$\sigma_1\tau_-\sigma_2$	0, −1	1,0	−1,0	0,1
15	$\sigma_1\tau_-\sigma_3$	0, −1	1,0	1, −1	−1,0
16	$\sigma_1\tau_-\sigma_4$	0, −1	1,0	−1,1	0, −1
17	$\sigma_1\tau_-\sigma_5$	0, −1	1,0	1,1	−1, −1
18	$\sigma_1\tau_-\sigma_6$	0, −1	1,0	−1, −1	0,0
19	$\sigma_1\tau_1\sigma_1$	−1, −1	0,0	0,0	1,1
20	$\sigma_1\tau_1\sigma_3$	−1, −1	0,0	0, −1	1,0
21	$\sigma_1\tau_1\sigma_5$	−1, −1	0,0	0, 1	1, −1
22	$\sigma_1\tau_2\sigma_2$	1, −1	−1,0	0,0	1,1
23	$\sigma_1\tau_2\sigma_4$	1, −1	−1,0	0,1	1, −1
24	$\sigma_1\tau_2\sigma_6$	1, −1	−1,0	0, −1	1,0
25	$\sigma_1\tau_3\sigma_1$	−1,0	0,1	0,1	1, −1
26	$\sigma_1\tau_3\sigma_3$	−1,0	0,1	0,0	1,1
27	$\sigma_1\tau_3\sigma_5$	−1,0	0,1	0, −1	1,0
28	$\sigma_1\tau_4\sigma_2$	1,1	−1, −1	0, −1	1,0
29	$\sigma_1\tau_4\sigma_4$	1,1	−1, −1	0,0	1,1
30	$\sigma_1\tau_4\sigma_6$	1,1	−1, −1	0,1	1, −1
31	$\sigma_1\tau_5\sigma_1$	−1,1	0, −1	0, −1	1,0
32	$\sigma_1\tau_5\sigma_3$	−1,1	0, −1	0,1	1, −1
33	$\sigma_1\tau_5\sigma_5$	−1,1	0, −1	0,0	1,1
34	$\sigma_1\tau_6\sigma_2$	1,0	−1,1	0,1	1, −1
35	$\sigma_1\tau_6\sigma_4$	1,0	−1,1	0, −1	1,0
36	$\sigma_1\tau_6\sigma_6$	1,0	−1,1	0,0	1,1

of polymorphism in kaolinite-type minerals has been made by Newnham (1961), who noted 36 different possible relative arrangements of two layers, differing in displacement and rotation of the layers. However, he limited his examination to structures with a repeat unit of not more than two layers, and also excluded 24 of the versions because of certain energy considerations.

The analytical method of representing a layer silicate crystal structure can be used to examine the entire range of two-storied layer minerals.

As already mentioned, the two-dimensional networks within a layer, and adjacent networks of neighboring layers, can be displaced relative to each other in six different ways. Displacements of both types (σ_i and τ_k) are, generally speaking, independent, so that any arbitrary structure will be defined by a sequence of values of σ and τ. From this sequence, the relative orientation of any two networks in different layers can be derived, and this information is essential in considering possible unit cells, deciding symmetry, and evaluating the uniformity and energetic stability of a structure.

Let the direction of the a axis of the whole structure coincide with the direction a_{2n+1} (e.g., a_3 in Fig. 11). Then the octahedral network of the upper layer can be displaced relative to the tetrahedral network of the lower layer by the amounts $\tau_0, \tau_+, \tau_-, \tau_1, \tau_3, \tau_5$ for the layer σ_{2n+1}, or by $\tau_0, \tau_+, \tau_-, \tau_2, \tau_4, \tau_6$ for the layer σ_{2n}. Whatever the preceding layer, six displacements are possible, three being the displacements τ_0, τ_+, τ_-, and three being odd if the next layer is odd and even if it is even.

In deriving the possible regular structures, we shall only consider sequences of networks and layers which possess the properties of periodicity and uniformity.

All the possible relative orientations of the layers can be derived by combining any arbitrary layer σ_i with all other σ_j using all possible values of τ_k. It will be seen that there are 36 such combinations possible.

Each such combination $\sigma_i \tau_k \sigma_j$ has its networks displaced by a particular amount, which we will call $\Delta_{M_1-M_2}$ (M_1 denotes cations of the upper layer, M_2 those of the lower layer):

$$\Delta_{Al-Si} = \tau_k, \quad \Delta_{Al-Al} = \sigma_i + \tau_k, \quad \Delta_{Si-Si} = \tau_k + \sigma_j,$$
$$\Delta_{Si-Al} = \sigma_i + \tau_k + \sigma_j. \tag{14}$$

It is easy to see that for a Δ with the components (0, 0) the cations of the layers in question will all superimpose one on the other in a normal projection on the ab plane; for $\Delta(0, \overset{\pm}{1})$ only half these cations will superimpose, and for a Δ, the first component of which differs from zero, there are no such superimpositions.

From Table 4, which lists all combinations when $\sigma_i = \sigma_1$, and the corresponding values of Δ, 12 groups of combinations can be distinguished, differing in the geometrical arrangement of their cations.

The condition of uniformity permits a structure to be made up only of combinations from the same group. Table 5 shows the distribution of these layer combinations among the groups. Symbols of the type 1(Si ↔ Si) and $\frac{1}{2}$(Al ↔ Si) show which cations are superimposed, and in what proportions, in a normal projection on the ab plane.

Transitions from one layer to another may be considered as equivalent, either if they are completely identical, i.e., they consist of identical relative displacements τ or relative layer rotations $\Delta\varphi$, or if the successive transitions are in opposite senses and the rotations $\Delta\varphi$ and $-\Delta\varphi$ alternate in the structure. As can be seen from Fig. 11, possible values of $\Delta\varphi$ are multiples of $2\pi/6$.

In addition, a sign of uniformity in a structure is when alternating layers σ and displacements τ are equivalent. This is the case when σ and τ are related by the operations of reflection in the axis a (i = 1, 5; 2, 4), b (i = 1, 2; 4, 5) or rotation by $\Delta\varphi = \pi$ (i = 1, 4; 2, 5; 3, 6).

All the regular structures which are obtained from the above examination are listed in Table 6. For each structure we give the sequence of combinations, its symbol in terms of σ, τ, the number of layers in the repeat unit, the projection of the c axis on the ab plane, the coordinates of the origin of the whole structure relative to the origin of the first layer, the space group, and a modification symbol for that particular structural modification. To avoid a long entry when the repeat unit is more than two layers, in structures with screw axes, only the first link of the chain is given, the others being obtainable by cyclic rearrangement.

From Table 6 it follows that some of the structures are enantiomorphic (I,1 and I,2; IV,1 and IV,2; etc.), in particular those differing only in the direction of rotation of the layers, and some of those with different positions of the rotation axis relative to the layers of the structure, as shown by the screw-axis term in the space group symbol and by the coordinates of the origin of the whole structure lying on this axis.

Table 5. Groups of Dioctahedral Layer Combinations

Diocta-hedral group	Group characteristics	Combination Nos.	Triocta-hedral group*
I	$1/2$ (Al $\leftrightarrow$ Si)	7 *, 9, 11, 13 *,15, 17	I'
II	$1/2$ (Si $\leftrightarrow$ Si)	23 *, 24, 28, 30, 34, 35 *	V'
III	1 (Si $\leftrightarrow$ Si)	22, 29 *, 36	VI'
IV	$1/2$ (Si $\leftrightarrow$ Si), $1/2$ (Al $\leftrightarrow$ Al)	25 *, 27, 31 *, 32	II'
V	$1/2$ (Al $\leftrightarrow$ Si), $1/2$ (Si $\leftrightarrow$ Al)	10 *, 12, 14, 16 *	III '
VI	1 (Al $\leftrightarrow$ Si)	1 *, 3, 5	I'
VII	$1/2$ (Al $\leftrightarrow$ Si), 1 (Si $\leftrightarrow$ Al)	8, 18	
VIII	$1/2$ (Si $\leftrightarrow$ Si), 1 (Al $\leftrightarrow$ Al)	20, 21	
IX	1 (Si $\leftrightarrow$ Si), $1/2$ (Al $\leftrightarrow$ Al)	26, 33	
X	1 (Al $\leftrightarrow$ Si), $1/2$ (Si $\leftrightarrow$ Al)	2,6	
XI	1 (Si $\leftrightarrow$ Si), 1 (Al $\leftrightarrow$ Al)	19 *	IV'
XII	1 (Al $\leftrightarrow$ Si), 1 (Si $\leftrightarrow$ Al)	4 *	III '

*See Tables 7 and 8 (p. 39).

The two-storied layer structure modifications can be assigned symbols analogous to those for the mica modifications (Smith, Yoder, 1956). The letters Tk, M, T, H, O indicate the symmetry class (triclinic, monoclinic, trigonal, hexagonal, or orthorhombic), and the number before the letter gives the number of layers in the repeat unit. The subscripts 1 and 2 on letters relate to structures with a monoclinic a or b axis, respectively (for uniformity of presentation and convenience in comparison, the same cell base ab is assumed for all structures). If the unit cell in these modifications is orthorhombic, a prime is added to its symbol. In addition, a monoclinic modification with an orthorhombic cell and with a twofold screw axis perpendicular to the face ab is denoted by the subscript 3. It should be explained that a symbol of the type used by Smith and Yoder (1956) does not define a unique structure in our usage. It can include a number of structures, each of which may be distinguished and defined uniquely by its symbol in terms of σ and τ.

It should be noted that some of the features of real layers by which they differ from the idealized scheme (Fig. 11) are not equally favorable to the formation of different structures.

Thus, when the tetrahedral networks are ditrigonal, corresponding to cubic packing of the anions in a layer (tetrahedron faces and upper faces of octahedra having opposite orientations), the hydrogen bonds are lengthened in some modifications and shortened in others. A check shows that for layers

Table 6. Regular Structures Made Up of Two-Storied Layers

Dioctahedral group	No.	Sequence of layer combinations	Structure symbol	No. of layers in repeat unit	c_x, c_y	x_0, y_0	Space group	Modification
I	1	7.7.7	$\sigma_2\tau_-\sigma_2$	1	$-1/3,0$	$0,0$	$P1$	$1T\kappa_1$
	2*	13.13	$\sigma_4\tau_+\sigma_4$					
	3*	7.13	$\sigma_3\tau_+\sigma_3\tau_-\sigma_3$	2	$-1/3,0$	$0,-1/3$	Cc	$2M_1$
	4	11.15	$\sigma_1\tau_+\sigma_3\tau_-\sigma_1$	2	$-1/3,0$	$0,-1/3$	Cc	$2M_1$
	5	17.9	$\sigma_1\tau_-\sigma_3\tau_+\sigma_1$	2	$-1/3,0$	$0,0$	Cc	$2M_1$
	6	9.9.9	$\sigma_3\tau_-\sigma_5\cdots$	3	$0,0$	$0,-1/9$	$P3_1$	$3T$
	7	15.15.15	$\sigma_3\tau_-\sigma_5\cdots$	3	$0,0$	$-1/6,1/18$	$P3_1$	$3T$
	8	11.11	$\sigma_3\tau_+\sigma_1\cdots$	3	$0,0$	$-1/6,-1/18$	$P3_2$	$3T$
	9	17.17	$\sigma_3\tau_-\sigma_1\cdots$	3	$0,0$	$0,1/9$	$P3_2$	$3T$
II	1	28.30	$\sigma_4\tau_1\sigma_5\tau_2\sigma_4$	2	$0,-1/3$	$-1/3,0$	Cc	$2M_2$
	2	34.24	$\sigma_1\tau_6\sigma_2\tau_3\sigma_1$	2	$0,-1/3$	$1/3,0$	Cc	$2M_2$
	3*	23.35	$\sigma_3\tau_4\sigma_6\tau_5\sigma_3$	2	$0,-1/3$	$1/3,0$	$Cc\ (R3c)$	$2M_2\,(6T)$
	4*	23.23	$\sigma_1\tau_5\sigma_4\tau_5\sigma_1$	2		$1/3,0$	$P2_1(P6_3)$	$2M_3\,(2H)$
	5*	35.35	$\sigma_5\tau_4\sigma_2\tau_1\sigma_5$	2	$0,0$	$1/3,0$	$P2_1(P6_3)$	$2M_3\,(2H)$
	6	28.28	$\sigma_3\tau_6\sigma_4\cdots$	6	$0,0$	$1/3,\ 1/3$	$P6_1$	$6H$
	7	34.34	$\sigma_2\tau_3\sigma_4\cdots$	6	$0,0$	$1/3,\ -1/3$	$P6_1$	$6H$
	8	24.24	$\sigma_3\tau_4\sigma_2\cdots$	6	$0,0$	$1/3,\ 1/3$	$P6_5$	$6H$
	9	30.30	$\sigma_3\tau_6\sigma_2\cdots$	6	$0,0$	$1/3,\ -1,3$	$P6_5$	$6H$
III	1	22.36	$\sigma_1\tau_9\sigma_2\tau_1\sigma_1$	2	$0,0$	$1/3,0$	Cc	$2M_2$
	2*	29.29	$\sigma_1\tau_6\sigma_6\tau_3\sigma_3$	2	$0,0$	$1/3,0$	$Ccm2_1\,(P6_3cm)$	$2O\,(2H)$
	3	22.22	$\sigma_3\tau_4\sigma_4\cdots$	6	$0,0$	$1/3,0$	$P6_1$	$6H$
	4	36.36	$\sigma_3\tau_2\sigma_2\cdots$	6	$0,0$	$1/3,0$	$P6_5$	$6H$
IV	1*	25.25	$\sigma_6\tau_2\sigma_6$	1	$0,-1/3$	$0,0$	$P1\,(R3)$	$1T\kappa_2(3T)$
	2*	31.31	$\sigma_3\tau_1\sigma_3$	1	$0,-13$	$0,0$	$P1\,(P3)$	$1T\kappa_2(3T)$
	3*	25.31	$\sigma_3\tau_5\sigma_3\tau_1\sigma_3$	2	$0,0$	$0,-1/3$	$Cc(P31c)$	$2M_1'(2T)$
	4	27.32	$\sigma_1\tau_3\sigma_5\tau_8\sigma_1$	2	$0,0$	$0,-1/3$	Cc	$2M_1$
	5	32.32	$\sigma_3\tau_1\sigma_5\cdots$	3	$0,0$	$-1/3,-1/3$	$P3_1$	$3T$
	6	27.27	$\sigma_3\tau_5\sigma_1\cdots$	3	$0,0$	$-1/3,\ 1/3$	$P3_2$	$3T$

Table 6 (continued)

Diocta-hedral group	No.	Sequence of layer com-binations	Structure symbol	No. of layers in repeat unit	c_x, c_y	x_0, y_0	Space group	Modifi-cation
V	1	14.12	$\sigma_1\tau_-\sigma_2\tau_-\sigma_1$	2	0,0	$-1/3, 0$	Cc	$2M'_2$
	2*	10.10	$\sigma_3\tau_-\sigma_6\tau_-\sigma_3$	2	0,0	$-1/3, -1/3$	$P2_1$	$2M_3$
	3**	16.16	$\sigma_3\tau_-\sigma_6\tau_+\sigma_3$	2	0,0	$-1/3, 1/3$	$P2_1$	$2M_3$
	4*	10.16	$\sigma_3\tau_+\sigma_6\tau_+\sigma_3$	2	$0, -1/3$	$-1/3, 0$	Cc	$2M_2$
	5	14.14	$\sigma_3\tau_-\sigma_4\cdots$	6	0,0	$-1/3, 0$	$P6_1$	$6H$
	6	12.12	$\sigma_3\tau_+\sigma_2\cdots$	6	0,0	$-1/3, 0$	$P6_5$	$6H$
VI	1*	1.1.1	$\sigma_6\tau\sigma_6$	1	$-1/3, 0$	$0, 0$	Cm	$1M$
	2	5.3	$\sigma_1\tau\sigma_5\tau_0\sigma_1$	2	$-1/3, 0$	$0, -1/3$	Cc	$2M_1$
	3	3.3.3	$\sigma_3\tau_0\sigma_5\cdots$	3	0,0	$-1/3, -1/9$	$P3_1$	$3T$
	4	5.5.5	$\sigma_3\tau_0\sigma_1\cdots$	3	0,0	$-1/3, 1/9$	$P3_2$	$3T$
VII	1	18.8	$\sigma_5\tau_-\sigma_4\tau_-\sigma_1$	2	$0, -1/3$	$0, -1/3$	Cc	$2M_2$
	2	8.8.8	$\sigma_3\tau_+\sigma_4\cdots$	6	0,0	$-1/3, 1/3$	$P6_1$	$6H$
	3	18.18	$\sigma_3\tau_-\sigma_2\cdots$	6	0,0	$-1/3, -1/3$	$P6_5$	$6H$
VIII	1	21.20	$\sigma_1\tau_1\sigma_5\tau_5\sigma_1$	2	0,0	$0, 0$	Cc	$2M_1'$
	2	20.20	$\sigma_3\tau_5\sigma_5\cdots$	3	0,0	$0, 0$	$P3_1$	$3T$
	3	21.21	$\sigma_3\tau_3\sigma_1\cdots$	3	0,0	$0, 0$	$P3_2$	$3T$
IX	1	33.26	$\sigma_1\tau_5\sigma_5\tau_1\sigma_1$	2	0,0	$0, 1/3$	Cc	$2M_1'$
	2	26.26	$\sigma_3\tau_5\sigma_5\cdots$	3	0,0	$1/3, -1/3$	$P3_1$	$3T$
	3	33.33	$\sigma_3\tau_1\sigma_1\cdots$	3	0,0	$1/3, 1/3$	$P3_2$	$3T$

Table 6 (continued)

Diocta-hedral group	No.	Sequence of layer com-binations	Structure symbol	No. of layers in repeat unit	c_x, c_y	x_0, y_0	Space group	Modifi-cation
X	1	2.6	$\sigma_1\tau_0\sigma_2\tau_0\sigma_1$	2	0,−1/3	−1/3, 0	Cc	$2M_2$
	2	2.2.2	$\sigma_3\tau_0\sigma_4\cdots$	6	0,0	−1/3, −1/3	$P6_1$	6H
	3	6.6.6	$\sigma_3\tau_0\sigma_2\cdots$	6	0,0	−1/3, 1/3	$P6_5$	6H
XI	1*	19.19	$\sigma_3\tau_3\sigma_3$	1	0,0	0,0(1/3,0)	$Cm(P31m)$	1M′ (1T)
XII	1*	4.4	$\sigma_3\tau_0\sigma_6\tau_0\sigma_3$	2	0,0	−1/3, 0	$Ccm2_1$	2O

Table 7. Trioctahedral Layer Combinations

No.	Layer combination	Dioctahedral combination Nos.	Dioctahedral group Nos.	Trioctahedral group No.
1	$\sigma_3 \tau_0 \sigma_3$	1, 9, 17	I, VI	I'
2	$\sigma_3 \tau_+ \sigma_3$	5, 7, 15		
3	$\sigma_3 \tau_- \sigma_3$	3, 11, 13		
4	$\sigma_3 \tau_1 \sigma_3$	31, 20, 27	IV, VIII	II'
5	$\sigma_3 \tau_3 \sigma_3$	33, 19, 26	IX, XI	IV'
6	$\sigma_3 \tau_5 \sigma_3$	32, 21, 25	IV, VIII	II'
7	$\sigma_3 \tau_0 \sigma_6$	4, 8, 18	V, VII, X, XII	III'
8	$\sigma_3 \tau_+ \sigma_6$	6, 10, 14		
9	$\sigma_3 \tau_- \sigma_6$	2, 12, 16		
10	$\sigma_3 \tau_2 \sigma_6$	35, 24, 28	II	V'
11	$\sigma_3 \tau_4 \sigma_6$	34, 23, 30		
12	$\sigma_3 \tau_6 \sigma_6$	36, 22, 29	III	VI'

Table 8. Groups of Trioctahedral Layer Combinations

Trioctahedral group No.	Combination Nos.	Group property
I'	1, 2, 3	1(Mg→Si)
II'	4, 6	$\frac{1}{2}$(Si↔Si), 1(Mg↔Mg)
III'	7, 8, 9	1(Mg↔Si), 1(Si↔Mg)
IV'	5	1(Mg↔Mg), 1(Si↔Si)
V'	10, 11	$\frac{1}{2}$(Si↔Si)
VI'	12	1(Si↔Si)

which are both either even or odd, only the displacements τ_0, τ_+, τ_- are favorable, and with one odd and one even layer, the only favorable displacements are those which are even if the upper layer is even, and odd if it is odd. These conditions rule out groups IV, V, and VII-XII in Table 6, i.e., half the total number of structures.

In addition, the O_{tetr} atoms and OH groups of the lower faces of the octahedra lying on the planes of symmetry of individual layers are somewhat depressed (Newnham, Brindley, 1956; Newnham, 1961). The requirement that the projections from the "bed" of the upper layer fit into depressions in the "roof" of the lower layer picks out, moreover, only the groups $\sigma_3 \tau_0 \sigma_3(1)$,

$\sigma_3 \tau_+ \sigma_1$ (11), $\sigma_3 \tau_- \sigma_5$ (15), $\sigma_3 \tau_6 \sigma_6$ (29), $\sigma_3 \tau_4 \sigma_2$ (24), $\sigma_3 \tau_2 \sigma_4$ (34) as being favorable, leaving eight structures in Table 6.

In agreement with these conditions, and as pointed out by Newnham (1961), the structures of dickite and nacrite correspond to the sequences I,4 and II,2 (Table 6). The latter condition is not obeyed in the structures of triclinic kaolinite (I,1 and I,2). In the opinion of Newnham (1961), a structure made up of the combinations (1) and (29) is energetically unfavorable because of the relative positions of the cations in this structure (Table 5).

When compared with the other minerals of the kaolinite group, the halloysites are a very special case, even though their structures contain the same two-storied dioctahedral layers. Their biggest difference is in the association of the two-storied layers with water molecules, which exist or existed in greater or lesser amounts in the interlayer spaces, a distinction being drawn between halloysites, and hydrated, anhydrous, and intermediate so-called metahalloysites. In addition, their crystals have a peculiar elongated tubular shape and distinctive optical constants and thermal characteristics.

As well as a variation in interlayer distance with water content, halloysite was accepted as having an extremely low degree of ordering in the arrangement of its layers, which were thought to be randomly displaced relative to each other by arbitrary distances along both the a and b axes, with simultaneous variations in their azimuthal orientations. This picture has, however, turned out to be inaccurate and in disagreement with new data.

Thus, electron-diffraction texture patterns (Zvyagin, 1954) have shown that the anhydrous halloysites, at any rate, have a high degree of ordering in their structures, as was indicated by their X-ray properties.

A qualitatively new result was obtained in a study of individual halloysite crystals by the microdiffraction method using an electron microscope (Honjo, Mihama, 1954; Kulbicki, 1954). The arrangement of the reflections on the patterns obtained by this method corresponded to a whole set of orientations of the crystal lattice around the direction of elongation of the particles, and indicated a true repeat unit of two layers. Specimens have been found in nature which have shown the structural features described, not only in electron-diffraction patterns from individual tubular crystals, but also in X-ray powder photographs; this was first pointed out by Japanese workers (Honjo, Kitamura, Mihama, 1954). It is true that these authors did not decide to relate the specimens that they studied to the halloysites, and cautiously called them tubular kaolins. There are completely satisfactory grounds for assuming that these specimens were, structurally, highly perfect halloysites, since, quite apart from the X-ray diffraction results, electron diffraction patterns of halloysites invariably possess the features noted above.

In their studies of "tubular kaolins," the Japanese authors (Honjo et al., 1954) arrived at a structure made up of two-storied layers, repeating every two layers, which theoretically approximated the actual diffraction data. In deriving this structure, the authors did not consider all possible azimuthal orientations of the layers, did not analyze the crystallochemical properties of their model, and were not convinced that their model was better than all others. The rules governing the construction of this model correspond to the two enantiomorphic structures $\sigma_4 \tau_- \sigma_4 \tau_0 \sigma_4$ and $\sigma_2 \tau_+ \sigma_2 \tau_0 \sigma_2$, with $\beta = 83°$. After rotating the system of coordinates by 180° about the normal to the layers, the two structures have the analytical formulas $\sigma_1 \tau_+ \sigma_1 \tau_0 \sigma_1$ and $\sigma_5 \tau_- \sigma_5 \tau_0 \sigma_5$, with $\beta = 97°$, and possess triclinic symmetry, since they have only translational symmetry elements. From their analytical formulas it follows also that the nonequivalent layer combinations $\sigma_1 \tau_+ \sigma_1$ and $\sigma_1 \tau_0 \sigma_1$ alternate in these structures, which is in conflict with the condition of uniformity. It is difficult to find justification for two σ_1 layers attaching themselves to the same layer (also σ_1) according to both the $\tau_+ \sigma_1$ and the $\tau_0 \sigma_1$ rule. This is even more incomprehensible in that the two types of combinations lead to different relative positions of the cation in adjacent layers. In the first combination, the cations are arranged as in kaolinite, so that half the Al atoms of a given layer are superimposed on Si atoms of the following layer in a projection on the ab plane. In the second combination, all Al atoms are superimposed on corresponding Si atoms, and so they experience a considerably higher electrostatic repulsion than in the first combination. Such a combination cannot be stable, and no kaolinite-type structure with this combination is known. If it were assumed that under some special circumstances such a structure could be formed, it would be impossible to produce a mechanism whereby, in the same structure, two combinations differing in geometry and energetic stability could arise and exist in regular alternation with one another. Because of the presence of unusual layer combinations in the structure, there is no justification for relating it to kaolinite or calling the material in question "tubular kaolin."

It will be demonstrated below (Chapter 4) that the structure $\sigma_3 \tau_+ \sigma_3 \tau_- \sigma_3$, derived from an examination of polymorphism and polytypism in kaolinite-type minerals, also satisfactorily agrees with the experimental data. This structure consists of the geometrically and energetically equivalent layer combinations $\sigma_3 \tau_+ \sigma_3$ and $\sigma_3 \tau_- \sigma_3$, which themselves represent two separate enantio-

Table 9. Regular Structures Made Up of Trioctahedral Two-Storied Layers, and Their Distinguishing Features

Structure No.	Structure symbol	Modification	Structure type*	Space group	Layer comb. seq.	c_x, c_y	$x_0 y_0$	Characteristics of corresponding dioctahedral groups and modifications		
								Group nos.	Modification	Space group
1	$\sigma_6\tau_0\sigma_6$	1M		Cm	1.1.1	$-\tfrac{1}{3},0$	$0,0$	VI,1	1M	Cm
2	$\sigma_3\tau_+\sigma_3\tau_-\sigma_3$	2M$_1$		Cc	2.3.2,3	$-\tfrac{1}{3},0$	$0,-\tfrac{1}{3}$	I,3	2M$_1$	Cc
3	$\sigma_3\tau_0\sigma_3\tau_+\sigma_3\tau_-\sigma_3$	3T	B	P3$_2$	1.2.3	$0,0$	$0,\tfrac{1}{9}$	–		P1
4	$\sigma_3\tau_0\sigma_3\tau_-\sigma_3\tau_+\sigma_3$	3T		P3$_1$	1.3.2	$0,0$	$0,-\tfrac{1}{9}$	–		P1
1	$\sigma_3\tau_1\sigma_3$	3T		R3	4.4.4	$0,0$	$0,0$	IV,2	1Tk$_2$	P1
2	$\sigma_6\tau_2\sigma_6$	3T	A	R3	6.6.6	$0,0$	$0,0$	IV,1	1Tk$_2$	P1
3	$\sigma_3\tau_1\sigma_3\tau_5\sigma_3$	2T		P31C	4.6.4.6	$0,0$	$\tfrac{1}{3},\tfrac{1}{3}$	IV,3	2M$_1$	Cc
1	$\sigma_3\tau_0\sigma_6\tau_0\sigma_3$	2O		Ccm2$_1$	7.7.7	$0,0$	$-\tfrac{1}{3},0$	XII,1	2O	Ccm2$_1$
2	$\sigma_3\tau_+\sigma_6\tau_-\sigma_3$	2M$_3$		P2$_1$	8.8.8	$0,0$	$-\tfrac{1}{3},-\tfrac{1}{3}$	V,2	2M$_3$	P2$_1$
3	$\sigma_3\tau_-\sigma_6\tau_+\sigma_3$	2M$_3$	C	P2$_1$	9.9.9	$0,0$	$-\tfrac{1}{3},-\tfrac{1}{3}$	V,3	2M$_3$	P2$_1$
4	$\sigma_3\tau_+\sigma_6\tau_+\sigma_3$	2M$_2$		Cc	8.9.8.9	$0,-\tfrac{1}{3}$	$-\tfrac{1}{3},-\tfrac{1}{3}$	V,4	2M$_2$	Cc
5	$\sigma_3\tau_0\sigma_6\tau_-\sigma_3\tau_-\sigma_6\tau_0\sigma_3\tau_+\sigma_3$	6H		P6$_5$	7.8.9	$0,0$	$-\tfrac{1}{3},-\tfrac{1}{3}$	–		
6	$\sigma_3\tau_0\sigma_6\tau_+\sigma_3\tau_+\sigma_6\tau_0\sigma_3\tau_-\sigma_6\tau_-\sigma_3$	6H		P6$_1$	7.9.8	$0,0$	$\tfrac{1}{3},\tfrac{1}{3}$	–		
1	$\sigma_3\tau_3\sigma_3$	1T	A	P31m	5.5.5	$0,0$	$\tfrac{1}{3},0$	XI,1	1M*	Cm
1	$\sigma_3\tau_4\sigma_6\tau_5\sigma_3$	6T		R3c	11.10	$0,0$	$\tfrac{1}{3},0)$	II,3	2M$_2$	Cc
2	$\sigma_3\tau_2\sigma_6\tau_5\sigma_3$	2H	D	P6$_3$	10.10	$0,0$	$\tfrac{1}{3},\tfrac{1}{3}$	II,5	2M$_3$	P2$_1$
3	$\sigma_3\tau_4\sigma_6\tau_1\sigma_3$	2H		P6$_3$	11.11	$0,0$	$\tfrac{1}{3},\tfrac{1}{3}$	II,4	2M$_3$	P2$_1$
1	$\sigma_3\tau_6\sigma_6\tau_3\sigma_3$	2H	D	P6$_3$cm	12.12	$0,0$	$\tfrac{1}{3},0$	III,2	2O	Ccm2$_1$

*Structures belonging to the same type give identical reflections for $k = 3k'$ (see Chapter 4).

morphic modifications of kaolinite. From the energetic-stability viewpoint, the existence of such a structure in nature would be justified just as much as would the existence of kaolinite, and the structure could, with some truth, be related to that of kaolinite, although as far as geometrical considerations and diffraction characteristics are concerned there would be no special point in this.

It might be foreseen that, because of the identical orientation of all layers and the high degree of equivalence of successive layer combinations, a breakdown in the regular alternation of the displacements τ_+ and τ_- would be highly probable. This is in good agreement with observations of imperfections in the halloysite structures, breakdown in these being stimulated by the effect of interlayer water.

It must be assumed that the limitations on structural varieties mentioned above may not be obeyed in the trioctahedral varieties, particularly where there is loss of ordering and formation of structure defects.

Trioctahedral layers have a plane of symmetry parallel to the a axis in all six azimuthal orientations, differing by rotations which are multiples of $2\pi/6$, for all layer sequences. If the origin of each layer is taken at the octahedron center lying on this plane of symmetry, it can therefore be considered that any sequence will be made up of the layers σ_3 and σ_6 only.

In order to obtain all combinations of layer pairs in this case, it is necessary in Table 4 to take the first layer as being σ_3, and substitute for the subscripts of τ and σ in the combinations $\sigma_3 \tau_j \sigma_k$. If, here, the layer σ_k is examined as if it were σ_3 or σ_6, depending on whether it is odd or even, then at the same time τ_j must be modified so that the relative displacements of the tetrahedral networks will be the same as they were before. The new values, τ_j', are obtained from the relationship

$$\tau_j + \sigma_{2n+1,2n} = \tau_j' + \sigma_{3,6}$$

or

$$\tau_j' = \tau_j + \Delta\tau, \quad \Delta\tau = \sigma_{2n+1,2n} - \sigma_{3,6}.$$

The quantity $\Delta\tau$ is dependent on the layer σ_k, and its components are obtained immediately from Table 3. For this transformation the layer pair combinations of Table 4 are altered to those listed in Table 7.

In this case, only 12 layer pair combinations are possible, and only six groups (Table 8), differing in the relative dispositions of their cations (these groups are marked with a prime). The decrease in the number of groups comes about because now a displacement of an octahedral network by $\pm b/3$ does not give rise to a different arrangement of its cations relative to the cations of another layer. By combining the combinations belonging to the same group we can derive all the possible regular trioctahedral structures. These structures are listed in Table 9.

The majority of the structures have analogs among the dioctahedral varieties (in Table 6 these are marked with an asterisk). In Table 9, the trioctahedral structures are compared with the dioctahedral structures which transform into the former on filling their vacant octahedra. It is clearly seen how an increase in layer symmetry leads, in some cases, to an increase in symmetry of the whole structure, accompanied by the transformation of modifications and formation of new modifications not present among the dioctahedral structures. The modification 1M' is transformed into 1T (lizardite), $1Tk_2$ into 3T, $2M_1'$ into 2T, some of the $2M_2$ structures into 6T, and some $2M_3$ into 2H. In addition, it appears possible to join together further combinations, giving new structures (the structures I', 3,4; III',5,6).

In a recently published article, Steadmen (1964) has given a derivation of trioctahedral kaolinite-type structures, based on somewhat different principles. The structures are described in terms of a sequence of operations made up of displacements S (measured between hexagon centers of tetrahedral networks of adjacent layers), and rotations of layers R by 180° about a normal to the layers. In the derivation of regular structures, certain empirical conditions were imposed on the sequence, these being in part more restrictive than those imposed by the uniformity condition, and in part not satisfying this condition. As a result, the structures derived in this way omit the six structures of type C, while eight structures contain layer pair combinations, belonging to groups II', and IV', V' and VI', which differ in the relative configurations of their cations. These structures do not possess the property of uniformity.

6. Regular Structures of Three-Storied Mica-Type Layers

Various mica modifications have been detected in nature and produced by artificial means. A number of experimental and theoretical studies have been devoted to their examination (Hendricks, Jefferson, 1939; Belov, 1949; Yoder, Eugster, 1955; Smith, Yoder, 1956; Radoslovich, 1958).

It has already been noted that the concept of a three-storied mica-type layer has more than one meaning. Each of its networks of tetrahedra or octahedra can have two opposite orientations, and each pair of adjacent networks can match up in three different ways. Thus, for a given chemical composition we can foresee 72 layers with different structures. However, the condition that the anions have cubic packing fixes the orientation of each network and reduces the possible number of different layers to nine. Under rare circumstances the cubic packing rule may be broken, and this is reflected in the reversal of orientation of an octahedral network. The number of possible layers is therefore equal to 18 exactly. But this is correct only for dioctahedral layers. For trioctahedral layers, with all octahedra occupied, the three different combinations of a tetrahedral network with an octahedral one are indistinguishable. The layers can differ only in the relative arrangement of the tetrahedral networks, and for a given octahedron orientation three layers are possible, which are, moreover, rotationally identical and can be transformed into one another through rotations of $\pm 120°$.

Of the nine dioctahedral layers (with the orientation of the octahedral network fixed), only three layers (with $\sigma_g = \sigma_i$) have the same symmetry as the trioctahedral layers, while the other six (with $\sigma_g \neq \sigma_i$) can be divided into two sets of three rotationally identical layers, which do not have a plane of symmetry and are polar both parallel and perpendicular to the layers. Moreover, as already noted, they do not satisfy the condition of structure uniformity.

It can be assumed that if such asymmetric layers actually exist in real structures, then either they are rare and accidental, or because of their nature they form disordered structures. At least in the cases where a structure has a strict and fairly small layer repeat unit, it can be assumed that the structure consists of symmetrical layers of one or more of the three sorts differing in the relative positions of their tetrahedral networks. Our subsequent examination will therefore be limited to those layer sequence varieties which may be defined in terms of layers having planes and centers of symmetry, but which differ in their relative orientations.

These layers are characterized by the relative displacements of the $2\sigma_i$ tetrahedral networks, and in Fig. 12 these are shown in the positions they can adopt relative to the axis a_1. In formation of a layer structure, the layers are stacked one upon the other in such a way that the upper loops of tetrahedra of the lower layer and the lower loops of the upper layer form hollows in which monovalent cations, for example K, can be distributed. Here, the corresponding hexagons formed by the Si atoms coincide on a projection on the plane of the layers. Thus, in these structures, the displacements $\tau = 0$, and any structure can be defined in terms of a sequence of values of σ_i alone. The conditions of periodic repetition and uniformity are satisfied by only six regular

Table 10. Regular Structures of Three-Storied Layers

No.	Layer sequence	n	c_x, c_y	x_0, y_0	Modification	Space group
1	$\cdots \sigma_3\sigma_3\sigma_3 \cdots$	1	−1/3,0	0,0	1M	$C2/m$
2	$\cdots \sigma_2\sigma_4\sigma_2\sigma_4 \cdots$	2	−1/3,0	0,0	2M$_1$	$C2/c$
3	$\cdots \sigma_3\sigma_5\sigma_1 \cdots$	} 3	0,0	0,1/9	3T	$P3_112$
4	$\cdots \sigma_3\sigma_1\sigma_5 \cdots$			0, −1/9		$P3_212$
5	$\cdots \sigma_3\sigma_6\sigma_3\sigma_6 \cdots$	2	0,0	0,0	2O	$Cc\,mm$
6	$\cdots \sigma_5\sigma_4\sigma_5\sigma_4 \cdots$	2	0, −1/3	0,0	2M$_2$	$C2/c$
7	$\cdots \sigma_3\sigma_4\sigma_5\sigma_6\sigma_1\sigma_2 \cdots$	} 6	0,0	0,1/3	6H	$P6_122$
8	$\cdots \sigma_3\sigma_2\sigma_1\sigma_6\sigma_5\sigma_4 \cdots$			0, −1/3		$P6_522$

structures from all the three-storied layer sequences, and these and their properties are listed in Table 10 on the same basis as that in Table 6. Of these sequences, 1-3 have cubic packing of the anions (...ccc...), and 4-6 have the approximate eight-layer packing ...ccchhccc... (Belov, 1947).

Up to the present time, all these varieties have been detected experimentally except for 2O and 6H. Radoslovich (1958) has pointed out that the variety 2M$_2$ can be formed if the tetrahedral networks have hexagonal symmetry, when their loops have the shape of regular hexagons, and he has noted that this is most probable for biotites and phlogopites, because of their large values of a and b, and for lepidolites, because of their low degree of replacement of Si by Al. In the majority of cases, the tetrahedral loops have a ditrigonal shape, which rules out a combination in which the octahedra of adjacent layers have opposite orientations.

Thus, the greatest departure from close-packing is shown by the modifications 2M$_2$, and 6H and 2O. The faces of adjacent tetrahedra of neighboring layers form trigonal prisms, bounded by hexagonal prisms with monovalent cations (K), and this arrangement can only be described as eight-layer stacking by a fair stretch of the imagination.

Structures can be envisioned in which the layers σ_3, σ_6 regularly play a special role in comparison to the layers $\sigma_1, \sigma_5, \sigma_2, \sigma_4$. In this case a structure of the type $\sigma_3\sigma_1\sigma_3\sigma_5\sigma_3\sigma_1\sigma_3\sigma_5$ might be formed, with a repeat unit c having the components $-\frac{1}{3}, 0$ in a normal projection on the ab plane (modification 4M). Since the structural control here extends to pairs of layers, this is conducive to the breakdown of such a sequence, and an eight-layer structure of type $\sigma_6\sigma_4\sigma_6\sigma_2\sigma_6\sigma_2\sigma_6\sigma_4...$, $c\left(-\frac{1}{3}, 0\right)$, may be formed. Such a structure may possibly have been met with experimentally, but may have been taken as a 24-layer structure (Hendricks, Jefferson, 1939). Moreover, very diverse sequences can arise through multiple and regular repetition of the twinning process described by Belov (1949).

The expressiveness of this approach to the examination of layer sequences in micas helps to avoid errors and confusion in their interpretation. For example, the monoclinic unit cell of the modification $2M_2$ can formally be obtained from the sequences $...\sigma_3\sigma_2...$ or $...\sigma_1\sigma_6....$ It is, however, not difficult to show that these are rotationally identical to $...\sigma_5\sigma_4...$, while only the arrangement $...\sigma_5\sigma_4...$ corresponds to the space symmetry group $C_{2h}^6 = C2/c$ which is possessed by this sequence.

When guided by the reasoning developed here, it is easy to detect and correct mistakes, for example, those made by Hendricks and Jefferson (1939). Thus, from Fig. 6 of their article (1939), it is clear that their "monoclinic hemihedral six-layer" structure is none other than the layer sequence $...\sigma_1\sigma_2...$ i.e., the modification $2M_2$. In the same way, their "triclinic six-layer" structure (Fig. 7a of their article) emerges as the layer sequence $\sigma_4\sigma_6$, i.e., the modification $2M_1$ rotated through 120°.

7. Regular Structures Made Up of Chlorite Packets

Chlorites are made up of alternating three-storied mica-type layers and one-storied brucite-type layers (Pauling, 1930b; McMurchy, 1934; Belov,1950) (see Fig. 6). Under these conditions, an enormous variety of structural modifications are possible, differing in both the structures of individual layers and in their relative arrangement, even for a single chemical composition.

To a fair approximation, the chlorites can be considered trioctahedral,* and their three-storied layers possess symmetry centers, half of which lie at the centers of one-third of all the octahedra. In the three-storied layers, $\sigma_g = \sigma_i$ from either of these conditions, and the displacements of the tetrahedral networks, and with them the layers themselves, are given by the values of $2\sigma_i$ (see Fig. 12 and Table 3).

---

*Dioctahedral chlorites have also been observed experimentally (Brown,1961; Müller, 1963), but they are very rare, and do not have a strict c repeat unit. For these reasons they can be neglected in a derivation of regular structures.

As we have already noted, for a given orientation of a mica-type layer there are, in all, four possible methods of attaching a brucite-type layer (see Fig. 13). In three of the methods the cations of the brucite-type layer superimpose, in a normal projection on the ab plane, on the cations of the nearest network of the mica-type layer, and in only one case does this superimposition not occur. It may be assumed that the latter method of arranging the layers is energetically more favorable. Here, the octahedra of the two layers have opposite orientations, and the lower corners of the brucite-type-layer polyhedra project onto the centers of the octahedra of the mica-type layer.

This arrangement of the layers is the one actually found in structural studies of chlorites (McMurchy, 1934; Steinfink, 1958), and the normally observed regular distribution of reflection intensities with $k = 3k'$ shows that this applies to the great majority of chlorites. It should, however, be noted that, in addition to chlorites consisting of σ packets, the existence of structures made up of σ', $|\sigma'$, and $|\sigma'|$ packets has recently been demonstrated, but these are met with only rarely (Brown, Bailey, 1962). It is therefore convenient to derive first the regular structures made up of σ packets.

In structures made up of σ packets, the tendency to have a reduced O−OH distance, because of hydrogen bonding, must lead to a rotation of tetrahedron faces such that, in forming a ditrigonal arrangement, they are oriented oppositely to the nearest octahedron faces, in accordance with a cubic (but not close-packed) packing of the anions of the three-storied layer. The cubic packing does not extend throughout the whole structure, because adjacent layers turn out to be sandwiched between deformed empty trigonal prisms.

In connection with the "insensitiveness" of the structure to displacement of the one-storied layer by multiples of $b/3$, the orientation of this being fixed relative to the a axis, one packet in the structure may differ from the previous one (working from bottom to top to avoid confusion) in the type and position of its three-storied layer σ_k. For a given position of the lower faces of the one-storied layer, its upper faces coincide, in a normal projection on the ab plane, with the spaces between the upper faces of the octahedra of the lower three-storied layer, and the corners of these faces of both layers are superimposed on one another. To these corners are attached the corners of the faces of the lower tetrahedral network of the upper three-storied layer. It is easy to see that there are six possible methods of attachment here, which can be distinguished in terms of the displacements τ_j between centers of hexagonal loops of the Si atoms of lower-layer upper tetrahedra and upper-layer lower tetrahedra, as measured on a projection on the ab plane. These displacements coincide with the analogous quantities τ_j (j = 1, 2, ..., 6) used earlier for kaolinite (Table 3).

In actual fact, the energetically most favorable arrangement of cations of adjacent layers is compatible with only three of the displacements τ_j, which have the same oddness or evenness as the three-storied layer σ_k lying above them. The strict observance of this condition leads to the result that the layers σ_i, σ_k together with the displacements τ_j must be all odd or all even throughout the whole structure. This is confirmed by the experimental general feature noted above in the majority of chlorites, that the distribution of reflection intensities (with k = 3k') implies a chlorite packet repeat unit of one packet, and in particular that the h0l reflections indicate that these chlorites give identical structure projections on the ac plane.

As with kaolinites and micas, in deriving the possible structural modifications of chlorites it is reasonable to impose the conditions of periodic repetition and uniformity on the sequences of chlorite packets, described in terms of a sequence of the symbols σ_i, τ_j.

The controlling structural factors which decide a given sequence of layers σ and displacements τ may be tied up with the interaction between networks and layers which are near to each other. In the case we are considering, the tetrahedral networks of adjacent three-storied layers can be so described. Since they are separated by one-storied layers, the interaction between these networks is weak, and the condition of uniformity must mean either a constant or a random alternation of different values of σ and τ, all either odd or even. There is correspondingly formed either an ordered or a disordered chlorite structure.

The weak structural control must favor a breakdown in periodicity of the sequence of values of σ and τ, and in practice chlorites do not usually have a strict packet repeat unit. As a rule, strict regular repetition is not observed in polycrystalline specimens, and is apparently only characteristic of individual single crystals, which hardly show up diffraction-wise in polycrystalline specimens. Also, the possibility cannot be ruled out that, during the dispersion of the specimens in the preparation of powder or texture samples, there occurs a loss in the ordering of layers which might originally have existed in the single crystals.

If we take the orientation of the three-storied layers as σ_3, the possible ordered sequences have a repeat unit of one packet and, depending on τ, may be represented as $\sigma_3 \tau_1 \sigma_3$, $\sigma_3 \tau_3 \sigma_3$, or $\sigma_3 \tau_5 \sigma_3$.

Table 11. Ordered Chlorite-Packet Sequences Obeying the Uniformity Condition

No.	$\sigma_3 \tau_j \sigma_3$	c_x, c_y	$\Delta\varphi$	$\sigma_i \tau_k \sigma_i$	c_x, c_y	Space group	Modification
1	$\sigma_3 \tau_1 \sigma_3$	1/3, —1/3	60°	$\sigma_4 \tau_2 \sigma_4$		C_i	1Тк$_i$
2	$\sigma_3 \tau_5 \sigma_3$	1/3, 1/3	—60°	$\sigma_2 \tau_4 \sigma_2$	—1/3,0		1M
3	$\sigma_3 \tau_3 \sigma_3$	1/3,0	180°	$\sigma_6 \tau_6 \sigma_6$		$C2/m$	

Table 12. Independent Single-Packet Regular Chlorite Structures Consisting of Arbitrary Packets

No.	Brown and Bailey's symbol	Analytical symbol	Space group	c_x, c_y	Number of specimens
1	Ia-2	$\|\sigma_3'\|\tau_0\|\sigma_3'\|$	$C2/m$	—1/3,0	10
2	Ia-4	$\|\sigma_1'\|\tau_+\|\sigma_1'\|$	$C\bar{1}$	—1/3,0	
3	Ib-1	$\sigma_6'\tau_3\sigma_6'$	$C2/m$	0,0	37
4	Ib-3	$\sigma_6'\tau_1\sigma_6'$	$C\bar{1}$	0, —1/3	
5	Ib-2	$\|\sigma_6'\tau_6\|\sigma_6'$	Cm	—1/3,0	13
6	Ib-4	$\|\sigma_2'\tau_4\|\sigma_2'$	$C1$	—1/3,0	
7	IIa-1	$\|\sigma_3\|\tau_0\|\sigma_3\|$	$C2/m$	—1/3,0	0
8	IIa-3	$\|\sigma_1\|\tau_+\|\sigma_1\|$	$C\bar{1}$	—1/3,0	
9	IIa-2	$\sigma_6\|\tau_3\sigma_6\|$	Cm	0,0	0
10	IIa-4	$\sigma_6\|\tau_1\sigma_6\|$	$C1$	0, —1/3	
11	IIb-2	$\sigma_6\tau_6\sigma_6$	$C2/m$	—1/3,0	243
12	IIb-4	$\sigma_2\tau_4\sigma_2$	$C\bar{1}$	—1/3,0	

In the general case of a sequence of n packets, the displacement of the center of symmetry of the n-th layer relative to the center of symmetry of the first layer is equal to the sum of displacements

$$\sigma^{(1)} + \sigma^{(n)} + \sum_1^{n-1} \tau^{(m)} + 2 \sum_2^{n-1} \sigma^{(m)}. \tag{15}$$

In the three ordered structures given, the normal projection of the c axis on the ab plane will thus be equal to $2\sigma_3 + \tau_j$ (j = 1, 3, 5), and its components along the axes a and b differ from zero or are greater than zero. After rotation of the systems of coordinates about the normal to the packets by angles of 60°, 180°, and 300°, respectively, the axis c has components $(-\frac{1}{3},0)$, and forms a monoclinic unit cell with the axes a and b, with an obtuse angle β. Under these circumstances, σ and τ change their subscripts as shown in Table 11.

Both the enantiomorphic triclinic structures and the third, monoclinic structure have been detected experimentally by Brindley et al. (1950), and have been investigated in detail by Steinfink (1958).

In addition to the polymorphic structures discussed above, real chlorite structures may contain other packet sequences. Their formation may be due to factors of a special nature, such as structural control exerted by a peculiar distribution of isomorphic cations, propagation of a chance stratification of packets through the spiral growth mechanism, and formation of a chlorite through degeneration of another silicate which possessed a particular layer sequence.

Thus, as well as the single-packet monoclinic and triclinic modifications made up of σ packets, which satisfy the condition of uniformity (structures A and B), Brindley et al. (1950) detected two-packet (structure C) and three-packet (structure D) modifications. In our analytical notation C and D can be represented as $\sigma_5\tau_3\sigma_3\tau_1\sigma_6$ and $\sigma_3\tau_1\sigma_1\tau_5\sigma_1\tau_5\sigma_3$. Neither structure appears at first sight to satisfy the uniformity condition. It is not clear, for example, why in structure D the layer σ_1 is followed in one case by $\tau_5\sigma_1$, and in another by $\tau_5\sigma_3$. It is true that the authors noted one small regular structural feature, which was that each three-storied layer had its centers of symmetry displaced relative to those of the preceding layer along the direction of its own axis a_4 by a distance of $-a_4/3$ (see Fig. 11). This will be appreciated on recalling that the components of the displacements of adjacent layers can be written as $\sigma_i + \tau_j + \sigma_k$, and then changing from a σ_k system to a σ_3 system. However, it is difficult at the moment to understand why this should happen, and why the relationship observed is not found in single-packet structures.

Even more surprising results were obtained by Brown and Bailey (1962). In an examination of regular chlorite structures with a repeat unit of one packet, they did not apply the condition of uniformity or consider energetic stability limitations. Under these conditions, packets of four types with six permissible displacements τ of adjacent packets can, of course, give 24 single-packet structures. Certain of these structures are enantiomorphs, and others are equivalent to one another, related by rotations about the axes a and b. Because of this, only 12 of the structures can be considered independent. These structures are listed in Table 12. The distinguishing features noted in the table for these structures can be derived from their analytical symbols.

Table 13. Distribution of Layer Silicate Structure Modifications

Modification	Space group	Number of structures 1 : 1 Di-octahedral	Number of structures 1 : 1 Tri-octahedral	2 : 1	2 : 2
1 Tκ₁	$P1$	2	—	—	—
1 Tκ₂ (3T)	$P1$, $R3c$	2	2	—	—
1 Tκ$_i$	C_i	—	—	—	2
1 M	Cm, $C2/m$	1	1	1	1
1 M'(1T)	Cm, $P31m$	1	1	—	—
2 M₁	Cc, $C2/c$	4	1	1	—
2 M₁' (2T)	Cc, $P31c$	4	1	—	—
2 M₂ (6T)	Cc, $C2/c$, $R3c$	6	2	1	—
2 M₂'	Cc	2	—	—	—
2 M₃(2 H)	$P2_1$, $P6_3$	4	4	—	—
2 O	$Ccm2_1$, $Ccmm$	2	2	1	—
3 T	$P3_1$, $P3_112$ / $P3_2$, $P3_212$	} 12	—	} 2	—
3 T	$P3_1$ / $P3_2$	—	} 2	—	—
6 H	$P6_1$, $P6_122$ / $P6_5$, $P6_522$	} 12	—	} 2	—
6 H	$P6_1$ / $P6_6$	—	} 2	—	—
Total		52	18	8	3

The structures made up of $\sigma_{3,6}$ layers and $\tau_{0,3,6}$ displacements have planes of symmetry parallel to the a axis. The presence or absence of a symmetry center in the packets depends on whether the one-storied layers on either side of a three-storied layer are attached in the same or in different ways. From Table 12 it can be seen that the packets of structures Nos. 5, 6, 9, and 10 are polar in this respect, and do not satisfy the uniformity condition. The analytical symbols indicate directly on which cations, and from what side of the three-storied layer, the cations of a one-storied layer are superimposed in a projection of the packet on the ab plane. It should be borne in mind that, in structures with displacements of τ_0, the tetrahedral networks and, consequently, also their cations in adjacent layers, are superimposed on one another in a projection on the ab plane and may repel each other in the corresponding structures. For the displacements $\tau_\pm$, the amount of superimposition is halved, and for τ_i ($i = 1,...,6$) it is completely absent.

Another factor affecting the energetic favorability of a structure is connected with the lengths of the O−OH bonds. Because of the rotation of the tetrahedra in formation of a ditrigonal pattern, these bonds are shortened in $\sigma, \sigma'|$ combinations of a one-storied and a three-storied layer, and are lengthened in $\sigma', \sigma|$ combinations.

In the last column of Table 12, the number of different chlorite specimens corresponding to a particular structure are listed, as determined by Brown and Bailey, partly from literature evidence, but mainly from their own investigations. The structures tied together with curly brackets mostly do not have strict c repeat units, and cannot be distinguished experimentally. From the viewpoint of the relative arrangement of cations and lengths of O−OH bonds, it is quite understandable why structures Nos. 11 and 12 predominate, since these correspond to structures given in Table 11. The second-highest number of specimens is shown for structures 3 and 4, which, although having a favorable cation distribution, have unfavorable O−OH bond lengths. The appearance of specimens with structures 1 and 2 is due to their short O−OH bonds, as this factor compensates, to some extent, for their unfavorable cation distribution. It is quite natural that no specimens with structures 7 and 8 are found, since both factors are unfavorable for these. Structures 9 and 10 apparently do not appear because they do not satisfy the uniformity condition. For the same reasons, it is remarkable that specimens with structures 5 and 6 have been detected.

Both these modifications, and any others which may be found in the future, can be examined and analyzed on the basis of the reasoning developed in the present account.

Generally speaking, none of the structural modifications of chlorites which we have examined, if their idealized packet models are accepted, contain grounds sufficient to decide which is to be preferred under given formation conditions, and none of them provide sufficient reasons for considering them as polymorphs. But the very fact that the regular structures described exist is an indication that deviations from the ideal layer forms and regular isomorphous-atom distributions exist in them, making different structures no longer equally probable under the same conditions, and thus transforming them into polymorphs.

The distribution of the layer silicate structure modifications derived above among the different layer types is shown in Table 13. This gives the modification symbols and the number of structures made up of 1:1 (dioctahedral and trioctahedral), 1:2, and 2:2 layers. The space groups given are slightly modified, depending on the presence or absence of symmetry centers in the layers.

CHAPTER 2

EXPERIMENTAL ELECTRON - DIFFRACTION
METHODS OF STUDYING MINERALS

The basis of the electron-diffraction method was the discovery, in 1927, of the phenomenon of electron diffraction. It was shown that electrons, like any other atomic particles, possess wave properties. A stream of electrons moving with a velocity v has the wavelength

$$\lambda = \frac{h}{mv},$$

(1)

where h is Planck's constant, and m is the mass of the electron.

The velocity of an electron and hence its corresponding wavelength depend on the accelerating voltage of the electric field applied to the electron. For an accelerating voltage of V, measured in volts, the wavelength is given by the formula

$$\lambda = \frac{12.225}{\sqrt{V}} \, A.$$

(2)

For a voltage of V $\sim$ 50 kV, $\lambda \sim$ 0.05 A.

In crystal structures, where the mass of the material is distributed in a periodic manner, with a periodicity of the order of angstroms or tens of angstroms, the structures form natural diffraction gratings for electrons, just as they do for X rays and neutrons. The wave-scattering patterns obtained from these gratings or lattices can be used to investigate the crystal structures.

In formal terms, the phenomena of diffraction of X rays, electrons, and neutrons are analogous, and they can all be described by a single geometrical theory of wave diffraction in a crystal lattice. However, electrons, neutrons, and X rays differ considerably in the nature of their interaction with substances when diffracted, and this gives rise to the individual features of the three methods.

The electron-diffraction method possesses the following distinguishing features:

1. Electrons interact with a substance about a million times more strongly than X rays and neutrons, so that their path within a substance must be very short. This determines the form of the specimen used and the investigational approach.

The simplest experimental method, and at the same time the one giving the most extensive structural information, is the study of transmission of electrons through thin films 10^{-5}-10^{-6} cm thick. These films may be prepared from a solution, or by coating or "sputtering" a film in a vacuum onto a surface from which it is subsequently separated, or they may be very fine cleavage flakes from a crystal having a perfect cleavage, or cross sections cut with a microtome. Of a different nature are specimens of finely crystalline substances prepared by precipitation onto extremely thin structureless support films from a solution or suspension. In such specimens, platy or fibrous particles align themselves parallel to the support, forming the so-called platy or fibrous textures.

An approach restricted to electron diffraction is the investigation of glancing-angle reflections from the surface zone of a specimen, which, because of oxidation, polishing, or other factors, may have a structure which differs from that deep inside the specimen. This is the method used to study massive specimens or finely crystalline materials precipitated on flat structureless surfaces.

A separate field is the study of molecules in gases by electron diffraction. The method has also been applied successfully to the analysis of liquid structures.

2. Electron-diffraction studies are carried out in a vacuum of the order of 10^{-4} mm Hg. This makes it possible to investigate specimens which are unstable under ordinary conditions because of oxidation or absorption of moisture. At the same time, the study of substances containing weakly bound water is made more difficult, since this water may be lost in a vacuum.

3. Electrons are scattered by the electric potential of the crystal lattice, which has its maxima at the positions of the atomic nuclei. One of the consequences of this is that the difference in scattering ability of light and heavy atoms is less than that for scattering of X rays.

This means that light atoms can be located more reliably and accurately in the presence of heavy atoms; for example, hydrogen atoms close to atoms of carbon.

4. The small wavelength of the electrons considerably improves the conditions for the appearance of diffraction reflections and simplifies geometrical analysis of the diffraction patterns, since these can be examined as direct representations of planar cross sections of the reciprocal lattice of the specimen.

5. Electron-diffraction patterns have individual features which depend both on the specimen being investigated and on which of the possible methods of controlling the electron trajectories is used in the particular instrument.

This is true of texture patterns (platy and fibrous textures), in which the reflections are dispersed over two dimensions and systematically grouped, according to their indices, along second-order curves (Pinsker, 1949; Vainshtein, 1956), and also of patterns from single crystals, including those of very small size ($\sim 1\,\mu$).

These patterns may show up features of a mineral which could never be recorded on an ordinary diffraction pattern either because the features applied only to a minor portion of the crystal or because they were tied up with minor structural variations in the substance.

6. Because of the more intense interaction of electrons with a substance, and the small wavelength of diffracted electrons, the boundary of the "crystalline state" is shifted toward smaller dimensions of particles. A substance which appears to be amorphous in X rays may show up as crystalline from electron-diffraction data, hence affecting its interpretation, classification, and investigational approach. In particular, this leads to a more fruitful approach to metamict minerals, if they posses crystalline regions of very small size.

7. The electron-diffraction method offers the following advantages: (a) it is possible to investigate exceedingly small amounts of a material ($\sim 10^{-5}$ g); (b) the diffraction pattern can be observed on a fluorescent screen during the diffraction process, allowing the quality of the diffraction pattern to be controlled and the most interesting portions of the specimen to be selected; (c) a diffraction pattern can be photographed with a very short exposure (of the order of a few seconds), allowing rapid structural changes taking place in a substance to be recorded.

1. Electron-Diffraction Apparatus

The diffraction of the electrons takes place in vacuum apparatus, either an electron-diffraction camera or an electron microscope. The electron-diffraction camera's basic function is to carry out studies involving diffraction of electrons, and it is therefore more highly adapted to this task than the microscope, permitting the most diverse specimens and crystals to be studied in various orientations. For the electron microscope, diffraction studies play a subsidiary role, supplementing the observation of minute objects at considerable enlargements, but only applicable over a narrow range of orientations of the crystal specimens.

A diagram of the diffraction of electrons in a camera is given in Fig. 15. The electron source is a curved incandescent tungsten filament, held at a negative potential relative to earth of 40-100 kV. The electric field arising between the cathode filament and the grounded anticathode accelerates the electrons emitted from the cathode, and these pass through a system of diaphragms to form a narrow electron beam. The streams of electrons making up the beam possess wave properties, and their wavelength is given by formulas (1) and (2).

As they pass through the specimen, which is secured in a special holder which may be moved about or rotated during the observation, the electrons are scattered in directions which depend on the structure of the specimen. The diffracted rays fall on a fluorescent screen to form a diffraction pattern which depends on the structure of the specimen and its position relative to the electron beam. To photograph the diffraction pattern, the screen is removed and its place taken by a photographic plate.

To increase the sharpness of the diffraction lines, an electromagnetic lens is used to focus both the central beam and the diffracted rays onto the screen of the diffraction camera.

The apparatus has electric power units to supply the lens windings and heat the filament. The accelerating voltage is derived from a high-voltage installation. A system of vacuum pumps

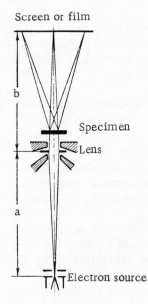

Fig. 15. Diagram of the diffraction of electrons.

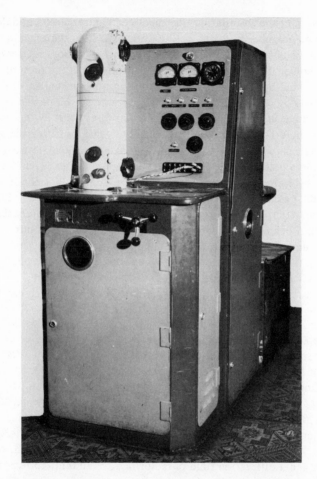

Fig. 16. General view of the EM-4 electron-diffraction
camera.

is used to bring about the vacuum required in the apparatus.

The different layouts, construction details, and individual features have
been described in detail in a review by Pinsker (1959). In the Soviet Union,
electron-diffraction cameras of several different types are being used, differing in their purposes and possible applications.

The EM-4 electron-diffraction camera (Bagdyk'yants, 1953) is a low-
height installation with a vertical ray path (Fig. 16). The electron source lies

in the bottom of the installation, the specimen camera in the middle, and the photographic chamber in the top. The distance L from the specimen to the screen may be 350 or 500 mm (by interposing an additional section). The specimen holder is designed to take seven specimens in the form of disks 3-4 mm across, held by the specimen film, for transmission studies, or one specimen for study by reflection. It allows the specimen to be moved by horizontal displacement in two mutually perpendicular directions, by rotation to any azimuthal angle about the axis of the apparatus, and by tilting the specimen about an axis perpendicular to the apparatus axis by any angle from 0 to 90°, measured on a graduated circle. The photographic camera contains 12 plates, either 4.5 by 6 cm or 6 by 9 cm, which can be exposed in succession without breaking the vacuum, which is particularly important in making multiple exposures. A system of two magnetic lenses results in an extremely sharply focused central beam and a high resolution of the diffraction points. The electric power units and the high-voltage apparatus are housed in a small cabinet attached to the camera column. The compactness of the unit does, however, limit the accelerating voltage, which cannot exceed 40-50 kV. The apparatus has an extra electron beam in the specimen chamber which can be used to remove the charge from a nonconducting specimen.

The EG electron-diffraction camera was evolved under the supervision of B. K. Vainshtein and Z. G. Pinsker (1958). Its constructional details were the fruit of many years of practical electron-diffraction studies in the Institute of Crystallography of the Academy of Sciences. The apparatus has a horizontal layout of its components from the electron gun to the photographic chamber, allowing easy access during operation, inspection, etc. A diagrammatic cross section of the device is shown in Fig. 17. The EG has accelerating voltages of 40, 50, 60, or 72 kV available, with a specimen−screen distance of L = 700 mm and a diffraction field of radius $r_{max} \simeq 70$ mm, corresponding to a value of $d_{min} \simeq 0.43$ A. The apparatus has provision for a camera with L = 250 mm. The specimen holder can be adjusted in any direction necessary without altering the value of L. The photographic chamber has a valve so that plates can be changed without admitting air to the apparatus. The plates, which measure 13 by 18 cm, can be exposed either completely or only in part, depending on the nature of the investigation.

The Sumy Electron-Microscope Factory produces several types of electron-diffraction camera (Boyandina, 1962). One of these, developed by Levkin (1957), is intended to work at voltages of 40, 60, 80, and 100 kV, and is suitable for the study of both solid and gaseous specimens. Another model, developed by a group of workers at the Vavilov Optical Institute (Alekseev et al., 1962) will give a direct record of the intensities of the diffracted rays.

For certain electron microscopes (EM-3, EM-5, EM-7), electron-diffraction adapters are available which allow the microscope to be used as a diffraction camera. There are two types of adapter, for transmission or reflection studies. Electron-microscope specimens which have been prepared on a metal grid, for transmission studies of textured specimens tilted 50-60° away from the perpendicular to the beam, are not suitable. For electron-diffraction work, the specimens must be prepared independently of the electron-microscope specimens, on discs with an inner diameter of about 2 mm.

Several other electron-diffraction cameras have been built in various laboratories in the Soviet Union. These include apparatus devised by Rumsh (1955), Ignatov, Bublik, and Pines (1954).

A particularly interesting ultrahigh-voltage apparatus has been devised by Popov (1959); in this the apparatus can be used as an electron microscope or an electron-diffraction camera by interchanging the units in the column. The high-voltage supply, the electron source, the photographic chamber, and the vacuum pumps are the same in both arrangements. The photographic chamber takes 12 plates, 6 by 9 cm. The relativistic speed of the electrons can be raised to 600 keV. Because of the high energy of the electrons, it is possible to carry out transmission studies on specimens up to 3 μ thick or more, depending on their nature.

Electron microscopes in which it is possible to obtain a diffraction pattern from a specimen, from a selected area a few microns across, are both very numerous and very diverse. There is no necessity or particular value in describing their constructional features in any detail. The only point of importance is an indication of the main methods used in these instruments to form the electron-microscope images and the diffraction patterns (Fig. 18). In these methods, the electron beam passes from the electron source through diaphragms and is focused by a condenser (or system of condensers) onto the specimen. All the electrons scattered in the same direction by the specimen are focused by the objective lens* onto a single point in its focal plane. If the

---

*In Fig. 18, the lenses are depicted symbolically as cross sections of glass optical lenses; the curvatures of the lenses indicate their refractive powers and their focal lengths.

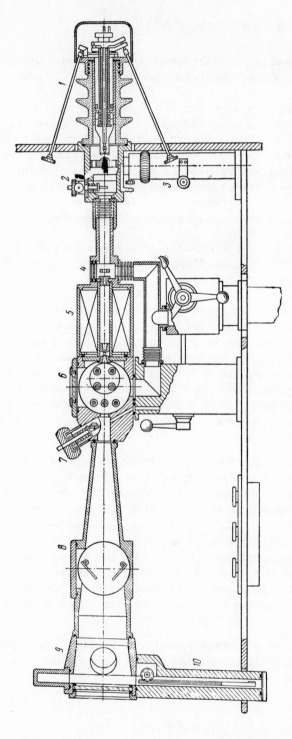

Fig. 17. The EG-75 electron-diffraction camera (Vainshtein, Pinsker, 1958; Pinsker, 1959). (1) Electron gun; (2) anode; (3) gun support; (4) intermediate chamber; (5) magnetic lens; (6) central chamber; (7) intermediate valve; (8) diffraction cone; (9,10) parts of the photographic chamber.

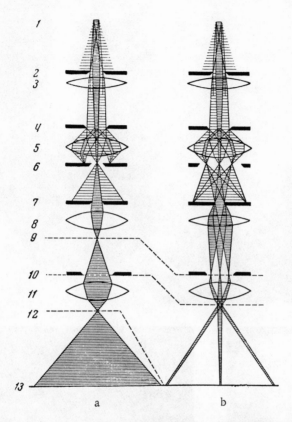

Fig. 18. Ray paths in selected-area or microdiffraction
method. (a) For direct image formation; (b) for diffraction-pattern formation (after Riecke, 1961). (1)
Electron source; (2) condenser diaphragm; (3) condenser; (4) specimen; (5) objective; (6) aperture diaphragm, or diffraction pattern; (7) first stage of image
formation, selector diaphragm; (8) intermediate lens;
(9) first stage in formation of diffraction pattern; (10)
second stage in image formation, limiting diaphragm
of projector lens; (11) projector lens; (12) second
stage in formation of diffraction pattern; (13) fluorescent screen or photographic emulsion.

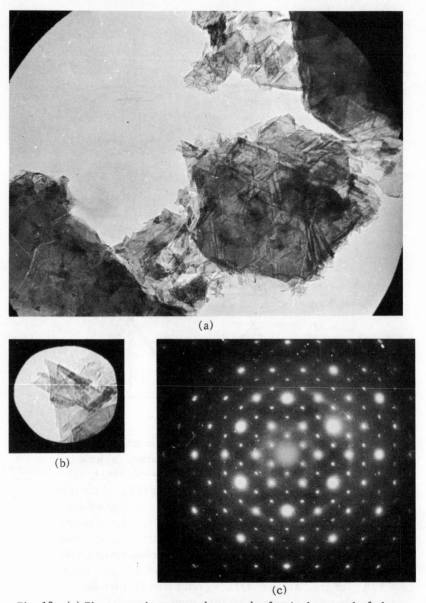

Fig. 19. (a) Electron-microscope photograph of a single crystal of platy serpentine; (b) a section of this isolated for diffraction with a diaphragm; (c) the diffraction pattern corresponding to this isolated section. Similar photographs have been obtained for lizardite (Zussman et al., 1957).

specimen possesses a crystalline lattice, then it will give rise to a discrete set of electron scattering directions, each of which will correspond to a point on the focal plane.

Thus, the diffraction pattern given by the specimen appears on this plane. As the electrons continue on their trajectories, they intersect the plane conjugate to the object in such a way that all the electrons passing through a given point of the object fall onto the same point on the conjugate plane. Consequently, the central nondiffracted beam forms a bright-field image in the conjugate plane, and each of the diffracted beams forms a dark-field image of the specimen. Any of these images may be separated off by placing an aperture diaphragm in the focal plane of the objective. However, even without such a diaphragm, the bright-field and dark-field patterns will not interfere, because of spherical aberration effects, and the conjugate plane of the objective will still contain a soft image of the specimen.

Through a change in its focal length, the intermediate lens after the conjugate plane can throw, on the plane in which the projector-lens diaphragm lies, either the image of the object in the conjugate plane of the objective or the diffraction pattern in the focal plane of the objective. Thus, the intermediate lens allows the electron microscope to be used for observing both a magnified image of the object and also its diffraction pattern. The projector lens throws on the final screen either a triply magnified image of the object or its doubly magnified diffraction pattern.

Diffraction from a small part of the specimen (microdiffraction or selected-area diffraction) can be achieved in two ways, by irradiating the specimen with a thin beam, or by using a selector diaphragm in the conjugate plane of the objective, through which pass only those electrons which have interacted with a selected area of the specimen and which have taken part in the formation of its image. Figure 19 shows a sequence of photographs of an enlarged image of an object, a portion of this selected by a selector diaphragm, and its corresponding diffraction pattern.

In practice, the second of the two methods is most often used. Microdiffraction studies have been carried out in the Soviet Union using UEMV-100, JEM-5Y, JEM-6A, EM-5, EM-7, HU-11, TESLA, and ELMI-D2 electron microscopes, among others.

The ultrahigh-voltage apparatus of Popov, set up as an electron microscope, has very considerable possibilities. It is possible to investigate a wide range of objects with this, and to study selected areas of about a tenth of the usual width.

2. Preparation of Specimens for Study

A brief outline has been given above of the requirements applicable to specimens for electron-diffraction study. These requirements indicate the types of object preferred in electron-diffraction investigations. Other types of specimen present complicated practical difficulties.

In mineral investigations, the best specimens for electron-diffraction studies are finely crystalline objects. The linear dimensions of the crystals, if they are isometric, must be roughly between 10^{-6} and 10^{-5} cm. If the crystals are not isometric, it is enough that their dimensions in at least one direction are quite small. This requirement is satisfied by the clay minerals, which consist of platy and, more rarely, fibrous particles.

In preparing clay mineral electron-diffraction specimens, the first stage is preparing the support. In this, "collars" or metal grids with apertures 2-5 mm across are placed on the bottom of a flat vessel (about 20 cm in diameter), and water is poured over them. On the water surface is placed one drop of a 0.5% solution of celluloid, collodion, or Parlodion in amyl acetate. The drop spreads out, and when the amyl acetate evaporates it leaves a thin film on the surface of the water. The water is removed from the vessel through a siphon, and the slowly settling film descends to cover the holes in the metal collars or grids. On the resulting ultrathin support is placed a drop of a previously prepared fine suspension of the material under study. When the drop dries out, it leaves a layer of particles on the film, of a suitable thickness for electron diffraction if the suspension was of the right dilution. Experience has shown that the finer are the crystals, the more the suspension should be diluted. Where the clay minerals are concerned, the greatest dilution is required for the montmorillonites, and the least for such minerals as dickite.

To find the optimum specimen thickness, it is necessary to prepare a series of specimens of the same material from various suspension concentrations, so that the series will contain both too-thin and too-thick specimens.

In the specimen preparation, the platy or fibrous particles of the clay minerals settle down on the surface of the films with the surfaces of the plates or the long directions of the fibers parallel to the support and, usually, randomly rotated about the normal to this plane, forming platy or fibrous textures. These are textured polycrystalline specimens, occupying an intermediate position between true polycrystalline specimens, in which there is no preferred orientation of the particles, and mosaic single crystals, in which the individual crystallites have all their corresponding crystallographic elements parallel.

It is extremely important that the sizes of the particles deposited do not exceed certain limits, so that they do not contain large particles which would be opaque to the electrons. Such particles are automatically eliminated in the sediment if the suspension is allowed to stand. To avoid coagulation of suspensions or aggregation of particles, and to obtain uniform distribution of the particles in the specimens, it is best to use ultrasonic dispersion and dry the specimens on a device vibrating at an ultrasonic frequency, or to use a peptizing agent (for example, NH_4OH). It is also possible to vary the rate of drying by adjusting the specimen heater.

Depending on the nature of the material and the details of the problem, it may also be possible to use another type of specimen in transmission studies. To prepare specimens without strongly preferred crystal orientation, or if the dispersed substance aggregates and precipitates too rapidly, it is possible to make up the suspension directly in a solution of the film material in amyl acetate. This means that the substance under study is introduced into the film which is formed on placing a drop of the solution onto the surface of the water. It can be assumed that in this case the particles are uniformly distributed within the specimen.

Specimens in the form of thin sections cut with a microtome are suitable for the study of substances, whatever their degree of dispersion and particle shape. Platy and fibrous minerals can be investigated in unusual orientations or orientations difficult to obtain in other ways, with the plates or fibers aligned parallel to the electron beam.

The sphere of application of electron-diffraction methods can be extended considerably through the use of reflection techniques. These allow diffraction patterns to be obtained from the faces of crystals, from the cleavage surfaces of rocks, and from deposits of finely dispersed minerals on massive flat surfaces. To avoid charge buildup on nonconducting specimens either an auxiliary electron gun can be used, or the specimen surface can be coated with a thin layer of metal or carbon by vacuum deposition. There are considerable possibilities for development and perfection of diffraction-specimen preparation techniques inherent in the application of the rich experience gained in the related electron-microscopy method. It is obvious that fruitful applications are likely to be found in electron diffraction for carbon and metal supports, selective extraction of mineral fragments from rocks through the stripping of different kinds of films from their surfaces, and so on.

3. Electron-Diffraction Patterns and Their Analysis

In electron-diffraction studies, the basic experimental material used consists of the diffraction patterns obtained from polycrystalline specimens, textures (Figs. 20, 21), and single crystals (Figs. 19c, 22) by transmission or reflection (Fig. 23); these are later subjected to further treatment and interpretation.

The different types of electron-diffraction pattern will be considered in more detail in subsequent sections. Here only their general features and methods of assessing the photographs will be described.

The amount and nature of the experimental work initially necessary, and the quality required in this, depend on the complexity and particular purpose of the work being carried out. For example, for a simple mineral diagnosis, a very scanty amount of data may suffice, while for a complete structure determination, extensive and varied data would be required.

The most important patterns for structure analysis are the texture diffraction patterns obtained in a transmission electron-diffraction arrangement, with the specimen inclined to the electron beam by an angle φ.

If the textured specimen is perpendicular to the electron beam ($\varphi = 0$), the diffraction pattern contains a series of ring reflections, the ratios of their

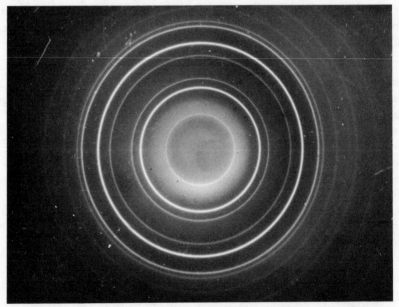

Fig. 20. Electron-diffraction texture pattern from a layer silicate ($\varphi = 0°$).

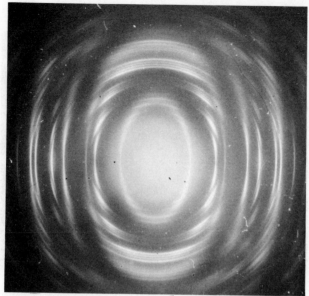

Fig. 21. Texture pattern from kaolinite with a triclinic
cell, at $\varphi = 55°$ (modification $1\,Tk_1$).

radii being characteristic for all layer silicates (see Fig. 20). When the specimen is tilted by an angle φ, each ring is replaced by a series of reflections,
which have the shapes of arcs of circles arranged in an ellipse, the minor axis
of which is equal to the radius of the original ring (see Fig. 21). The number
of these reflections increases as the angle φ increases. Thus, electron-diffraction patterns from inclined textured specimens or, as they are often called,
oblique texture patterns, possess a full, and yet regularly distributed, set of reflections in their diffraction fields. At the same time, they are comparatively
simple to obtain experimentally. In the particular case of layer silicates, the
diffraction patterns contain hkl reflections grouped around the ellipses according to their values of $3h^2 + k^2$, as will be shown below (see Fig. 21). Thus,
the reflections are spread out in two dimensions and have a systematic arrangement. Compared with X-ray powder photographs of the same materials, texture patterns show less superimposition of reflections. This all makes their indexing and intensity determination easier. As a result, it is possible not only
to determine the unit cell constants directly, but also, in a number of cases,
to apply Fourier methods to obtain a complete structure determination. These
new possibilities for the study and determination of mineral structures are also

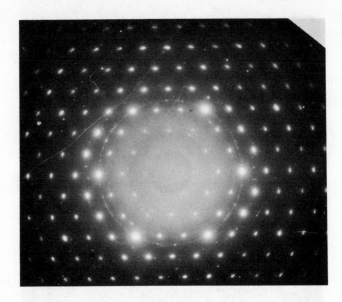

Fig. 22. Point electron-diffraction pattern from a dickite
single crystal.

of paramount importance in other practical problems such as the diagnosis of
minerals, phase analysis of rocks, etc., since the more accurately the structures are known, the better will be the performance of these tasks.

These diffraction patterns do not, however, include reflections with all
types of indices. Reflections with large values of l do not appear. For them
to register, the specimen must be tilted at a large angle to the electron beam
($\varphi > 60°$), but in this case, because the path of the electrons in the substance
is lengthened, the general quality of the diffraction pattern deteriorates; the
reflections become more diffuse and the background more pronounced. The
basal reflections $00l$ are completely lacking. For them to register it would
theoretically be necessary to rotate the specimen through $\varphi = 90°$, but in practice it is not possible to obtain a transmission diffraction pattern at this value
of φ. Another factor is that oblique texture patterns from layer silicates may
show overlap of reflections with the same value of $3h^2 + k^2$, since these fall on
the same ellipse. So, in spite of the features favorable to structural analysis
noted above, these diffraction patterns need to be worked over carefully, and
additional patterns of another type must be obtained.

It is above all necessary to strive for the highest possible quality in the texture patterns, so that the diffuseness and elongation of their reflections and the contrast shown by the patterns are governed solely by the nature of the structures and morphologies of the crystals, and are not affected by poor specimen preparation. To increase the sharpness of the reflections, the use of two focusing lenses is very important. To ensure good contrast in the diffraction patterns, it is desirable to use voltages over 60,000 V.

For layer silicate structure investigations the diffraction field must include reflections with indices at least up to h = 6 (12th ellipse) or up to k = 12 (15th ellipse). These conditions are met by the 6 by 9 cm photographs produced by the EM-4 electron-diffraction camera, with L = 350 mm and V = 40,000 V. There is very little point in using the more distant ellipses, since, due to their great lengths and low intensities, and because of distortions in the diffraction pattern tied up with the curvature of the reflecting sphere, the determination of their positions and indices becomes uncertain.

To reveal reflections with large values of l, it is necessary not only to prepare diffraction patterns at the optimum angle $\varphi \sim 55°$, at which the pat-

Fig. 23. Electron-diffraction pattern obtained from celadonite by the reflection method.

terns retain their sharpness and contrast, but also to obtain patterns at angles of $\varphi \approx 56\text{-}70°$, even though the reflections on these become considerably more diffuse. In this case, it is most desirable to use the highest accelerating voltages.

To obtain the basal reflection $00l$ from layer silicates, a reflection phtograph may be produced from an oriented coating of an ideally polished stainless steel surface (Mitra, Rao, 1955). Here also it is desirable to use the highest possible accelerating voltage.

Where exact results for lattice symmetry and a,b unit cell constants are required, single-crystal electron-diffraction patterns are of considerable value, since they reveal the symmetry of the mineral lattice directly, and separate off reflections which, under other conditions, are superimposed. In particular, they are better than anything else for showing up superlattice repeat units and other fine structural features. Moreover, they allow a number of other ancillary problems to be dealt with.

These patterns can be obtained in normal electron-diffraction cameras if their condenser-lens systems produce a sufficiently fine electron beam (Cowley, 1953; Cowley, Rees, 1953), and this is played on isolated microcrystals in the specimen, or if the accelerating voltage is high enough (V > 100 kV) to render accessible to diffraction the comparatively large crystallites which occupy a considerable proportion of the specimen area and predominantly control the diffraction pattern (see Fig. 22). In this way it would be possible in principle, through the movement of the specimen holder, to obtain point diffraction patterns from specimens in any arbitrary orientation with respect to the electron beam, thus producing images of various cross sections of the reciprocal lattice and revealing its symmetry directly. It is obvious that, for the single-crystal patterns obtained in this manner to have any practical value, strict control of the orientation of the single crystals with respect to the electron beam is necessary. Only then can the values of the l indices and the signs of the hk indices be taken as definite for the reflections on the pattern produced by a crystal lying perpendicular to the electron beam. Then, by rotating the crystal azimuthally about its normal, and by tilting it through various angles, it is possible to obtain the desired three-dimensional sections of the reciprocal lattice and the corresponding diffraction reflections.

Using Popov's ultrahigh-voltage apparatus, set up as a diffraction camera, studies were made of two kaolinite specimens (Nos. 553a and 593) and one dickite specimen (No. 20) from Vikulova's collection. To prepare diffraction specimens, the minerals were ground up and made up into suspensions. Drops of freshly shaken suspension were highly diluted with distilled water and placed on a celluloid support. This gave specimens made up of widely scattered particles, and if the latter were large in size, they gave point diffraction patterns typical of single crystals. An auxiliary electron gun was used to remove electrostatic charges.

Point diffraction patterns were obtained both with and without a focusing (condenser) lens. In the first case, the patterns were found with great ease, but the reflections were sharper in the second. Diffraction patterns were obtained from specimens held perpendicular and inclined at various angles to the electron beam.

It was also possible to carry out microdiffraction studies using four-lens electron microscopes, by selecting only those electrons which had been scattered by small selected areas of the specimens.

Microdiffraction combines the features and uses of both electron microscopy and electron diffraction. It allows an area about 1 μ across in the field of view of the electron microscope to be selected (a considerably smaller area in Popov's apparatus) and a diffraction pattern of the area to be obtained immediately afterwards or, conversely, it is possible to select a given reflection on the diffraction pattern and then obtain a dark-field electron-microscope image of the parts of the specimen diffracting in the direction of the reflection.

Thus, the microdiffraction method can be used in several ways: (1) to study the structure of very small single crystals; (2) to carry out a detailed mineralogical analysis, as it is possible to locate the finest inclusions and insignificant impurities in a complex specimen; (3) to investigate twinned and intergrown crystals, and intermediate zones of transformation of one mineral into another; (4) to determine crystallographic directions and distribution of different reflecting faces; and (5) to establish the nature of defects and other imperfections in the regular arrangement of crystals of small dimensions.

In studies of layer silicates by the microdiffraction method, diffraction patterns were obtained from crystals lying approximately perpendicular to the electron beam and having the shapes of individual plates, either isometric, elongated, or rolled into a tube. Here, the platy particles gave a hexagonal network of point reflections (see Fig. 19), corresponding to the pseudohexagonal basis of layer silicate lattices, while particles rolled up into tubes gave more complex patterns, reminiscent of X-ray rotation photographs (see Fig.24).

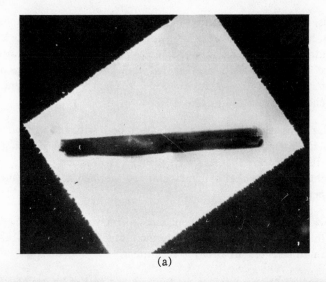

(a)

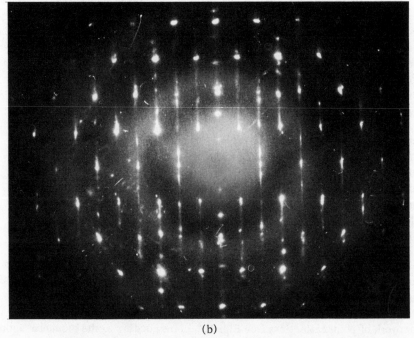

(b)

Fig. 24. (a) Electron-microscope photograph (b) diffraction pattern from a halloysite tube elongated along the [010] direction.

Analysis of a diffraction pattern is divided into two stages, first a geometrical analysis and then analysis of intensities (see Chapters 3 and 4). The measurements made on the diffraction pattern thus consist of the determination of the relative arrangement of the reflections, and the estimation of their intensities.

The most important characteristic of a diffraction pattern is the quantity $L\lambda$, its individual scale, where L is the specimen–screen distance, and λ is the wavelength. The product $L\lambda$ is usually found from photographs of standard substances. It is convenient to add the standard substance to the specimen after photographing the specimen, and then repeat the exposure with both present. A useful standard is a 0.5% solution of NaCl. If drops of this solution are placed on the specimen and dried at about 60°C, they form a fine polycrystalline precipitate which gives sharp ring reflections. Slower drying leads to crystallization of a textured precipitate, which gives a diffraction pattern containing layer lines, and the separation of these lines can be used to work out the angle of inclination of the specimen, φ. Drying in air may give mosaic single crystals or large single crystals, not very suitable for diffraction purposes.

Formulas for determining $L\lambda$ from photographs of standard substances will be given in Chapter 3.

Electron-diffraction patterns of single crystals consist of two-dimensional rectilinear networks of point reflections. Their geometry is determined by the distances between reflections and by the angles between rows of reflections. Purely geometrically (not considering intensities), the network is two-dimensionally periodic, being defined by its "unit cell" parallelogram, fixed by two cell sides and the angle between them. These quantities may be measured extremely simply (with a ruler, measure, protractor, etc.). The cell constants on a diffraction pattern can be found very accurately by measuring the distance along a row of reflections and then dividing this by the number of cell sides in the row. An IZA-2 optical length gauge can also be used, and with this the lengths of individual cell sides can be determined accurately.

In texture patterns, the position of a reflection for a given specimen inclination angle φ is completely fixed by the following factors: (1) the ellipse on which it lies, and (2) the particular point of the ellipse which it occupies. The former can be designated by the length of the minor axis of the ellipse, and the latter by the distance between the reflection and this axis (the "height" of the reflection). In addition, an important characteristic of the position of a reflection is its distance from the center of the diffraction pattern. We will run ahead here and note that the first two quantities form the basis for determining the unit cell of the crystal lattice, while the third quantity is directly related to the interplanar distance corresponding to the particular reflection.

It will be shown in Chapter 3 that for layer silicates the minor axes of ellipses on texture patterns depend only on the indices hk, and may be denoted by b_{hk}. The major axes of the ellipses vary with the angle of tilt φ of the specimen, so that $a_{hk} = b_{hk}/\cos \varphi$. When $\varphi = \pi/2$, the ellipses degenerate into pairs of straight lines. In this case, the distances of the reflections from the minor axes are given by

$$D_{\pi/2} = \sqrt{R_{hkl}^2 - b_{hk}^2}, \qquad (3)$$

where R_{hkl} is the distance of the reflection from the center of the diffraction pattern. For an arbitrary angle φ,

$$D_\varphi = \frac{D_{\pi/2}}{\sin \varphi}. \qquad (4)$$

Using these relationships, both $D_{\pi/2}$ and D_φ can be determined from measurements of R_{hkl} and b_{hk}. This is especially convenient when the reflections have the shapes of long arcs. Here, the distance R can be measured much more accurately than D, since it is necessary to locate the midpoint of the arc for the latter. Moreover, by measuring R and converting it to $D_{\pi/2}$, it is possible to avoid any inaccuracy in reading φ. It should, however, be noted that direct measurement of D_φ does not introduce any appreciable error into the final result, even with long-arc reflections. The point is that the radii R alter to a lesser extent than the heights D on going from one reflection to another, so that R is thus less sensitive to the position of a reflection. The greater is the difference, the larger is φ, and the further the corresponding ellipse lies from the center, i.e., the greater is b_{hk}. In fact, bearing in mind that D is one of the coordinates of the ellipse with axes b_{hk} and $b_{hk}/\cos \varphi$, it is not difficult to show that

$$\frac{\Delta R}{\Delta D} = \frac{D}{R} \sin^2 \varphi. \qquad (5)$$

If there are no reflections on the minor axis of the ellipse, and b_{hk} cannot be measured directly, particularly when the reflections are thinly scattered, then Vainshtein's geometrical construction can be used (1956, Fig. 53); this corresponds to conversion of an oblique texture pattern ($0 < \varphi < 90°$), with a given angle φ and reflection radii R, to a "direct" texture pattern ($\varphi = 90°$). Also, expression (3) can be used in reverse, so that, knowing the radius R of any

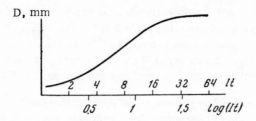

Fig. 25. Blackening curve (after Vainshtein,
1956).

reflection and the value converted to $\varphi = \pi/2$ of its height, $D_{\pi/2}$, it is possible
to find the corresponding value of b_{hk}. At the same time, these values give
an indication of which reflections belong to the same ellipses.

In these texture patterns it is much simpler to measure the distances,
not from the center (which is usually difficult to distinguish in the dense background), nor from the imaginary line of the minor axis, but instead to find the
distances between reflections related by a center or a plane of symmetry, respectively. These measurements give the values $2b_{hk}$, 2D, 2R, and also $2L\lambda$.

With a fairly large diffraction pattern ($2L\lambda > 60$), all the distances can
be measured accurately enough using a ruler. They can, however, be measured much more accurately with an IZA-2 optical length gauge. The values
of $2b_{hk}$ and 2D can be measured very simply in this case, because the corresponding sections are usually parallel to the edges of the photographic plate
containing the diffraction pattern. The diffraction pattern is fixed against the
freely adjustable stop of the movable stage of the length gauge, by one or
other of its edges, so that the stage moves along the direction of the section to
be measured. Through the left-hand tube of the gauge the extremities of the
section being measured can be viewed, while the right-hand tube shows positional readings to an accuracy of 0.001 mm. The difference between the
readings for the two extremities gives the required distance directly. It is usually worth measuring the reflection heights 2D to an accuracy of 0.1 mm, and
$2b_{hk}$ to an accuracy of 0.01 mm. In measuring 2R, the photographic plate is
fixed by a spring clip at a single point, about which it can rotate, the point
depending on the direction of the reflection diameter being measured. The
value of 2R can be measured to an accuracy of 0.01 to 0.001 mm, according
to the sharpness of the reflections. If the reflections are very closely positioned, or if it is necessary to scan all the positions where, theoretically, reflec-

tions ought to lie, then the reflection heights can be measured directly on the
IZA-2 comparator as follows: the stage is set at the reading "100.000" the
minor axis of any given ellipse is brought up to the measuring line, and then
on moving the stage, the values of D are read directly in the tube as the difference between their comparator readings and the reading "100.000."

Electron-diffraction patterns of polycrystalline specimens consist of reflections in the form of concentric rings, with positions defined by their radii
R, or rather diameters 2R, since the latter are measured more simply. These
values, like those in texture patterns, are directly related to the interplanar
spacings corresponding to the reflections. The ring diameter can also be measured with a ruler, measure, or IZA-2 length gauge.

The methods which can be used to measure the reflection intensities depend on the type of diffraction pattern and the method of recording it. Direct
electrometer measurements are undoubtedly the most accurate. These, however, come up against experimental difficulties and apparatus complications.
Although we may shortly expect the appearance of electron-diffraction
cameras able to register intensities directly (Alekseev et al., 1962), the method of photographing diffraction patterns and then evaluating their intensities
will still retain its importance, at least so far as clay minerals are concerned.
In this method, the intensity of a reflection is quoted in terms of its density of
blackening D on the photographic plate. The value of D will be a function of
the beam intensity I and the time t that the beam acts on the plate. Over a
given range of values of It, the following relationship holds quite well:

$$D = k\log (It).$$

This expression can therefore be used to find the relative intensities of
reflections on multiple-exposure diffraction patterns. Let us suppose that we
have a set of diffraction patterns taken with exposures forming a geometrical
progression. For example, the exposures might be of 2, 4, 8, 16 sec, etc., or
they might be in the ratio of 1, $\sqrt{2}$, 2, $2\sqrt{2}$, 4, etc. Lines of equal density on
different photographs correspond to the condition $I_1 t_1 = I_2 t_2$, from which it follows that

$$\frac{I_1}{I_2} = \frac{t_2}{t_1}.$$

Blackening values can be evaluated with a microphotometer. The method has been described in detail by Yamzin and Pinsker (1949). From photometer measurements of multiple-exposure photographs, a curve of blackening

against log (It) can be constructed (Fig. 25), and this can be used to determine different intensities on the same diffraction pattern. It is essential here that the blackening values fall on the straight-line part of the curve.

A rough estimate of reflection intensities on a particular pattern can be obtained visually. Such an estimate is useful in superficial diagnosis of minerals and their structural varieties. Here, Pinsker's nine-point scale (1949) can be used; this puts subjective visual estimates of the "strong" (s.), "medium" (m.), or "very weak" (v.w.) type in correspondence with definite numbers based on objective measurements, according to the following sequence:

v.v.s.	100	m.w.	8
v.s.	70	w.	5
m.s.	36	v.w.	2
m.	18	v.v.w.	1

The simplest and most accurate intensity measurements are those of polycrystalline diffraction patterns. On moving a photometer along any straight line passing through the center of the pattern, the instrument registers the complete sequence of ring reflections.

Point diffraction patterns are less suitable for this method, because, in these, quite a large proportion of the blackened areas fall in the saturated part of the scale (the horizontal part of the curve in Fig. 25). Moreover, the intensities of point reflections depend on how closely the positions of the crystals correspond to reflection conditions. They are also affected by secondary scattering effects suffered by the diffracted rays, etc. To make the first factor less significant, Cowley (1953) has suggested subjecting the diffracted rays to the action of a transversely varying electric field, which would spread the point reflections out into rectangular spots with degrees of blackening which would mostly fall on the straight-line portion of the curve in Fig. 25.

With texture patterns, photometric measurements, particularly direct electrometer measurements of intensities, are greatly complicated by the complexity of the reflection distribution rules. This is particularly true of layer silicate diffraction patterns. Because the lattices of these minerals are not rectangular, the reflections are very often distributed at varying intervals along the ellipses. Some of the ellipses partly overlap. Under these conditions, only visual estimates of intensities are practicable. In practice the intensity of the strongest reflection appearing on the weakest diffraction pattern is taken as 10,000 (this is the factor by which the intensity of the strongest reflection usually exceeds that of the weakest). Working from this, on other diffraction pat-

terns there will be reflections which will be roughly as many times intrinsical-
ly weaker as the exposure of their pattern is stronger, so that their actual intensity will be roughly equal to that of the chosen reflection. Intensities of
intermediate reflections are found by interpolation. It is also possible in these
diffraction patterns to make photometer measurements of reflections lying
along a single straight line, for example, the line of minor axes of the ellipses, and
to use these reflections as reference standards in estimating the intensities of the
other reflections.

Multiple-exposure photographs also help the least intense reflections to
show up, on the overexposed photographs. To obtain the largest possible set
of reflections, an examination should be made of all points on the diffraction
pattern at which the lattice dimensions theoretically indicate that reflections
should appear; this helps to pick up reflections which might otherwise be over-
looked.

CHAPTER 3

THE GEOMETRICAL THEORY OF ELECTRON DIFFRACTION AND ANALYSIS OF CLAY MINERAL DIFFRACTION PATTERNS

1. Theoretical Basis of the Relationship Between the Crystal Lattice and the Diffraction Pattern: The Reciprocal Lattice

In a crystalline substance, the material particles (atoms, ions, or molecules) are distributed in a regular manner, forming the crystal structure. In this structure, both the individual particles and their groupings alternate in three dimensions in such a way that their relative environments are the same throughout the structure. Thus, any arbitrary point in the structure corresponds to a multitude of other points, identical to it and arranged in a definite sequence which varies with the particular structure. Any of these sets of identical points having the same relative spatial arrangement is a characteristic feature for a given structure and, under the name of a space lattice, serves as a means of describing the structure.

This lattice is periodic in three dimensions. It can be considered as made up of parallelepipeds, attached to one another at the faces in such a way as to fill completely the space available. In magnitude and direction, the edges or these parallelepipeds correspond with the period or repeat distances of the identical points, or lattice points as they are called, which coincide with the corners of the parallelepipeds.

To describe the lattice, it is sufficient to define its unit parallelepiped, taking the edges of this as the axes of coordinates. To describe the structure, the coordinates of atoms within the parallelepiped must be given. These parallelepipeds, which are repeated in the directions of their three noncoplanar axes to form the whole lattice, may be chosen in numerous ways. The unit cell of the lattice is chosen to be that parallelepiped which has the highest symmetry and the smallest volume.

81

The unit cell, and with it the lattice, is characterized by the lengths of its sides (the cell constants) **a, b, c** and the angles between these, α, β, γ. In vector form, the lattice and the cell are defined by three vectors, **a, b, c.**

Depending on their shapes, the cells are divided up into seven symmetry classes. The cells with smallest volume (called primitive cells, with lattice points only at the corners) may not possess the highest symmetry. The cells with the highest symmetry may contain lattice points not only at the cell corners, but also at the centers of one set of faces, of all faces, or at the center of volume of the cell. Fourteen different types of crystal lattice exist (the Bravais lattices), each with a different maximum-symmetry unit cell.

The combination of the symmetry and internal structure of the contents of the unit cell with translational displacements equal to the repeat distances creates new symmetry elements (planes of gliding reflection, screw axes). A set of these elements defines the spatial symmetry of a crystal structure. E.S. Fedorov has shown that 230 of these space symmetry groups exist.

A crystal lattice may also be considered as a set of parallel lines of lattice points or parallel planes of lattice points. It is possible to select innumerable different systems of parallel lines of points or planes of points. Each system differs in its orientation relative to the coordinate axes of the lattice. These orientations are designated by the numerical indices hk$l$ (indices of lines of points are written in square brackets, those of planes of points in round brackets). The indices of a line are directly proportional to the projections of any section of it onto the axes of coordinates, while the indices of a plane are inversely proportional to the intercepts cut off by the plane on the axes of coordinates. These intercepts and projections are measured in units equal to the repeat distances in the lattice.

The basis of crystal-structure analysis is the phenomenon of diffraction of waves by a crystal lattice. The phenomenon can be described in the following terms. Each lattice point, as a representative of a repeating group of material particles, has a certain scattering ability. Under the influence of a beam of primary waves falling on the object, the lattice points become sources of secondary waves. In many cases, it can be assumed that the lattice points are excited only by primary waves, excitation by scattered waves being neglected. In addition, the differences in phase of waves scattered by different lattice sites are taken to be governed solely by the geometry of the lattice. These assumptions distinguish the kinematic theory of wave diffraction from the dynamic theory. The secondary waves are scattered in all directions, and inter-

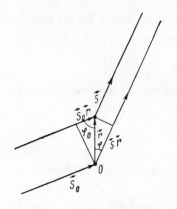

Fig. 26. Relative arrangement of primary (s_0) and secondary (s) rays for scattering by the origin of coordinates and the point r.

fere with others in various ways, according to the arrangements and scattering abilities of their sources.

A plane monochromatic wave of wavelength λ and frequency ν, propagating through a three-dimensional space, the points of which are given by the vector r, in a direction given by the unit vector s forming an angle φ with the vector r, is described by the exponential function $Ae^{2\pi i\left(\nu t - \frac{r\cos\varphi}{\lambda}\right)}$. In actual fact, the exponential factor with the imaginary index expresses the periodic dependence of the function on the argument. The argument in the index has the form such that the vibration state at the origin of coordinates $r = 0$ at time t will be observed at the point r after the interval of time

$$\Delta t = \frac{r\cos\varphi}{\nu\lambda} = \frac{r\cos\varphi}{v},$$

necessary for the wave front to move with velocity v toward the point r through a distance equal to the projection of r on the direction s. To avoid the use of the angle φ, this projection is best written in the form of the scalar product (sr), and this notation will be used in what follows.

In considering the result of mutual interference of secondary waves scattered by lattice points, it is necessary to focus attention on the phase relationships imposed by the lattice points when acting as the wave sources.

From Fig. 26 it can be seen that the incident ray s_0 traverses a path toward the arbitrary scattering point r which is greater by the distance $r\cos\varphi_0 = rs_0$ than the path taken toward the origin; the path of the scattered ray s from the point r is less by the distance $r\cos\varphi = rs$ than its path from the point 0. In all, the path travelled by the ray scattered by the lattice point r differs from the path of the ray scattered by the lattice point at the origin by a distance

$$s_0 r - s r = (s_0 - s, \ r).$$

Accordingly, the ray scattered at $\mathbf{r}$ lags in phase behind the ray scattered at the origin by a quantity

$$2\pi\,(s_0 - s,\ \boldsymbol{r})/\lambda.$$

Thus, if at the observation point $\boldsymbol{r}$ the ray coming from the origin is.s described by the factor

$$e^{2\pi i\left(\nu t - \frac{sR}{\lambda}\right)},$$

then for the ray coming from the point $\boldsymbol{r}$ the factor will have the form

$$e^{2\pi i\left(\nu t - \frac{sR}{\lambda} - \frac{(s_0 - s,\ \boldsymbol{r})}{\lambda}\right)} = e^{2\pi i\left(\nu t - \frac{sR}{\lambda}\right)}\,e^{2\pi i\left(\frac{s - s_0}{\lambda},\ r\right)}. \tag{1}$$

To explain the features of wave scattering by an object, we use an instantaneous representation of the scattering at an arbitrary moment of time t and an arbitrary observation point R. Here we can exclude the first factor in (1) from consideration and direct attention to the second factor only, which is the one concerned with the geometrical distribution of the scattering centers in the object.

For a set of point centers $\mathbf{r}_j$ with scattering power f_j, the scattering amplitude will be expressed by the sum

$$S\,(\mathbf{s}) = \Sigma f_j e^{2\pi i\left(\frac{s - s_0}{\lambda},\ \boldsymbol{r}_j\right)}. \tag{2}$$

To a large degree, the actual distribution of scattering material throughout the object is continuous. It can be characterized by the continuous scattering-ability function $\varphi(\mathbf{r})$ which, of course, has its maxima at the positions of atoms and takes a value close to zero in the intervening spaces. According to this picture of the object, the summation over the discrete scattering centers f_j in (2) must be replaced by an integration over continuously varying values of the function $\varphi(\mathbf{r})$, and the scattering amplitude will be given by the integral

$$S(\mathbf{s}) = K\int \varphi\,(\mathbf{r})\,e^{2\pi i\left(\frac{s - s_0}{\lambda},\ \mathbf{r}\right)}\,d\tau, \tag{3}$$

where the integration is carried out over the whole of the scattering volume, and the function $\varphi(\mathbf{r})$ and the multiplier K both depend on the nature of the waves. In the particular case of electrons $K = (2\pi m e)/h^2$, and $\varphi(\mathbf{r})$ is the electric potential in the lattice.

In the case of a crystalline object, $\mathbf{r} = \mathbf{r}_m + \mathbf{r}'$, where $\mathbf{r}_m$ gives the position of the origin of the m-th cell, $\mathbf{r}'$ gives the positions of points within the cell, and expression (3) can be broken down into a sum of integrals over the volume Ω of the individual unit cells:

$$S(s) = K \int_{\Omega} \varphi(\mathbf{r}') \, e^{2\pi i \left(\frac{\mathbf{s}-\mathbf{s}_0}{\lambda}, \, \mathbf{r}'\right)} \, d\tau \cdot \sum_{m} e^{2\pi i \left(\frac{\mathbf{s}-\mathbf{s}_0}{\lambda}, \, \mathbf{r}_m\right)}.$$

(4)

Bearing in mind that the function $\varphi(\mathbf{r})$ in the unit cell is made up of the scattering abilities of the individual atoms present in the cell, i.e.,

$$\varphi(\mathbf{r}') = \sum_{j} \varphi_j(\mathbf{r}' - \mathbf{r}_j),$$

the integral over the cell in (4) can be written as

$$\Phi(s) = \sum_{j} f_j(s) \, e^{2\pi i \left(\mathbf{r}_j, \frac{\mathbf{s}-\mathbf{s}_0}{\lambda}\right)},$$

(5)

where

$$f_j(s) = K \int_{\Omega} \varphi_j(\mathbf{r}' - \mathbf{r}_j) \, e^{2\pi i \left[(\mathbf{r}' - \mathbf{r}_j), \frac{\mathbf{s}-\mathbf{s}_0}{\lambda}\right]} d\tau.$$

(6)

Therefore,

$$S(s) = \Phi(s) \, D(s),$$

(7)

where $\Phi(\mathbf{s})$ and

$$D(s) = \sum_{m} e^{2\pi i \left(\frac{\mathbf{s}-\mathbf{s}_0}{\lambda}, \, \mathbf{r}_m\right)}$$

are the expressions for the structure amplitude and the interference function, respectively.

Since formulas (3) and (4) are Fourier integrals of $\varphi(\mathbf{r})$, then the function $S(\mathbf{s})$ can be considered as the result of a Fourier transformation of the crystal lattice.

The system of coordinates of the crystal lattice is given by the set of three vectors a, b, c, and within it the vectors r_m, r_j are determined by their contravariant coordinates m_1, m_2, m_3; x_j, y_j, z_j as $r_m = m_1 a + m_2 b + m_3 c$, $r_j = x_j a + y_j b + z_j c$ (m_1, m_2, m_3 are integral). If, in this system of coordinates the second multiplier of the scalar products of indices [the vector $P = (s - s_0)/\lambda$] is given by its covariant coordinates p_1, p_2, p_3 defined by the expressions

$$p_1 = (Pa), \quad p_2 = (Pb), \quad p_3 = (Pc), \tag{8}$$

then

$$\Phi(P) = \Sigma_j f_j\, e^{2\pi i\,(p_1 x_j + p_2 y_j + p_3 z_j)}, \tag{9}$$

$$D(P) = \sum_m e^{2\pi i\,(p_1 m_1 + p_2 m_2 + p_3 m_3)}. \tag{10}$$

Considering the amplitude S as a function of the vector P, it is convenient to use a system of coordinates a^*, b^*, c^* in which p_1, p_2, p_3 are contravariant coordinates, i.e., where

$$P = p_1 a^* + p_2 b^* + p_3 c^*. \tag{11}$$

From ordinary analytical geometry we know that such a trio of vectors a^*, b^*, c^* is reciprocal to the trio a, b, c, i.e., they satisfy the conditions

$$aa^* = bb^* = cc^* = 1; \quad ab^* = ac^* = ba^* = bc^* = ca^* = cb^* = 0 \tag{12}$$

and consequently also the equations

$$a^* = \frac{[bc]}{(a\,[bc])} = \frac{[bc]}{\Omega}, \quad b^* = \frac{[ca]}{\Omega}, \quad c^* = \frac{[ab]}{\Omega}. \tag{13}$$

Equations (12) and (13) may also be derived directly from the substitution of (11) in (8).

The interference function $D(P)$ has modulo maxima at the points H for the integral values $p_1 = h$, $p_2 = k$, $p_3 = l$, and the regions in which it differs appreciably from zero are less, the greater the dimensions of the crystal.

Thus, for example, if the crystal were in the shape of a parallelepiped with edges in the direction of the crystallographic axes a, b, c, and with N_1, N_2, and N_3 cells, respectively, along these, then considering (10) as the product of three geometrical progressions with common ratios $e^{2\pi i p_j}$, this can be brought to the form

$$D = \prod_{j=1,\,2,\,3} \frac{\sin \pi N_j p_j}{\sin \pi p_j} e^{\pi i (N_j - 1) p_j}. \tag{14}$$

The square of the modulus of D,

$$|D|^2 = \frac{\sin^2 \pi N_1 p_1}{\sin^2 \pi p_1} \cdot \frac{\sin^2 \pi N_2 p_2}{\sin^2 \pi p_2} \cdot \frac{\sin^2 \pi N_3 p_3}{\sin^2 \pi p_3} \tag{15}$$

has principal maxima equal to $N_1^2 N_2^2 N_3^2$ at points with integral values of $p_j = h_j$ and much weaker side maxima at the points

$$p_j = h_j \pm \frac{n + 1/2}{N_j} [n = 1, 2, \ldots, (N_j - 2)].$$

From the positions of the principal maxima at the integral-value points $p_j = h_j$, $|D|^2$ falls to zero at the points $p_j = h_j \pm 1/N_j$, so neglecting the side maxima we can take it that the dimensions of the regions we are concerned with, in the directions $a*$, $b*$, $c*$, are equal to $a*/N_1$, $b*/N_2$, $c*/N_3$.

For normal-sized crystals these regions are very small, and $D(\mathbf{P})$ cuts, from the three-dimensional distribution $\Phi(\mathbf{P})$, a discrete set of $\Phi(hkl)$. In its turn, the distribution $\Phi(\mathbf{H})$ is superimposed on the points of $D(\mathbf{H})$, extinguishing certain of these with its zero values, and imparting to the remainder a density $|S(\mathbf{H})|$ and the phase pertaining to $\Phi(\mathbf{H})$ to form an infinite figure with a singular point at the origin (Vainshtein, 1960).

Thus, the function $S(\mathbf{H})$ emerges as an analytical method of assigning a unique lattice, in which only the geometry of the distribution of lattice points, and not the distribution of values of $\Phi(\mathbf{H})$ pertaining to these points, is characterized by the repeat units $\mathbf{a}*$, $\mathbf{b}*$, $\mathbf{c}*$. Its appearance is a natural consequence of our consideration of the diffraction phenomenon and in form, as we have already noted, it is a Fourier transformation of the crystal lattice. In structural analysis it is known as the reciprocal lattice, and serves as an intermediate link between the crystal lattice and its corresponding diffraction pattern, where it is a most valuable tool (Ewald, 1921).

The geometrical positions of the crystal lattice points which for a given $\mathbf{H}$ scatter in the same phase, are given by the equation

$$\mathbf{rH} = hx + ky + lz = \text{const}. \tag{16}$$

i.e., by the equation of the plane which has the vector $\mathbf{H}$ normal to it. The planes connecting the points with integral coordinates are the ordinary lattice-point planes (hk$l$); according to (16), they are perpendicular to the reciprocal-lattice vector $\mathbf{H}$(hk$l$).

The distance between these lattice-point planes, d_{hkl}, is equal to the distance from the origin to the closest plane of the (hk$l$) system. It is obtained by substituting in the equation of the plane, written in the normal form,

$$\mathbf{r}\mathbf{H}/\mid \mathbf{H} \mid = 1/\mid \mathbf{H} \mid,$$

the value $\mathbf{r} = 0$. This means that

$$d_{hkl} = 1/\mid \mathbf{H} \mid. \tag{17}$$

To sum up, let us list the most important properties of the reciprocal lattice. The integral-valued reciprocal-lattice vector $\mathbf{H}$ is perpendicular to the direct-lattice planes of indices hk$l$ equal to the components of the vector $\mathbf{H}$, and in length it is equal to the reciprocal of the corresponding interplanar distance d_{hkl}.

In its essence, the function $S(\mathbf{H})$ expresses the diffraction properties of the crystal lattice. The directions of coherent scattering are determined by the values of $\mathbf{H}$ at the maxima of the function $D(\mathbf{H})$ or, in other words, by the equation

$$\mathbf{H} = \frac{\mathbf{s} - \mathbf{s}_0}{\lambda} = h\mathbf{a}^* + k\mathbf{b}^* + l\mathbf{c}^*. \tag{18}$$

If (18) is multiplied in turn by $\boldsymbol{a}$, $\boldsymbol{b}$, $\boldsymbol{c}$, we obtain the three Laue rules for the directions $\mathbf{s}$ of the diffracted rays

$$\mathbf{a}\left(\boldsymbol{s} - \boldsymbol{s}_0\right) = h\lambda, \quad \mathbf{b}(\boldsymbol{s} - \boldsymbol{s}_0) = k\lambda, \quad \mathbf{c}\left(\boldsymbol{s} - \boldsymbol{s}_0\right) = l\lambda. \tag{19}$$

From Fig. 27, in agreement with (17), the Bragg reflection condition may also be derived, so that we have

$$1/\boldsymbol{d} = \mid (\boldsymbol{s} - \boldsymbol{s}_0)/\lambda \mid = 2\sin\vartheta/\lambda, \quad 2\boldsymbol{d}\sin\theta = \lambda. \tag{20}$$

The Laue and Bragg conditions impose limitations on the wavelengths λ which are suitable for diffraction, for given values of $\mathbf{a}$, $\mathbf{b}$, $\mathbf{c}$, and $\mathbf{d}$. In particular, it is clear that $\lambda \leq 2\mathbf{d}$.

As can be seen from Fig. 27, which is a geometrical illustration of condition (18), the directions of all diffracted rays coincide with the directions of

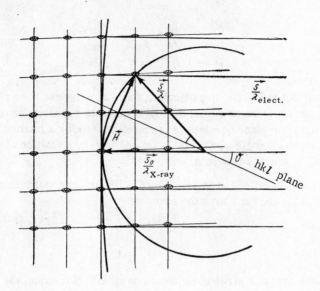

Fig. 27. Reflecting sphere and "plane" in the reciprocal lattice.

those radii of the so-called reflecting or Ewald sphere which pass through points of intersection of the sphere with reciprocal lattice points. The reflecting sphere has a radius of $1/\lambda$ and its center lies at a distance of $1/\lambda$ in the direction s_0 from the origin of the reciprocal lattice.

It is a very important point that the function $S(\mathbf{H})$ allows us to establish, in a concrete form, a relationship between points of direct and reciprocal space which are related by those symmetry elements which transform one into the other on going from the direct to the reciprocal lattice (Vainshtein, Zvyagin, 1963). As proof of this, it may be noted that on substitution in the power index of $S(\mathbf{H})$ of the vector $\mathbf{r}$ by another vector $\mathbf{r}'$, the components of which, x', y', z', are linearly and uniformly dependent on x, y, z in the form

$$x' = \beta_{11}x + \beta_{12}y + \beta_{13}z,$$
$$y' = \beta_{21}x + \beta_{22}y + \beta_{23}z, \qquad (21)$$
$$z' = \beta_{31}x + \beta_{32}y + \beta_{33}z,$$

then the scalar product $(\mathbf{r}'\mathbf{H})$, if x', y', z' are expressed in terms of x, y, z, turns out to be equivalent to the scalar product $(\mathbf{r}\mathbf{H}')$, so that

$$h' = \beta_{11}h + \beta_{21}k + \beta_{31}l,$$
$$k' = \beta_{12}h + \beta_{22}k + \beta_{32}l, \qquad (21')$$
$$l' = \beta_{13}h + \beta_{23}k + \beta_{33}l.$$

From this it follows, in particular, that if $\mathbf{r}'$ is derived from $\mathbf{r}$ by reflection in a certain plane or origin, then $\mathbf{H}'$ is obtained by reflection of the vector $\mathbf{H}$ in the same plane or origin. If $\mathbf{r}'$ is related to $\mathbf{r}$ by a rotation about a certain axis by an angle φ, then $\mathbf{H}'$ is related to $\mathbf{H}$ by a rotation about the same axis, by the same angle φ.

In fact, for reflection in a center of symmetry, the matrices of the transformations (21) and (21') have the form

$$\begin{pmatrix} -1 & 0 & 0 \\ 0 & -1 & 0 \\ 0 & 0 & -1 \end{pmatrix}.$$

In the case of a plane of reflection or an axis of rotation, the relationships between points emerge clearly, these being invariant for any choice of coordinate system, if the coordinate axes are chosen so that b is perpendicular to the plane of reflection, and c* coincides with the axis of rotation. Then reflection and rotation are described by the matrices

$$\begin{pmatrix} 1 & 0 & 0 \\ 0 & -1 & 0 \\ 0 & 0 & 1 \end{pmatrix}, \quad \begin{pmatrix} \cos\varphi & \sin\varphi & 0 \\ -\sin\varphi & \cos\varphi & 0 \\ 0 & 0 & 1 \end{pmatrix},$$

and by interchanging rows and columns in these matrices, we arrive at the conclusions given above.

That these conclusions are correct can be easily verified from the fact that the relative positions of the vectors $\mathbf{r}'$ and $\mathbf{H}$, and $\mathbf{r}$ and $\mathbf{H}'$ are geometrically identical (Fig. 28). The numerical values of the corresponding coordinates are wholly determined by equations (21) and (21').

If the axis, plane, and origin of coordinates in question are symmetry elements, then they form closed sets of points in direct or reciprocal space. If, in each term of $S(\mathbf{H})$, the vector $\mathbf{r}$ is replaced by the corresponding vector $\mathbf{r}'$, this leads only to an interchange of terms and does not affect either their complete sum or the multiplying factors $\Phi(\mathbf{H})$ and $D(\mathbf{H})$.

If there is a center of symmetry at the origin, the $\Phi(\mathbf{H})$ terms can be grouped with respect to the points $\mathbf{r}_j$ and $-\mathbf{r}_j$:

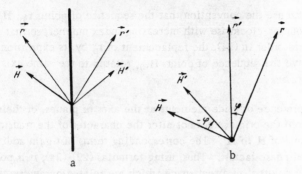

Fig. 28. Relative positions of the vectors **r** and **r'**
and **H** and **H'** under circumstances where they are
related by reflection in a plane (a), and by rotation about an axis (b). The plane and axis are perpendicular to the plane of the drawing, while the
vectors are shown in projection on this plane.

$$\Phi(H) = \sum_j f_j [e^{2\pi i(\mathbf{H},\ \mathbf{r}_j)} + e^{2\pi i(\mathbf{H},\ -\mathbf{r}_j)}] = \sum_j f_j [e^{2\pi i(\mathbf{H},\ \mathbf{r}_j)} + e^{2\pi i(-\mathbf{H},\ \mathbf{r}_j)}];$$

$$\Phi(\mathbf{H}) = \Phi(-\mathbf{H}). \qquad (22)$$

A plane of symmetry passing through the origin allows the $\Phi(\mathbf{H})$ terms
to be grouped with respect to the reflectionally identical points **r** and **r'**:

$$\Phi(\mathbf{H}) = \sum_j f_j [e^{2\pi i(\mathbf{H}, \mathbf{r}_j)} + e^{2\pi i(\mathbf{H}, \mathbf{r}_j')}] =$$

$$= \sum_j f_j [e^{2\pi i(\mathbf{H}, \mathbf{r}_j)} + e^{2\pi i(\mathbf{H}', \mathbf{r}_j)}]; \quad \Phi(\mathbf{H}) = \Phi(\mathbf{H}'), \qquad (23)$$

where **H'** is derived by replacing $\mathbf{r}_j'$ in (23) by its expression in terms of $\mathbf{r}_j$.

If there is an n-fold rotation axis passing through the origin, the $\Phi(\mathbf{H})$
terms can be grouped with respect to the rotationally identical points $\mathbf{r}_{j0}$,
$\mathbf{r}_{j1}, \ldots, \mathbf{r}_{j,n-1}$, and

$$\Phi(\mathbf{H}_0) = \sum_j f_j \sum_{p=0}^{n-1} e^{2\pi i(\mathbf{H}_0, \mathbf{r}_{jp})} = \sum_j f_j \sum_{p=0}^{n-1} e^{2\pi i(\mathbf{H}_{n-p}, \mathbf{r}_{j0})}$$

$$\Phi(\mathbf{H}_0) = \Phi(\mathbf{H}_m) \quad (m = 0, 1, \ldots, n-1). \qquad (24)$$

We can use the convention that the sequence of points r_p, H_m are numbered off counter-clockwise with increasing index number, so that with an unchanged term order in (24), the replacement of r_p by its expression in terms of r_0 will give the sequence of points H_m, related to the same axis but numbered clockwise.

The presence of glide elements at the axes or planes, or their positioning away from the origin, does not alter the character of the transformations of H_0 to H_m, and of H to H'. The corresponding terms just gain additional, fully defined phase factors. Thus, using formulas (22)-(24), it is possible to derive, for any points r' in direct space which are related by any symmetry element, a group of phase-identical points in reciprocal space which are related by the corresponding symmetry element, by transformation of the scalar product $r'H$ to rH'.

Because of the distribution of its lattice points $|S(H)|$, the reciprocal lattice does not have translations, and possesses a singular point at its origin. The spatial character of a crystal lattice may show up in it in terms of particular phase relationships for symmetrically related lattice points. An important feature of the function $S(H)$ is that it enables us to derive the basic relationships between the phase properties corresponding to a particular spatial symmetry in the crystal lattice and the point or diffraction symmetry of the reciprocal lattice (Vainshtein, Zvyagin, 1963). The possible special features of crystal lattice symmetry include absence of any symmetry center, presence of translational symmetry elements, and distribution of symmetry elements outside the limits of the unit cell. If these features are not present, then from expressions (22)-(24) it follows that the phases of symmetrical lattice points coincide. Thus, for example, according to (24), the structure factor for any of the rotationally identical lattice points H_m is the same as the sum in (24), but with cyclic permutation of the terms, taken in the reverse direction, i.e.,

$$\Phi\left(H_m\right) = \sum_j f_j \sum_{p=-m}^{n-m-1} e^{2\pi i (H_{n-p}, \, r_{j0})}.$$

$$(25)$$

In (24), the term with $p = -m$ is identical to the term with $p = n - m$ ($m = 0, 1, \ldots, n-1$), since n is the periodicity of the cyclic permutation. When the crystal lattice lacks a center of symmetry, then

$$\Phi\left(H\right) = \sum_j f_j \, e^{2\pi i Hr_j}; \quad \Phi\left(-H\right) = \sum_j f_j e^{-2\pi i Hr_j},$$

and in the reciprocal lattice, $\Phi(-H) = \Phi^*(H)$.

If the symmetry of the crystal lattice includes translational elements, then instead of the set of identical $\mathbf{r'}$ points which would correspond to the symmetry elements without translations, the whole set of elements forms a set of $\mathbf{r''} = \mathbf{r'} \pm (m\mathbf{T})/n$ points, where $\mathbf{T}$ is the repeat unit in the direction of translation, and m,n are integers, the values of which depend on the nature of the symmetry element. The corresponding transformation to $\Phi(\mathbf{H''})$ is linked not only with the permutation of terms in expressions (23) and (24), but also with the appearance of an additional phase multiplier

$$e^{\pm \frac{2\pi i m \mathbf{H} \mathbf{T}}{n}}. \tag{26}$$

Thus, in the particular case of an n_1 screw axis,

$$\Phi(\mathbf{H}_0) = \sum_j f_j \sum_{p=0}^{n-1} e^{2\pi i \left[(\mathbf{H}_{n-p}, \mathbf{r}_{j0}) + p\frac{\mathbf{H}\mathbf{T}}{n} \right]} \tag{27}$$

$$\Phi(\mathbf{H}_m) = \sum_j f_j \sum_{-m}^{n-m-1} e^{2\pi i \left[(\mathbf{H}_{n-p}, \mathbf{r}_{j0}) + \frac{(p+m)\mathbf{H}\mathbf{T}}{n} \right]} = \Phi(\mathbf{H}_0) e^{\frac{2\pi i m \mathbf{H}\mathbf{T}}{n}}. \tag{28}$$

In these formulas, the scalar product $\mathbf{H}\mathbf{T}$ has the same value for all $\mathbf{H}_m$ vectors related by the n-fold rotation axis, and so no index is attached to the vector $\mathbf{H}$ in formulas (26)-(28). With the axis n_{n-1},

$$\Phi(\mathbf{H}_m) = \Phi(\mathbf{H}_0) e^{-\frac{2\pi i m \mathbf{H}\mathbf{T}}{n}}. \tag{29}$$

If $\mathbf{T} = \mathbf{c}$, then

$$\Phi(\mathbf{H}_m) = \Phi(\mathbf{H}_0) e^{\pm \frac{2\pi i m l}{n}}. \tag{30}$$

Similarly, with the planes c, a, n,

$$\Phi(\mathbf{H'}) = \Phi(\mathbf{H}) e^{2\pi i \frac{M}{2}}, \tag{31}$$

where $M = l$, h, h + l, respectively.

In the general case,

$$\mathbf{T} = m_1\mathbf{a} + m_2\mathbf{b} + m_3\mathbf{c}, \quad \mathbf{H}\mathbf{T} = m_1 h + m_2 k + m_3 l. \tag{32}$$

The positioning of lattice symmetry elements away from the origin leads to the appearance in equations (22)-(24) of a further phase multiplier $e^{2\pi i (\mathbf{H}\Delta)}$, where Δ is the vector giving the displacement of the symmetry center from the origin.

2. Geometrical Analysis of Electron-Diffraction Patterns

It has been shown above that the diffraction directions are given by the intersections of the reflecting sphere with the reciprocal lattice. Because of the very small wavelength λ in electron diffraction, the radius $1/\lambda$ of the reflecting sphere is considerably greater (about 100 times greater) than the distance between reciprocal lattice points. Therefore, the intersecting part of the reflecting sphere can be assumed to be approximately planar (Fig. 27). The arrangement of the primary and diffracted rays and the reflecting sphere in reciprocal space is analogous to the arrangement of the same rays and the screen in the diffraction camera. If L is the specimen—screen distance, and R is the distance from the zeroth reflection to a reflection on the diffraction pattern, then from Fig. 27,

$$L : 1/\lambda = R : H, \ R = L\lambda H. \tag{33}$$

Since $H = 1/d$, then for calculation of the interplanar distances we have the extremely simple formula

$$d = L\lambda/R. \tag{34}$$

The intensity I of a diffracted ray, which is governed by the intensity I_0 of the primary ray and the diffractional properties $\Phi(\mathbf{H})$, $D(\mathbf{H})$ of the specimen, is inversely proportional to the square of the distance L from the object to the point of observation, since a real diffracted ray is divergent. It is given by the equation*

$$I = \frac{I_0}{L^2} |\Phi(\mathbf{H})|^2 |D(\mathbf{H})|^2. \tag{35}$$

From the above it is apparent that an electron-diffraction pattern is an approximate representation of the plane cross section of the reciprocal lattice

---

*For dynamic scattering, I is directly proportional to $|\Phi(\mathbf{H})|$.

which passes through the zero lattice point and is perpendicular to the electron beam. Here, the reciprocal lattice distances appear on the diffraction pattern on the scale $L\lambda$, with a lattice-point density coefficient of I_0/L^2.

This property will govern the method used to analyze any type of diffraction pattern, taking into account in each case the particular character of the specimen's reciprocal lattice and its orientation relative to the electron beam. Here, the indexing and measurement of the pattern will be based on the direct relationship noted above between the pattern and the reciprocal lattice, according to which the radius of any reflection hkl represents the reciprocal lattice vector $\mathbf{H} = h\mathbf{a}^* + k\mathbf{b}^* + l\mathbf{c}^*$.

Diffraction Patterns from Single Crystals. Only a single crystal with a perfect structure, or a mosaic single crystal consisting of identically oriented blocks (angular spread not greater than 2-3°), will have a reciprocal lattice of the type described above. In the ideal case of a precise s_0 direction and a reciprocal lattice made up of ideal points, the reflecting sphere may pass through only three lattice points, including the origin. Consequently, the diffraction pattern may include, besides the primary beam, only two diffracted rays. Under real conditions, however, it is possible to form a large number of reflections, corresponding quite closely to the appropriate plane section of the reciprocal lattice. This is assisted by the following factors, which operate as a result of the negligible curvature of the reflecting sphere.

1. The crystallites are of small dimensions, particularly in the direction of the primary beam. This feature transforms the reciprocal lattice points into regions of finite size, here called nodes, so that they come into contact with the reflecting sphere if it passes close to their centers.

2. The electron beam diverges, which transforms the reflecting sphere into a zone of finite thickness, and increases the probability of its touching certain reciprocal lattice points.

3. The crystal may have a mosaic character, with its fragments oriented in slightly different directions. Independently of the first factor, this again leads to enlargement of the lattice points into finite volumes or nodes, which will intersect a reflecting sphere which passes nearby.

Single-crystal diffraction patterns which depict reciprocal lattice planes, in particular, coordinate planes, give direct values of the constants a^*, b^*, c^*, and from these the constants of the direct lattice, a, b, and c. The hk$l$ indices of reflections are simply equal in this case to their coordinates in the system a *, b *, c *, measured on the scale Lλ.

The processing and analysis of single-crystal patterns can be carried out particularly simply when it is known which crystal plane is perpendicular to the electron beam and, therefore, which reciprocal lattice plane the pattern represents. A single diffraction pattern may then yield the indices of reflections present and reciprocal lattice dimensions along the series of points lying on it. In particular, if the diffraction pattern depicts a coordinate plane, this gives directly two lattice constants (for example, a* and b*) and the corresponding reflection indices (hk0).

By comparing different point diffraction patterns from the same crystal, its three-dimensional axis system in reciprocal space can be found; all the reflections can be indexed, the constants a*, b*, and c* found, and the angles between their directions established.

The general features of the technique of analyzing electron diffraction patterns have been described in Vainshtein's book (1956). He has shown that if the angle of tilt of the reciprocal lattice planes depicted is known, then two diffraction patterns are sufficient for a solution. If the angle of tilt is not known, then for a complete analysis three patterns are necessary. The task is made easier if the crystal can be rotated azimuthally about the direction of the primary beam, so that the axis of inclination of the crystal in the plane perpendicular to the beam coincides in direction with a reciprocal lattice axis (for example, b*). Then a series of photographs of the crystal, at different inclinations to the original perpendicular position, will establish the type of reciprocal lattice plane which is perpendicular to the axis of inclination (for example, the plane a*c*). With two reciprocal lattice coordinate planes, the coordinate vectors **a** *, **b** *, **c** * can be found directly. Their absolute values are found from the radii R of the corresponding reflections:

$$a^* = \frac{R_{h00}}{L\lambda h}, \qquad b^* = \frac{R_{0k0}}{L\lambda k}, \qquad c^* = \frac{R_{00l}}{L\lambda l}. \tag{36}$$

The angles between the axes are measured directly on the images of the a *b* and a *c* planes.

For the case where it is not possible to measure the lengths of all three constants directly on the patterns (for example, if c* cannot be measured), Vainshtein (1956) has derived the following formulas:

If $\beta = \pi/2$, then

$$c^* = \frac{\sqrt{R_{hkl}^2 - R_{hk0}^2}}{L\lambda l}.$$ (37)

If $\beta \neq \pi/2$, then

$$c^* = \frac{1}{L\lambda l} \sqrt{\frac{R_{hkl}^2 + R_{hk\bar{l}}^2 - 2R_{hk0}^2}{2}}, \qquad \cos\beta^* = \frac{R_{h0l}^2 - R_{h0\bar{l}}^2}{4R_{h00}R_{00l}}$$ (38)

Both the geometry and the intensity distribution of single-crystal diffraction patterns may be affected by certain additional effects. The first of these is two-dimensional diffraction. This arises through the enlargement of reciprocal lattice points which takes place when the crystal size is reduced. It can lead to partial overlap, or even complete merging, of the lattice points to form reciprocal lattice nodes consisting of straight lines or rods. Thus, the pattern may show so-called forbidden reflections, which correspond to points not lying in the reciprocal lattice plane section depicted by the diffraction pattern.

Another effect is secondary extinction, which occurs when intense diffracted rays themselves play the part of a primary beam and on propagating through the lattice form their own diffraction patterns. These patterns are superimposed on the original, since the geometry of the reflections is unaltered, causing a redistribution of the intensities observed; strong reflections become somewhat weaker, and weaker ones stronger. If one part of the crystal is rotated relative to another part through a certain angle, the secondary diffraction patterns will not coincide with the primary ones. Stabenow (1959) has shown that if the rotation brings certain reciprocal lattice angles of the two crystal parts into coincidence, the secondary patterns combine with the primary ones to form a network of reflections which possesses constants characteristic of the "superlattice" formed by the combination of both structures, projected on their common plane. Secondary diffraction shows up most prominently when it is due to crystals of another structure lying on the crystal giving the basic pattern.

These effects must be borne in mind in the analysis of systematic absences to determine space symmetry groups, and in the analysis of reflection intensities.

Single crystals of layer silicates and clay minerals have a platy or fibrous shape. The most highly developed faces are those parallel to the layers and networks of the structures, so that if these are perpendicular to the electron beam, the diffraction pattern is an image of the reciprocal lattice plane

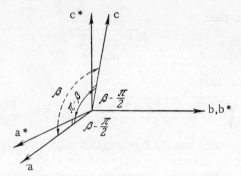

Fig. 29. Direct and reciprocal axes in a
monoclinic lattice.

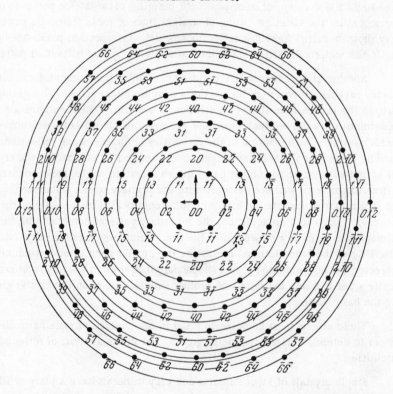

Fig. 30. The hk network of a layer silicate with a face-centered unit
cell with b = $a\sqrt{3}$. The circles join up series of reciprocal lattice
points hk which are all parallel to the c* axis and equidistant from it.

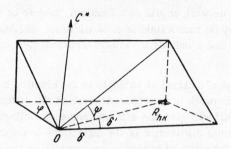

Fig. 31. Diagram of the relationships between the angles φ, ψ, δ, δ'.

which passes through the origin and is parallel to the ab face of the crystal lattice unit cell. In the pseudolayer silicates, i.e., the sepiolites and palygorskites, the axis which is the analog of the a axis in layer silicates is usually called the c axis, but this obviously does not affect the essential general features of their diffraction patterns, these features being equally applicable to layer and pseudolayer silicates.

The layer silicates are characterized by a centered rectangular ab unit cell face, with b = $a\sqrt{3}$. This, together with the equations numbered (13), determines the relative arrangement of the axes of the direct and reciprocal lattices, as shown in Fig. 29. If the c axis forms angles of β and α, not being right angles, with the axes a and b, respectively, then the c* axis is perpendicular to the ab plane, the axes $a*$ and b* form angles of $\beta*$ and $\alpha*$ with the c* axis, and the ab plane does not coincide with the $a*b*$ plane. Since the angle between the axes a and c is $\gamma = \pi/2$, the axes $a*$ and b* lie in the $ac*$ and $bc*$ planes, respectively, forming angles of $(\pi/2)-\beta*$ and $(\pi/2)-\alpha*$ with the a and b axes. Therefore, the projections of the vectors $a*$ and b* on the ab plane are equal to $a* \sin \beta*$ and b* $\sin \alpha$ *.

From formulas (13) it follows, in agreement with Fig. 29, that

$$a^* = \frac{bc \sin \alpha}{abc \sin \alpha \sin 3^*} = \frac{1}{a \sin \beta^*}, \quad b^* = \frac{ca \sin \beta}{bca \sin \beta \sin \alpha^*} = \frac{1}{b \sin \alpha^*}. \quad (39)$$

Thus, when the vectors $a*$ and b* are projected on the plane ab, they mark off lengths of $1/a$ and $1/b$ on the plane, with an angle $\gamma = \pi/2$ between these lengths, in spite of the fact that α, β differ from $\pi/2$. The ratio of the projected lengths is $1/a : 1/b = \sqrt{3}$, so that electron diffraction patterns of single crystals held with their ab plane perpendicular to the beam typically show a hexagonal network or grid of reflections, in spite of the fact that the crystal lattice may be monoclinic or even triclinic. Because the ab cell face is centered, the h and k indices are always unmixed, i.e., both even or both odd (Fig. 30).

If the crystal is inclined at an angle to the electron beam, the regularity of the hexagonal network is distorted. Analysis of the distortion observed in the hexagonal motif will give the axis and angle of inclination. Under these conditions, the actual distribution of the reflections corresponds to an oblique planar section of the series of lattice points, parallel to the c^* axis. Let the angle between the intersecting plane of reflection and the ab plane, lying perpendicular to c^*, be φ, the angle between an arbitrary straight line in the inclined plane and the line of intersection of both planes be δ, and the angle between the projection of the straight line on the ab plane and the same line of intersection be δ'. Then, from Fig. 31, it can be seen that the angle ψ between any straight line and its projection on the ab plane will be given in terms of δ or δ' by the formulas

$$\sin \psi = \sin \varphi \sin \delta, \quad \tan \psi = \tan \varphi \sin \delta'. \qquad (40)$$

It is quite obvious that the distance between any two lines of lattice points lying parallel to c^* will be $1/\cos \psi$ times greater along an inclined line than it will along the perpendicular between the pair of lines.

If a reflection in a "perpendicular" or "direct" photograph lies R_{hk} away from the center, then the same reflection on an oblique photograph will be $R_{hk}/\cos \psi$ away from the center. Changes in distances due to specimen tilting are easy to establish if a diffraction pattern is available for the same specimen, taken with the same value of $L\lambda$, but with the specimen perpendicular to the beam. The observed distances can also be compared with the theoretical values calculated for the appropriate value of $L\lambda$, if the unit cell constants of the mineral are known. It is, however, better to analyze the distortion in the regular network using a single diffraction pattern. By comparing the different distances along symmetrically equivalent directions on the pattern, it is possible to find the direction nearest to the axis of inclination ($\delta = 0$, $\psi = 0$) and the direction of maximum tilt ($\delta = \pi/2$, $\psi = \varphi$). The extra distances between the reflections in the direction with $\delta = \pi/2$ give directly the angle of inclination of the crystal, φ.

The distortion of the originally regular hexagonal network due to crystal tilt also shows up in changes in the angles between rows of reflections. If an arbitrary straight line in a "perpendicular" pattern lies at an angle δ' to

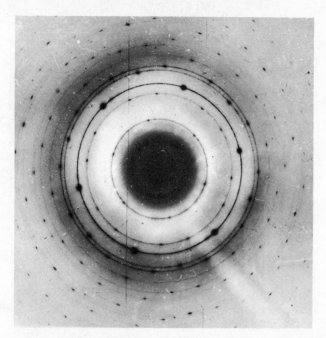

Fig. 32. Electron-diffraction pattern from a platy kaolinite
crystal perpendicular to the electron beam.

the axis of inclination, then from Fig. 31 it can be seen that the corresponding
straight line in an inclined pattern is at an angle δ, where

$$\tan \delta = \tan\delta'/\cos \varphi. \qquad (41)$$

The angle between any two straight lines, $\delta_2' - \delta_1'$, becomes the angle
$\delta_2 - \delta_1$, where δ_2 and δ_1 are given by equation (41).

The above technique was used in the analysis of single-crystal patterns
prepared on a 400-kV Popov electron-diffraction camera. By making use of
known unit cell data, it was found possible to index not only the hk indices,
but also the l indices of those reciprocal lattice points which appeared on the
diffraction-pattern cross section.

Figure 32 shows a diffraction pattern from a single crystal of kaolinite
lying with its ab plane perpendicular to the electron beam. The hexagonal

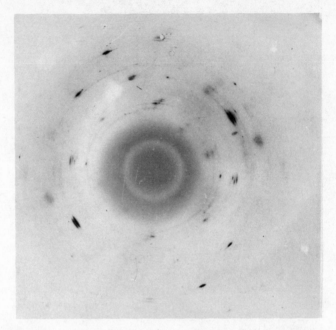

Fig. 33. Electron-diffraction pattern from a platy kaolinite
single crystal inclined at 60° to the electron beam.

network of reflections illustrates the structure of the silicate layers, with their
hexagonal motifs. A consequence of the pseudohexagonal symmetry of the
pattern is that the choice of the directions of the a^*- and b^*-axis projections
can be ambiguous. They can be tied down more exactly by rotating the specimen about each of three directions mutually inclined at 120° in the diffraction pattern, and these might be directions $[0k0]^*$, $[hh0]^*$, and $[h\bar{h}0]^*$. For a
known b^* direction, the indices of the reflections are determined directly by
the coordinate network. The point reflections arrange themselves on circles,
a consequence of diffraction from the multiplicity of fine crystals with random azimuthal orientations which forms the platy texture.

The point reflections thus directly determine the hk indices of the
circles or ellipses on normal or oblique layer silicate texture patterns, respectively. Because the a^* axis of the monoclinic reciprocal-lattice unit cell of
kaolinite is inclined to the ab plane at an angle of $\beta - (\pi/2) \simeq 15°$, the l
indices are obtained by rounding off the values of -0.525 h to the nearest
whole number.

Figure 33 shows a diffraction pattern from a single crystal rotated about 60° away from the perpendicular to the electron beam. Because the axis of rotation did not coincide with any one of the pairs of mutually perpendicular lattice-point lines lying in the ab plane, and because the axis is inclined to the projection of a^* on the ab plane by about 10°, the previously perpendicular projections of the a^* and b^* directions now form an angle of 108°. The hk indices are determined from the distorted coordinate network (see Fig. 30), and the l indices are worked out by rounding off the values of $0.6\,k - 0.1\,h$ to whole numbers.

Figure 34 shows a diffraction pattern from dickite. This pattern does not include the continuous circles which indicate that a mass of very fine crystals is taking part in diffraction in addition to the main crystal. It appears that secondary scattering has affected this diffraction pattern, leading to a certain amount of equalization of intensities and increase in their distribution symmetry.

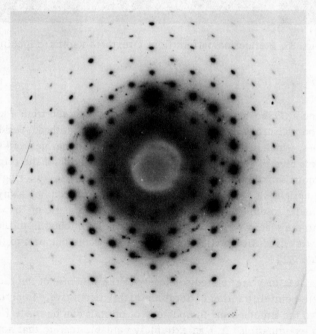

Fig. 34. Electron-diffraction pattern from a dickite single crystal.

Fig. 35. Reflection pattern from a textured kaolinite specimen.

Figure 35 shows a point diffraction pattern obtained from a kaolinite coating on polished stainless steel, by the reflection method. Besides the obvious lines, which are $00l$ reflections, a two-dimensional network of point reflections is clearly visible. An analysis of the distortion of the original hexagonal motif showed that this point pattern was produced by a single crystal standing upright on its a edge and with its surface rotated 35° away from the perpendicular to the beam, about an axis making an angle of 15° with the normal to the specimen surface. The hk indices are found from the coordinate network, and the l indices are the nearest whole numbers to the values of $0.55\,h - 0.15\,k$.

Because they are related so closely to the structure of the reciprocal lattice, these point diffraction patterns are highly expressive. Important deductions as to the structure and morphology of crystals can be made from a purely visual examination. It is an extremely valuable feature that not only can the geometry of the pattern be considered, but any regularity in distribution of the reciprocal lattice-point densities shown can also be taken into account.

The possibilities inherent in visual treatment of layer silicate single-crystal diffraction patterns can be illustrated using some examples of different types of pattern obtained by the microdiffraction method.

Figure 36 shows a diffraction pattern from a kaolinite single crystal; it has a regular hexagonal network of reflections, indicating that the crystal is roughly perpendicular to the electron beam. This is also shown by the equal intensities of the strong reflections $0\overset{\pm}{6}0$, $3\overset{\pm}{3}1$, $\overset{-\pm}{3}31$. In the central sextet the weakest pair are apparently the reflections $\overset{\pm\pm}{1}10$, since theoretically they must be the weakest (see Table 32 of Chapter 4), and moreover they are most distant from the ab plane depicted. The next pair of reflections (working anti-clockwise) must have the indices $1\overset{\pm}{1}0$. Theoretically, these should have the greatest intensity, but the corresponding lattice points are somewhat more dis-tant from the ab plane than the points of the third pair, $0\overset{\pm}{2}0$. Because of this the $0\overset{\pm}{2}0$ reflections are hardly any weaker than the $\overset{\pm\mp}{1}10$ reflections. It is apparent here, as in other cases, that secondary scattering has contributed to

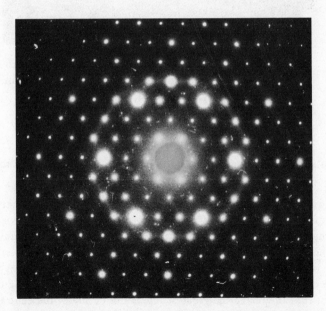

Fig. 36. Diffraction pattern from a kaolinite single crystal showing the regular hexagonal grid and symmetry of the in-tensity distribution possessed by the reciprocal lattice plane perpendicular to the c* axis.

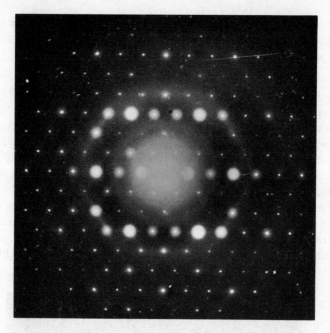

Fig. 37. Kaolinite single-crystal diffraction pattern. The loss of regularity in the grid and intensity distribution indicates that the crystal is tilted.

an equalization of intensities and an increase in the symmetry of their distribution.

In the diffraction pattern shown in Fig. 37, the hexagonal network or grid is distorted. The crystal is rotated by about 12° around an axis perpendicular to the rows of spots. Of the central sextet, the points which may approach closest to the plane inclined at this angle to the ab plane are those with the indices $\overset{\pm\mp}{1}10$. It is clear that it is the reflections with these indices which also lie on the line of maximum tilt ($\psi = \varphi$) on the diffraction pattern, since, theoretically, they have the greatest intensities of any of the central sextet, and they are in fact much stronger than the other four reflections. From this we can determine the directions of the $a*$ and b* axis projections on the pattern without any ambiguity, and thus determine the indices of all the reflections.

a

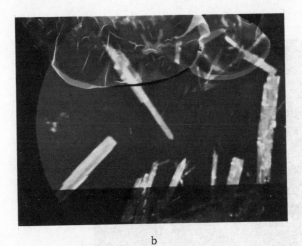

b

Fig. 38. (a) Bright-field image and (b) dark-field image of the same sector of a celadonite specimen.

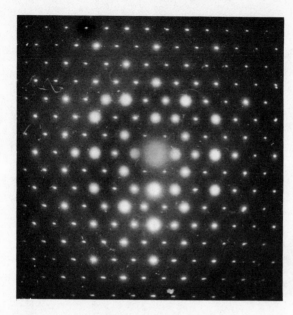

Fig. 39. Diffraction pattern from celadonite single crystal without superlattice.

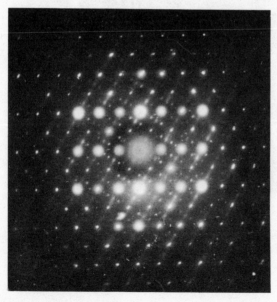

Fig. 40. Diffraction pattern from celadonite single crystal with superlattice.

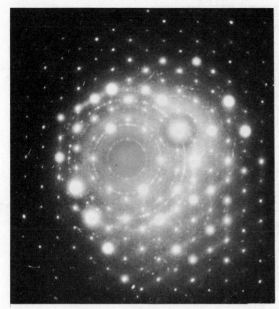

Fig. 41. Diffraction pattern from a large hydromica single crystal and the mass of fine crystallites lying on it. The effects of secondary diffraction are clearly seen, with the intense diffracted rays scattered by the single crystal playing the part of primary beams as they pass through the polycrystalline part, each giving a ring reflection.

Fig. 42. Diffraction pattern from an antigorite crystal.

Fig. 43. Diffraction pattern from a regular combination of two kaolinite crystals rotated relative to one another by 22°.

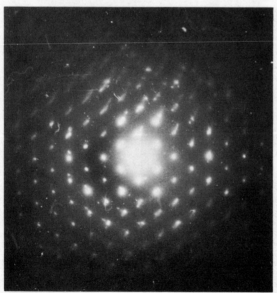

Fig. 44. Diffraction pattern from a deformed kaolinite crystal.

In the microdiffraction method it is necessary to have both bright-field and dark-field images of the crystals as well as diffraction patterns. Figure 38(a) shows a bright-field image of celadonite crystals, and Fig. 38(b) a dark-field image of the same crystals. The variations in contrast here are not due to varying crystal thickness, but to crystal fragmentation, as a result of which different parts do not all end up in equally favorable positions for diffraction. This crystal fragmentation, with disorientation of the blocks, helps the development of the diffraction patterns. These patterns are shown in Figs. 39 and 40. Figure 39 relates to a crystal held roughly perpendicular to the electron beam (the network of reflections is not distorted). The most intense reflections, according to Table 39 of Chapter 4, must be those with 020 indices. Moreover, the corresponding lattice points lie in the ab plane section of the reciprocal lattice. This allows the complete diffraction pattern to be indexed without ambiguity. The other pattern from celadonite (Fig. 40) shows weaker subsidiary spots between those of the basic network. They lie in rows in the hk directions, and divide the distances between the main reflections into six equal parts. Since the main network has unmixed hk indices, the subsidiary reflections correspond to a superlattice constant, equal to three times the absolute value of the celadonite a constant. It is highly probable that the pattern does not correspond to a real superlattice, but is only due to secondary diffraction or to the superimposition of two celadonite crystals, rotated relative to one another.

Secondary diffraction shows up very clearly in the diffraction pattern of a hydromica (Fig. 41). This shows discrete rings, not only around the central spot, but also around the most intense diffraction reflections of the main pattern. These rings are due to secondary diffraction from tiny plates covering the large crystal which gives the basic point pattern.

A genuine superlattice constant is very clearly seen in the diffraction patterns obtained from antigorite single crystals. Antigorite is made up of two-storied corrugated silicate layers arranged in such a way that the waves are parallel to the b axis, and they alternate along the a axis with a superlattice constant of A = na. In most cases, A has a value of about 45 Å, but isolated specimens have been found with smaller (~16 Å) and greater (~100 Å) values of A. In the corresponding diffraction patterns (see Fig. 42), the hexagonal network of point reflections is replaced by a pattern of closely spaced reflection groups, each group running along the a* direction. The distances between adjacent reflections in a group give the value of A directly. From Fig. 42 it can be seen that there are five group reflections in a distance of 4 mm, so that KA* = 0.8 mm (where K is a proportional factor). Over a dis-

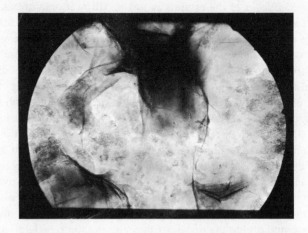

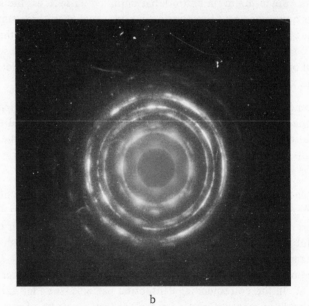

b

Fig. 45. (a) Electron-microscope photograph and (b)
diffraction pattern from a montmorillonite flake.

tance of 55 mm in the b*-axis direction there are 14 gaps between the rows of reflections parallel to $a*$, so KB* = 3.93 mm. Since the distance Ka * is difficult to measure directly, and we can assume that Ka * = $3.93\sqrt{3}$ = 6.8 mm, then from the relationship $a*/A*$ = 8.5 it follows that A = 8.5 a. If $a \approx 5.3$ Å, then A = 45 Å.

A diffraction pattern from a combination of two kaolinite single crystals, not showing secondary diffraction effects or subsidiary reflections, is shown in Fig. 43. The two crystals are rotated relative to one another in such a way that some of the reflections from one crystal are superimposed on reflections from the other, giving an unusual honeycomb pattern. The diffraction pattern shows that the reflection $3\bar{1}0$ is superimposed on 310, 240 on 150, etc., which corresponds to a relative rotation of about 22°. Combinations of kaolinite crystals with this angle of rotation have been observed frequently.

Diffraction patterns from single crystals will also show up deformation of the crystal. If the crystal is bent, the reciprocal-lattice point rows, in particular those parallel to the c* axis, trace out a particular surface (cylinder, cone, hyperboloid). The plane cross section of such a reciprocal lattice will give groups of reflections arranged on second-order curves in the diffraction pattern. Figure 44 shows a diffraction pattern obtained from the curved edge of a kaolinite crystal, with the reflections grouped on arcs of hyperbolas. A detailed and quantitative treatment of this type of pattern has been reported by Ehlers (1954).

Figure 45 shows an electron micrograph of a thin plate of montmorillonite which is very imperfect in its crystallographic morphology. The reflections appear as swellings on continuous circles, indicating that the plate is deformed and that its parts have different relative orientations.

A special case of particular interest is when a layer silicate crystal is curved around an axis lying in the ab plane, and has the morphology of a tube. Crystals of the two-storied layer minerals halloysite and chrysotile have tubular shapes, and they give very distinctive electron-diffraction patterns (see Figs. 24, 46, 47, 48, 49).

Diffraction from cylindrical lattices has been considered from the theoretical standpoint by Fok and Kolpinskii (1940), Waser (1954), Whittaker (1954, 1955), and Jagodzinski and Kunze (1954). The substance of these studies, and their results, may be summarized as follows.

The diffraction properties of tubular objects are described by the general expression for scattering amplitudes (3), if scattering-center coordinates

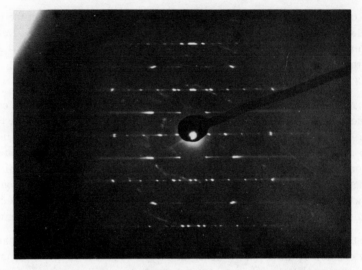

Fig. 46. Diffraction pattern from a clino-chrysotile tube, showing the h0l ($\bar{2}0l$) reflections which fix the c repeat unit at one layer.

characteristic of this type of object are inserted. In the case we are considering it is convenient to use a cylindrical system of coordinates; the positions of points will be defined in terms of the cylinder radius r, the azimuthal angle φ measured from a given reference plane passing through the axis of the cylinder, and the coordinate z measured along the cylinder.

If the scattering centers are arranged on each cylinder on a series of circles, repeating over a distance d, then $z = z_0 + md$. The coordinates φ_j of points arranged in regular fashion on a circle can be represented in the form $\varphi_j = \varphi_0 + 2\pi j/N$, where N is the number of centers on a circle of radius r, so that the distance between adjacent points measured along the arc of the circle is $b = 2\pi r/N$.

A tubular lattice may be formed by making a roll of either a single layer or several superimposed layers. A somewhat approximate approach which is convenient for calculation is to consider the structure as a set of co-axial cylindrical lattices, each of which has its value of r and its own value of r_0, φ_0, but all of which have the same value of the distance b. If the distance between successive cylinders is t, then for cylinder number g we have

$$r_g = r + gt, \quad 2\pi r_g \approx N_g b, \quad N_g \approx N + 2\pi gt/b.$$

The points of the reciprocal lattice can be defined in their turn using their own cylindrical coordinates R, Φ, Z.

In the power indices of the S terms, the scalar products of the reciprocal and direct space vectors should be expressed in terms of these coordinates, remembering that they are equal to the product of the absolute values of these vectors and the cosine of the angle between them. Without going into the details of all the transformations involved, we will just quote the final result obtained by Waser (1954):

$$S = \frac{\sin Q\pi dz}{\sin \pi \, dz} \sum_g N_g e^{2\pi i z_{og} z} \sum_{n=-\infty}^{+\infty} J_{nN_g}(2\pi r_g R) \, e^{inN_g(\Phi + \pi/2 - \varphi_{og})} , \quad (42)$$

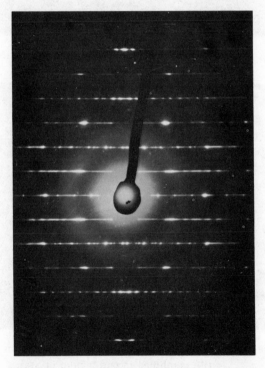

Fig. 47. Diffraction pattern from a clino-chryso-tile tube, showing subsidiary $\overset{+}{2}0l$ reflections which double the c unit.

Fig. 48. Diffraction pattern from an ortho-chrysotile tube.

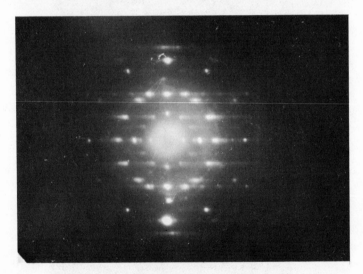

Fig. 49. Diffraction pattern from ortho and para forms of chrysotile combined in one tubular crystal.

where Q is the number of circles on each cylinder, and J_{nN_g} is a Bessel function of order nN_g. Then, just as in the simple case of a parallelepiped crystal with a rectilinear lattice, the square of the modulus of S determines the shape and repeat distances of the reciprocal lattice points. In this particular case, $|S|^2$ must be averaged over its local oscillations with changes in R and Φ.

The first factor in (42) is analogous to the factors in (14). According to the factor, the reciprocal lattice points are arranged on planes, perpendicular to the tube axis, and a distance 1/d apart. Within each of these planes, the arrangements and shapes of the lattice points are determined by Bessel functions of different orders. In particular, the periodicity of the scattering centers along the circles is expressed by the averaged Bessel functions $<J_{nN_g} (2\pi r_g R) >_{av}$. These functions have the property that their highest values occur in the vicinity of values of the argument which are equal to the order of the function, i.e., for $2\pi r R = nN$ or $R = (nN)/(2\pi R) = n/b$. Thus, as R increases, there arises a sequence of lattice points repeating over an interval of about 1/b. The form of the n-th lattice point is determined by the corresponding Bessel function. For n > 0, it shows a steep rise where R < n/b and a gradual fall where R > n/b. Its maximum value falls as n increases and is somewhat displaced from the position $R = n/b_0$ by an amount which depends on n and N. On the point of the periodicity of the cylinders, this is shown up in the structure of the lattice points described by the functions $J_0 (2\pi r_g R)$. In particular, when $z_{0g} = 0$, this periodicity corresponds to the lattice points repeating over a distance of 1/t. If z_{0g} values are randomly distributed, then this arrangement will only hold in the equatorial plane (z = 0). If z_{0g} varies linearly from cylinder to cylinder, this leads to splitting of the lattice points found when $z_{0g} = 0$.

It is natural that the electron-diffraction pattern from a single tube shows a plane section of the reciprocal lattice we have described. The pattern is rather similar to that which would be obtained by rotating a rectilinear (noncurved) crystal about the tube axis. In chrysotiles, the a constant usually runs along the generatrix of the cylinder, and the b constant along the circles. The reciprocal lattice planes lying perpendicular to the tube axis correspond to layer lines of constant h index and separated by distances of $1/d = 1/a \sin \beta$, the lattice points being described by Bessel functions of order nN, i.e., hk reflections with k = n. On the diffraction pattern (see Figs. 46, 47), the h0 reflections are point reflections, while the reflections with k ≠ 0 have a rayed appearance, with sharp inner edges and a gradually fading tail. These reflections are arranged in a hexagonal pattern like the point patterns given by flat platy crystals. The reflections corresponding to the regular repetition of co-axial cylinders separated by distances of t can, for the case where $z_{0g} = 0$, be considered as images of the intersection of a reflecting plane with the h0l lattice points when they are rotated about the c axis. These reflections are arranged along layer lines with h = 0, 2, 4, etc. The distances of these reflections from the image of the a axis are proportional to $ha^* \cos\beta + lc^*$, which can be seen by rotating Fig. 29 about the a axis. Thus, when $\beta \neq \pi/2$, the h0l reflections split up into pairs as the sign of h is changed, the pairs lying on either side of vertical lines parallel to the a axis and passing the 00l points at distances proportional to $\pm |h| a \cos\beta$, where the amount of splitting decides the value of the angle β. Zussman, Brindley, and Comer (1957) have shown that this feature serves to distinguish between ortho-chrysotile and clino-chrysotile. If z_{0g} has a random distribution, then only the 00l reflections can be considered separately, these repeating over a distance proportional to $1/t$ along the equatorial layer line of the diffraction pattern.

In the simplest case, the reflections with k = 0 give the cell and its constant c from the first layer (Fig. 46). This is actually a pseudo-repeat unit. Since there is no strict repeat unit here, embracing a given number of layers, then the clearly represented basal reflections have the successive indices 001, 002, 003, etc. A check shows that these particular diffraction patterns are of clino-chrysotile.

According, however, to work carried out by the present author in conjunction with V. A. Shitov, the electron-diffraction patterns of most clino-chrysotile crystals show subsidiary reflections of lower intensity as well as the usual h0l reflections, and these, in accordance with the formula D = hp + lq, show good agreement with the values q' = q/2, p' = q' − p, and determine the constants of the two-layer monoclinic cell. As well as the h0l subsidiary reflections, there are observed on the rays, or instead of the hk rays with k ≠ 0, rows of closely spaced discrete point reflections, these being more clearly separate, the clearer and more intense are the subsidiary h0l reflections [see Fig. 47]. The points with k ≠ 0 unfortunately do not appear in sufficiently clear or reproducible form on any of the rays hk, $\overset{\pm\ \pm}{hk}$, $\overset{\pm\mp}{hk}$ for them to be used in checking the validity of the double-height cell. It is very probable that at least a portion of these points have a different cause, and are due to interference of waves scattered by cells lying on the same diameter of the tube but on different sides of the tube axis. According to Whittaker (1963), this fine structure in the hk bands is dependent on the shape of the transverse tube cross section and on the diameter of the chrysotile tube.

According to the theory of serpentine polytypism, the observed $20l$ reflections, which correspond to a one-layer cell, are typical of a certain group of polytypal (polymorphic) modifications and also of disordered-structure serpentines belonging to the B structure type (see Chapter 4). Whatever the number of layers included in the c repeat distance of these modifications, and whether or not there is a strict c repeat unit, the arrangement and intensities of these reflections must remain unchanged. The polytypism theory, which was constructed for plane-faced crystals and rectilinear lattices, does not allow for the possible existence of structures in which the $h0l$ reflections are determined by a two-layer monoclinic cell, the more so when half these reflections are characteristic of a one-layer cell. It is necessary to bear in mind that the $h0l$ reflections are not sensitive to any displacements whatsoever of fragments of the structure along the b axis. This leads to the suggestion that the subsidiary $h0l$ reflections are due to distortions in the ideal B-type structures which are such that corresponding atoms in adjacent layers have different x coordinates. It is quite obvious that the bending of the layers and the differences in the radii and arcs of curvature over which the b axis is bent results in the distances between identical atoms in different layers becoming different, and this cannot but lead to breakdown of the positional correspondence existing between atoms of successive layers. However, the appearance of discrete $h0l$ reflections, indicating a repeat unit of two layers, indicates that this breakdown is accompanied by regular displacements of the layers along the x axis by amounts $+\Delta x$, $-\Delta x$, and these displacements occur more regularly, the higher the degree of perfection of the crystal. Crystal perfection favors both intra-tube interference (Whittaker, 1963) and also the formation of fine structure in the hk bands.

Under these circumstances, it is natural that the basal reflections have the sequence of indices 002, 004, 006, . . . , 002n.

Chrysotile electron-diffraction patterns quite often show splitting of the hk rays into pairs of rays, lying on either side of the positions of the nonsplit reflections in the patterns of particles elongated along their a axis. Here, the magnitude of the splitting is proportional to the value of the index k, so that neither $h0l$ nor $00l$ reflections undergo splitting. This effect is without doubt evidence that the tube axis is somewhat tilted away from the [100] direction, the angle of tilt being measurable directly on the diffraction pattern, since the angle ψ between the lines hk and $h\bar{k}$ with the same h is equal to twice ε, the

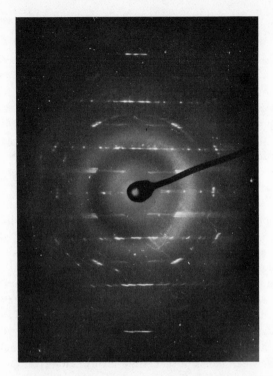

Fig. 50. Diffraction pattern from a clino-chrysotile tube in which layers form coaxial zones
through rotation about [100] and another close-
lying direction ($\varepsilon \sim 2°$).

angle of inclination of the axis of elongation to the axis a, that is, $\psi = 2\varepsilon$.
This angle can be measured most simply from the 0k and $0\overline{k}$ rays. If ξ is the
distance between reflections which are symmetrical relative to the image of
the rotation axis, ζ is the distance between reflections symmetrical relative
to the image of the c^*-axis, and R is the radius of a reflection, then

$$\cos \varepsilon = \xi/2R, \sin \varepsilon = \eta/2R, \tan \varepsilon = \eta/\xi.$$

Under experimental conditions, the value of ε has been found to vary
widely between 0 and 2°. Some diffraction patterns do not show pairs of hk
rays, instead they have triplets, with the middle rays lying in positions
exactly corresponding to the direction of elongation, [100] (the a axis in Fig.
50). This shows that the layers making up a given single tube may have several possible orientations relative to the tube axis.

Diffraction patterns are often found showing increased distances between the layer lines lying parallel to the 00l series of reflections, caused by the axis of the chrysotile tube being tilted relative to the electron beam. From changes in the ratios of the distances between 0k0 reflections and layer lines, it is possible to work out the angle of tilt by a similar method to that described above for single crystals. In addition, these patterns characteristically show reductions in the distances between h0l reflections and hk rays disposed symmetrically with respect to the rotation axis, just as in texture patterns the distances between reflections disposed symmetrically with respect to the major axis of an ellipse are less than the corresponding distances in direct texture patterns, where the ellipses have degenerated into pairs of parallel lines. In particular, the closer hk rays and h0l reflections, normally lying on either side of the rotation axis because of a nonrectangular cell, run together into a single reflection lying on the rotation axis.

Fig. 51. Diffraction pattern from a prismatic-tube crystallite of halloysite, clearly showing the unusual relative arrangement of the 00l and 02l reflections.

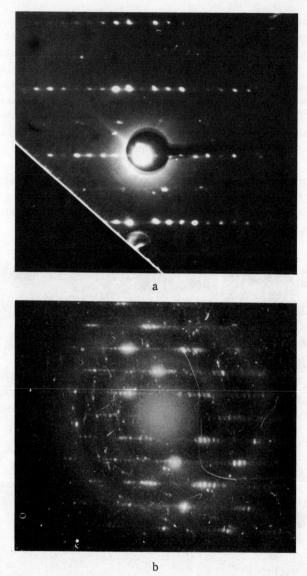

a

b

Fig. 52. Diffraction patterns from sepiolite fibers. (a) With the bc plane normal to the electron beam; (b) in an inclined position. The c axis runs along the fiber.

A regular c repeat unit, in the radial plane of a cylinder, is in some degree of contradiction with a constant b repeat unit along circles of different radii r. Even if the scattering centers of different cylinders in one radial plane were arranged in regular fashion, this ordering could not be retained in another radial plane unless the b repeat units changed in proportion to the radii of the cylinders.

It is more likely that an ordered arrangement of scattering centers along radii exists in more or less plane-parallel blocks, formed through the creation of cracks and fissures in the tubes to relieve stresses due to lattice deformations. In this connection, the diffraction patterns obtained from halloysite single crystals have considerable significance.

In halloysites, the long direction of the tube usually coincides with the b axis. The reflections lie in layer lines, each of which has a given value of the index k. As a rule, practically all the reflections are separate. This is an indication that the concept of a cylindrical halloysite lattice does not correspond exactly with reality. More precisely, halloysite crystals are prismatic, with approximately plane faces. The diffraction patterns of halloysite crystals often appear as two patterns superimposed, containing hk and $0kl$ reflections (Fig. 51). It is worth noting that the $02l$ reflections lie on intermediate levels compared to the $00l$ reflections.

Thus, the $00l$ reflections have only even indices, with $l = 2l'$, while these indices are odd in the $0kl$ reflections, i.e., $l = 2l' + 1$. Because of this, diffraction patterns from halloysite single crystals give a cell constant of $c \simeq$ 14 A, which embraces two two-storied layers. This result, which was first obtained by some Japanese workers (Honjo, Mihama, 1954; Honjo, Kitamura, Mihama, 1954), has also been confirmed in a diffraction pattern obtained by Popov and the present author (1959) (see Fig. 24). In Chapter 4 we shall see which structural modification corresponds to a given observed intensity pattern. It is of great importance to determine to what extent the above result applies to the mineral halloysite in general, rather than to certain rare individual crystals only.

It will be appreciated that such behavior must not be taken as an indication that we were dealing with ortho-chrysotile in the case under consideration, because the remaining reflections were unequally distributed, as with clino-chrysotile. It is true that the distances between these reflections were altered, which had to be taken into account in working out the cell constants.

In contrast to chrysotile and halloysite crystals, fibrous particles which are not in the form of tubes give patterns typical of platy crystals, although admittedly rather elongated ones. In the particular case of a sepiolite fiber diffraction pattern (see Fig. 52), this shows a b constant three times as large as that in ordinary layer silicates.

The fact that the single-crystal diffraction patterns of different minerals are different means that we can use the method to investigate mixtures of minerals and aggregates of particles. Figure 53 shows an electron-microscope photograph of a clump of particles, and diffraction patterns from its isometric and elongated particles, respectively. The diffraction patterns show that the isometric particles are plates of the single-layer platy serpentine lizardite, these being covered by randomly oriented fine chrysotile fibers, while the elongated particle is a chrysotile tube.

Electron-Diffraction Patterns from Disordered Poly-crystalline and Textured Specimens. If the specimen contains a number of crystals, then their reciprocal lattices, in orientations related to those of their direct lattices, are combined into a single reciprocal lattice with a common origin of coordinates.

If the crystals forming a true polycrystalline specimen do not possess any shared general orientation, their combined reciprocal lattice will be formed from the reciprocal lattice of an individual single crystal by rotating it about a single fixed point, the origin of coordinates. Under these conditions, each lattice point describes a sphere, and the reciprocal lattice nodes consist of a set of concentric spheres, the radii of which are equal to the lengths of the corresponding $\mathbf{H}$ vectors. The diffraction pattern from such a polycrystalline specimen, being the image of a plane cross section of its reciprocal lattice, consists of a set of concentric circles of radii $R = L\lambda H$. By the use of equation (34), these patterns give the values of the interplanar lattice distances d_{hkl} directly.

An electron-diffraction pattern of this type can only be indexed, and the unit cell determined, through an analysis of the relationship $1/d^2 = |\mathbf{H}|^2$; if this is expanded as the product of the vector $\mathbf{H}$ by itself, this gives, in the general case of a triclinic lattice, a quadratic formula of the type

$$1/d^2 = h^2 a^{*2} + k^2 b^{*2} + l^2 c^{*2} + 2hka^*b^* \cos \gamma^* +$$
$$+ 2klb^*c^* \cos \alpha^* + 2lhc^*a^* \cos \beta^* =$$
$$= (1 - \cos^2 \alpha - \cos^2 \beta - \cos^2 \gamma + 2 \cos \alpha \cos \beta \cos \gamma)^{-1} \times$$

$$\times \left\{ \frac{h^2}{a^2} \sin^2 \alpha + \frac{k^2}{b^2} \sin^2 \beta + \frac{l^2}{c^2} \sin^2 \gamma + \frac{2hk}{ab} (\cos \alpha \cos \beta - \cos \gamma) + \right.$$

$$\left. + \frac{2kl}{bc} (\cos \beta \cos \gamma - \cos \alpha) + \frac{2lh}{ca} (\cos \gamma \cos \alpha - \cos \beta) \right\}.$$

If the symmetry group is higher than triclinic, this expression is considerably simplified. Nonetheless, the indexing of a diffraction pattern of this type is quite a complicated and tedious business, whether carried out by either the analytical or the graphical methods available (Kitaigorodskii, 1952).

It is much easier to index diffraction patterns if the polycrystalline specimens have any sort of preferred orientation. In view of the features of the specimen-preparation methods discussed above, such orientation is very probable. If the crystallites have nonisometric shapes, some type of preferred orientation is virtually inevitable.

The most common and practically important arrangement is one in which the crystallites have one face parallel to the supporting film, but without any shared azimuthal orientation relative to the plane of the support, and so form a platy texture. This texture is particularly characteristic of layer silicate electron-diffraction specimens, because their crystallites usually have a platy shape, but it can also occur when the crystallites are not obviously platy, but fibrous or isometric instead.

As might be imagined, the reciprocal lattice which corresponds to a platy texture is formed by rotating the single-crystal reciprocal lattice about a normal to the film, i.e., about the texture axis. In layer silicates, each crystallite lies with an ab plane on the film, so this plane forms the texture plane, and the texture axis lies along the $c*$ reciprocal lattice axis. When the lattice is rotated about the texture axis, lattice points with like hk indices, lying on lines parallel to $c*$, form nodes having the shapes of sets of rings arranged on coaxial cylinders (Fig. 54).

In such a lattice, the position of a ring-node is given by the radius of the cylinder b'_{hk}, which is equal to the projection of the corresponding **H** vector on the ab plane, and by the distance D'_{hkl} measured along the cylinder surface from the ring to the ab plane, which is equal to the projection of the **H** vector on the $c*$ axis.

According to equation (39), the projection of **H** on the ab plane is equal

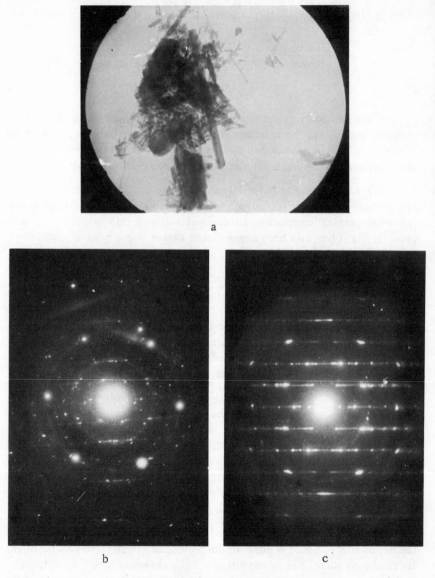

Fig. 53. Illustration of phase analysis by the microdiffraction method. (a) General view of an aggregate specimen; (b) diffraction pattern from a platy particle (lizardite); (c) diffraction pattern from a tubular particle (chrysotile).

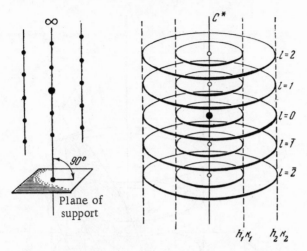

Fig. 54. Formation of the reciprocal lattice of a textured polycrystalline specimen by rotation of a single-crystal reciprocal lattice about the texture axis (Vainshtein, 1956).

to

$$\mathbf{b}'_{hk} = \frac{h}{a}\,\mathbf{i} + \frac{k}{b}\,\mathbf{j}, \tag{43}$$

where $\mathbf{i},\mathbf{j}$ are the unit vectors along the directions of the axes $\mathbf{a},\mathbf{b}$, and

$$b'_{hk} = \sqrt{\frac{h^2}{a^2} + \frac{k^2}{b^2}}. \tag{44}$$

Since $a \simeq b\sqrt{3}$ for layer silicates, then

$$b'_{hk} = \frac{\sqrt{3h^2 + k^2}}{b}. \tag{45}$$

The projection of the vector $\mathbf{H}$ on the c^* axis is

$$D'_{hkl} = ha^* \cos\beta^* + kb^* \cos\alpha^* + lc^*. \tag{46}$$

As the diffraction pattern from a textured polycrystalline specimen represents a reciprocal-lattice cross section, it consists of reflections lying on a plane section through the ring-nodes, arranged in accordance with the disposition of these rings relative to the reflecting plane.

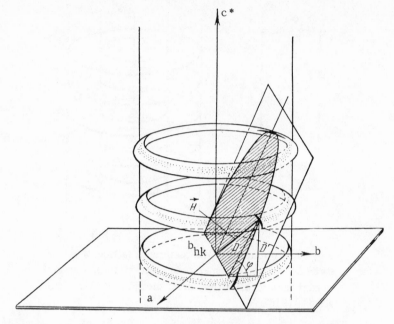

Fig. 55. Oblique section through a ring-node cylinder. The vector **H**
is projected on the plane ab and the axis $c*$.

Because the ab crystal faces are not all exactly parallel to the plane of
the support, the reciprocal lattice nodes actually have the shape of spherical
bands. The reflections on the diffraction patterns therefore have the shapes of
arcs of circles, and these are arranged, according to their hk indices, on plane
sections of cylinders, i.e., on ellipses with shapes depending on the angle of
inclination φ of the specimen to the electron beam (Pinsker, 1949).

If the specimen is perpendicular to the electron beam ($\varphi = 0$), the pattern represents a normal section of the reciprocal lattice (perpendicular to the
$c*$ axis), the ellipses are transformed into circles, and the diffraction pattern
only shows those reflections corresponding to rings lying in the ab plane of the
reciprocal lattice. These reflections, being normal sections of spherical bands,
appear as continuous circles (Fig. 20). At a given angle $\varphi > 0$, the diffraction
pattern shows an oblique section of the reciprocal lattice (Figs. 21, 55), containing ellipses with minor axes b_{hk} equal to the corresponding circle diam-
eters on the perpendicular photograph, and with major axes equal to $a_{hk} = b_{hk}/\cos \varphi$. So the greater the value of φ, the more elongated the ellipses,

the shorter the reflection arcs, and the more reflections are contained on the pattern (see Fig. 78). However, the $00l$ lattice points, which lie along the c^* axis, only intersect the reflecting plane, and thus only appear on the diffraction pattern, when $\varphi = 90°$, which only applies in practice during studies of textured specimens by reflection. In this case, the diffraction pattern shows a direct section of the reciprocal lattice, with the ellipses degenerated into pairs of straight lines (see Fig. 23).

With the reflection positions found in these diffraction patterns, the cylinder radius b_{hk}^* in the reciprocal lattice corresponds to the minor axis b_{hk} of the hk ellipse, and the distance D_{hkl}^* of the hkl node from the ab plane corresponds to the distance D_{hkl} of the reflection from the minor axis of the ellipse. It is convenient to call this latter quantity the height of the reflection.

It is quite obvious (see Fig. 55) that

$$b_{hk} = L\lambda \sqrt{h^2/a^2 + k^2/b^2}, \tag{47}$$

so that for layer silicates

$$b_{hk} = L\lambda \sqrt{3h^2 + k^2}/b, \tag{48}$$

$$D_{hkl} = \frac{L\lambda}{\sin \varphi}(ha^* \cos \beta^* + kb^* \cos \alpha^* + lc^*) = hp + ks + lq. \tag{49}$$

From the properties of the reciprocal lattice, we know that when $\gamma = \pi/2$,

$$\cos \beta^* = -\frac{\cos \beta}{\sin \alpha}, \ \cos \alpha^* = -\frac{\cos \alpha}{\sin \beta}, \ \cos(\widehat{cc^*}) = \sqrt{1 - \cos^2\alpha - \cos^2\beta}.$$

If we substitute these values in (49) and apply (39), we get

$$p = -\frac{L\lambda}{\sin \varphi} \cdot \frac{\cos \beta}{a \sqrt{1 - \cos^2\alpha - \cos^2\beta}},$$

$$s = -\frac{L\lambda}{\sin \varphi} \cdot \frac{\cos \alpha}{b \sqrt{1 - \cos^2\alpha - \cos^2\beta}}, \tag{50}$$

$$q = \frac{L\lambda}{\sin \varphi} \cdot \frac{1}{c \sqrt{1 - \cos^2\alpha - \cos^2\beta}}.$$

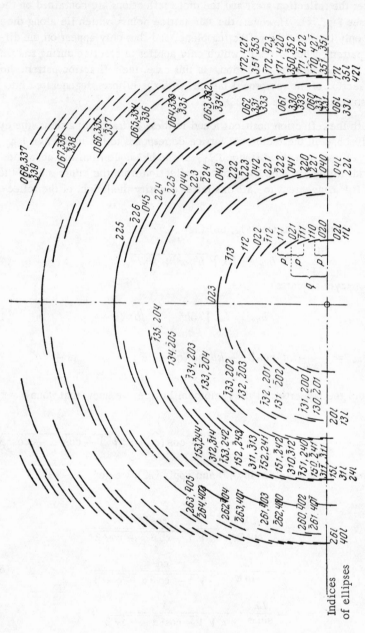

Indices
of ellipses

Fig. 56. Diagram of the geometrical arrangement of hk*l* reflections in an oblique texture pattern from a mineral with three-storied layers, modification 1M (p = q/3; only half the pattern is shown).

Table 14. Sequence of Ellipses on Oblique Electron-Diffraction Texture
Patterns from Layer Silicates

Ellipse no.	hk	$3h^2 + k^2$	Ellipse no.	hk	$3h^2 + k^2$
1	02, 11	4	11	0.10, 55	100
2	13, 20	12	12	39, 60	108
3	04, 22	16	13	2.10, 48, 62	112
4	15, 24, 31	28	14	1.11, 57, 64	124
5	06, 33	36	15	0.12, 66	144
6	26, 40	48	16	3.11, 4.10, 71	148
7	17, 35, 42	52	17	2.12, 59, 73	156
8	08, 44	64	18	1.13, 68, 75	172
9	28, 37, 51	76	19	4.12, 80	192
10	19, 46, 53	84	20	0.14, 3.13, 5.11, 77, 82	198

The equations (47)-(49) express the geometrical rules which govern the
positions of the reflections. An analysis of these positions with the use of these
equations will give the hk indices of the ellipses, the quantities p, s, and q,
and then the l indices of the reflections. It is therefore possible to compare
different approaches to a structure analysis problem right at the diffraction
pattern examination stage, before any structural data have been derived, by
using the simple quantities p, s, and q, which are directly and graphically related to the arrangement of the reflections and the unit cell. As soon as the
choice of p, q, and s has been shown to be well-founded by the geometry of
the diffraction pattern, equations (47)-(50) will give the unit cell directly.

According to (48), a given ellipse will contain not only reflections with
the same hk indices, but also reflections which have the same value of $3h^2 +
k^2$. Because of this, one ellipse may contain several hk combinations. Due to
the base-centering of layer silicate cells, the indices hk are unmixed, i.e.,
all even or all odd. The sequence of ellipses and hk indices corresponding to
these is easy to establish by setting out all possible combinations of pairs of
unmixed numbers in order of increasing value of $3h^2 + k^2$ (see Table 14).

The fact that there is a single set of ellipses for all layer silicates is an
expression of their common features. Their dissimilar features are shown up
individually through the arrangement and qualitative properties of the reflections in each ellipse. Figure 56 is a diagram of the geometrical layout of re-

flections in diffraction patterns from mica-type minerals with a repeat unit of one layer.

After measuring the minor axis of any ellipse, formula (48) can be used to determine b, and then $a = b/\sqrt{3}$. These constants can be found most simply from the minor axes of the first two ellipses,

$$b = 2L\lambda/b_{02}, \quad a = 2L\lambda/b_{20}, \tag{51}$$

but it is more accurate to use the minor axes b_{06} of the fifth ellipse and b_{40} of the sixth ellipse.

To find c, α, and β, the other quantities which characterize the unit cell of a layer silicate, it is necessary to analyze the distribution of reflections on the ellipses to find the values of p, s, and q.

In an oblique texture pattern, the whole set of diffraction points is represented by the reflections in one quadrant. The arcs which are symmetrically placed relative to the major axis of the ellipse are, in fact, sections of the same reciprocal lattice ring-node. The reflections on the lower half of the pattern are sections of reciprocal lattice rings which are centrosymmetric and equivalent to those giving the reflections of the upper half. We can therefore consider only those reflections with D, p, q, s $\geq$ 0. According to (50), these will yield a cell with $\alpha, \beta \geq \pi/2$.

If we examine the sequence of reflections on any ellipse, the following features may be noted. Since $D \geq 0$, then for any ellipse the indices $l \leq$ (hp + ks)/q can only be positive; the hk indices can take a plus or minus sign. If $l \leq$ (hp + ks)/q, the hk indices take a plus sign, and the l index a plus or minus sign (see the diffraction pattern diagram in Fig. 56, where the arrangement of the reflections obeys the condition that p = q/3).

To sum up, then, for a given absolute value of the index l, there exists on a given ellipse two reflections for each index h, k $\neq$ 0, or one reflection for each index h, k = 0. No matter what hk$l$ index is chosen, the next reflection to it with the same hk indices, but with its third index equal to l + 1 or $l-1$, must be separated from it along the height D by a distance of $\Delta D = q$. Thus, in any interval $\Delta D = q$, there fall reflections of all possible hk combinations for the given ellipse, and none of these can be met with in the given interval more than once, or else the difference in height of reflections with the same hk and successive l would be greater than or less than q. Consequently, q is equal to the interval of height occupied on the given ellipse by that number of successive reflections which is equal to the number of differ-

ent hk combinations for the ellipse, these combinations differing either in absolute values or in hk values, or in both simultaneously.

The absolute values of hk are determined in the first stage of analysis
of the diffraction pattern, from the values of b_{hk}. For the layer silicates they
are already known, having the values indicated in Table 14; the signs can be
established from an examination of the way the reflections deviate from the
orthogonal law $D = l\,q$ and the monoclinic law

$$D = hp + lq. \tag{52}$$

Actually, if the cell is orthorhombic, the reflections hkl, $\bar{h}kl$, $h\bar{k}l$, $h\bar{k}\bar{l}$
coincide, and the reflections are arranged on layer lines parallel to the minor
axes of the ellipses. If the cell is monoclinic, the reflections hkl, $h\bar{k}l$ coincide, and the reflections hkl and $\bar{h}kl$ lie on either side of the line $D = l\,q$,
at distances of $\Delta D = \pm hp$.

If the cell is triclinic, i.e., $s \neq 0$ in addition, then the changes in sign
of the index k lead to a splitting of the reflections which would correspond to
a monoclinic lattice, these here forming pairs lying away from the $D = hp +$
$l\,q$ level at distances of $\Delta D = \pm ks$. By considering the mode of arrangement
of the reflections, it is possible to draw some preliminary conclusions on the
number of variables governing the heights of reflections on the patterns and
on the shape of the crystal lattice unit cell; the first move here is to compare
the heights of the $0kl$ and $h0l$ reflections, which show the largest of the split-
tings which are due to the cell being monoclinic or triclinic, particularly if p
or s is small.

After it has been established, even tentatively, that the quantities p and
s differ from zero, and thus that the number of different reflections on each
ellipse having a given value of l is known, then q is measured, this being the
difference in height of successive reflections on the ellipse having the same
hk values and differing only in their value of the index l. These reflections
can be easily distinguished according to the methods given above, as long as
they are all separately represented on the diffraction pattern. If then the number of different reflections for each l is equal to n, no matter which reflection
is chosen, at a distance of n + 1 reflections from it on the particular ellipse
will be a reflection having the same hk indices and an l index which differs
by one. Therefore, q is equal to the difference in height between the first and
(n + 1)-th reflections. So when p = s = 0 (orthorhombic cell), D does not depend on hk, and q is equal to the difference in height between successive reflections. If only s = 0 ($\alpha = \pi/2$, monoclinic cell), then for layer silicates q
is equal to the difference in height between every second reflection for the
first and third ellipses, between every third reflection for the second ellipse,

every fifth reflection for the fourth ellipse, etc. If, in addition, s ≠ 0, then the number of reflections entering into the calculation is increased according-ly. After determining q, the values of p and s can be found from the differ-ence in height of reflections at the levels D = lq.

In practice not all the reflections appear on the diffraction pattern, some being weak, and others being closely spaced and overlapping. For ex-ample, for minerals made up of three-storied layers, the fifth ellipse usually does not show separately the trio of reflections 06l, $\overline{3}3(l+1)$, 33($l-1$); in-stead, these appear as one (Fig. 56). To find the true values of p, q, and s, it is therefore necessary to compare the heights of reflections on different el-lipses carefully, to make an effort to avoid overlooking weak reflections, and not to take several close-lying reflections as a single reflection. The criteri-on for a correct choice of values of p, q, and s is whether or not it is possible, using (49), to express D for all reflections in terms of the whole numbers hk$l$. Examples of distributions of reflections on ellipses, with their heights D ex-pressed in terms of p, q, and s, will be found below in Tables 16-27, and also in the section dealing with the description of electron-diffraction patterns used for structural studies in Chapter 5.

It should be noted that because there is some freedom of choice in se-lecting the cell in the direct lattice, and the corresponding cell in the recip-rocal lattice, the quantities p, q, and s can be chosen in innumerable ways, such that the heights of reflections will be expressed by equation (49) in the form of whole numbers hk$l$. Naturally, in different cases these indices will be different. If we know the relationship between the values of p, q, and s for the different cells, then from formula (49) we can not only index a diffrac-tion pattern, we can also find in a simple manner the reflection-index trans-formation corresponding to a given coordinate-system or cell transformation.

In the case of a monoclinic cell, the whole set of values p', q' which satisfy the equation D = hp + lq for integral h, l for a given direction of the texture axis c* (or, which comes to the same thing, for a fixed texture plane ab) can be expressed by the equations

$$p' = \frac{mq}{n} \pm p, \quad q' = \frac{q}{n} \quad (m = 0, \pm 1, \pm 2, \ldots; \quad n = 1, 2 \ldots). \quad (53)$$

In actual fact, q' can be equal to any fractional part of the height inter-val q, and p' can be taken as the distance along the height from the base level to any reflection not lying on any of the levels D = lq'.

Since

$$D = hp + lq = \pm hp' + (nl \mp mh)\, q',\tag{54}$$

then

$$h' = \pm h, \quad l' = nl \mp mh.$$

It will be apparent that since the repeat distance of reflections along the height is q' < q, then not all the heights D' = h'p' + l'q' will satisfy the equation D = hp + lq.

Intermediate reflections, if they exist, have the heights

$$D = h'p' + l'q' = \pm h'p + \frac{h'm + l'}{n}\, q.\tag{54'}$$

When used for indexing the values of p' and q', the reflections may have arbitrary h' l' indices, and to these there must correspond, for the quantities p and q, the index h = ±h' and the fractional index $l = (h'm + l')/n$. Out of the vast set of reflections which give lower values of q' = q/n, the reflections which also belong to the set with maximum q' = q (when n = 1) are distinguished by the fact that they obey a relationship derived from (54'):

$$l' + mh' = nl.\tag{55}$$

For the other reflections, of course,

$$l' + mh' = nl + n'\, (n' < n).\tag{56}$$

To satisfy the requirements for a unit cell (minimum c and β as close to a right angle as possible), we must select from all the possible values the lowest value of p and the highest value of q.

We will consider some special cases.

1. p' = q − p, q' = q. Correspondingly, h' = −h, l' = l + h. Since p = q − p', then p and p' are equally correct, and from them the lowest should be chosen, i.e., the one which is less than q/2.

2. p' = p, q' = q/2. Then for the reflections satisfying the values of p, q, h' = h, l' = 2l; for the others, h' = h, l' = 2l + 1. However, if the diffraction pattern contains no reflection with D = hp + (2l + 1)q', the choice of these values of p' and q' and the corresponding doubling of c would not be justified. If such reflections are observed, then this choice would be necessary.

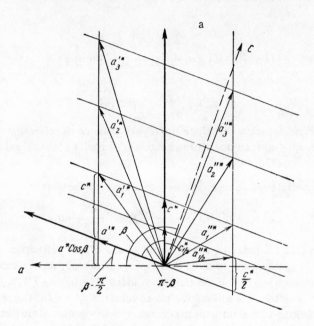

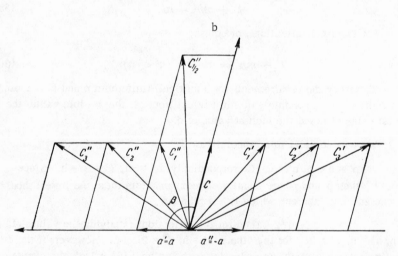

Fig. 57. Relative arrangement of axes: (a) $a*$, $c*$ in the reciprocal lattice; (b) a, c in the direct lattice.

If here $p > q/2$, then we should take $p" = q/2 - p$. The corresponding new indices would be $h" = -h$, $l" = l' + h$. The reflections with heights satisfying equation (52) would take the indices $h" = -h$, $l" = 2l + h$. The value $q' = q$ would be satisfied by reflections with $l" + h" = 2l$.

3. $p' = p$, $q' = q/n$, $h' = h$, $l' = nl + n'$ (where $n' < n$). This transformation would also only be just justified if the corresponding intermediate reflections were present. If, here, $p > q/2n$, it would be necessary also to select another, minimum p' in accordance with formula (53).

Since, according to equations (46) and (49), p and q are proportional to the projections on the $c*$ axis of the translations $a*$ and $c*$, respectively, then their different values, as in formula (53), correspond to transformation of the reciprocal lattice axes:

$$a'^* = \pm a^* + \frac{m}{n} c^*, \qquad (57)$$

$$c'^* = \frac{c^*}{n}. \qquad (57')$$

We know that for any linear transformation of the axes of a direct lattice, the axes of the corresponding reciprocal lattice are transformed using a matrix derived from the matrix of the inverse transformation of the direct lattice axes by replacing rows by columns.

Since, geometrically, the direct and reciprocal lattices are mutually covariant, and the direct lattice is "inverted" with respect to the reciprocal lattice, the transformations (57) for the axes of a reciprocal lattice correspond to the following transformation of a direct lattice:

$$a' = \pm a, \qquad c' = \mp ma + nc. \qquad (58)$$

Figures 57(a) and 57(b) represent different directions of $a*$ and $c*$, and the corresponding directions a, c, including the second special case examined above where $m = 1$ and $n = 2$. The value corresponding to this case is $p' = (q/2) - p < p$, and the angle formed by the axes $a' = -a$, $c' = 2c + a$ is closer to $\pi/2$ than is the original angle β.

In the general case corresponding to transformation (58), the relationship between the monoclinic angles β' and β can be derived in analytical form by comparing the corresponding p, p' values which are related by equation (53).

Table 15. Scheme of Monoclinic Heights

h	....	$\bar{2}$	$\bar{1}$	0	1	2	3···	···	l	···
0				0	q	$2q$	$3q$	···	lq	···
$\bar{1}$					$-p+q$	$-p+2q$	$-p+3q$	···	$-p+lq$	···
1				p	$p+q$	$p+2q$	$p+3q$	···	$p+lq$	···
$\bar{2}$					$-2p+q$	$-2p+2q$	$-2p+3q$	···	$-2p+lq$	···
2				$2p$	$2p+q$	$2p+2q$	$2p+3q$	···	$2p+lq$	···
$\bar{3}$						$-3p+2q$	$-3p+3q$	···	$-3p+lq$	···
3			$3p-q$	$3p$	$3p+q$	$3p+2q$	$3p+3q$	···	$3p+lq$	···
....										
$\bar{h}$								···	$-hp+lq$	···
$\bar{h}$	.... $hp-2q$	$hp-q$	hp	$hp+q$	$hp+2q$	$hp+3q$	···	$hp+lq$	···

For a monoclinic lattice ($\alpha = \pi/2$), according to (50),

$$p = -\frac{L\lambda}{a\tan\beta\,\sin\varphi}, \qquad q = \frac{L\lambda}{c\sin\beta\,\sin\varphi}; \tag{59}$$

$$\frac{\tan\beta}{\tan\beta'} = \frac{p'}{p} = \frac{m}{n}\frac{q}{p} \pm 1, \qquad \cot\beta' = \frac{-\dfrac{m}{n}\dfrac{a}{c} \pm \cos\beta}{\sin\beta}. \tag{60}$$

After obtaining reliable values of p, q, and s, with which the determination of the unit cell constants is finalized, the final indexing of the diffraction pattern is carried out.

As already noted, the indexing is carried out by means of equation (49). It will be helpful here to compile a table of theoretical reflection heights. It is convenient to first take the "monoclinic" heights as D = hp + lq (52), putting down the heights of the reflections on each ellipse, and then "split" each "monoclinic" reflection into two "triclinic" reflections with heights of

$$D_{\text{tri}} = D_{\text{mon}} \pm ks.$$

The table of "monoclinic" heights should include values of $D \geq 0$. We end up with the following type of scheme (for clarity it has been assumed that $q \approx 3p$) (Table 15).

If we compare the theoretical heights with the experimental ones, this will provide a final check on how well these are satisfied by the values of p,q and s found, and will show whether or not there are extraneous reflections present which do not fit in with the lattice adopted (if there are, it must be considered whether these can be related to impurities or to combination with the dominant structure of another structure or structure modification). If there is good agreement, then the indices of the reflections are found, and it can be seen which reflections coincide, partly overlap, or just touch; this is of the utmost importance in the subsequent evaluation of intensities.

If platy crystals have a planar, nonisometric shape, for example, if they exist as elongated plates, then the orientation scatter which exists in a textured specimen will be different in the different crystallographic directions. The crystals show a departure from strict parallelism with the texture plane (by an angle $\Delta\alpha$) which is greater, the smaller the degree of elongation in the particular direction. For the same difference t in level between two extreme points on a crystal, $\tan(\Delta\alpha) = t/l$, where l is the distance between these points.

Correspondingly, the transverse dimensions of the spherical reciprocal lattice bands, and thus also the lengths of the arc reflections on the diffraction pattern, will be variable and dependent on the hk$l$ indices. The more compact reflections can be treated by the method of calculation which uses the height D; for the others, the radius R must be measured, and this converted to D using formulas (3)-(4) of Chapter 2.

If the set of values obtained for D is sufficient for determination of the unit cell, then, using the theoretically possible values of b_{hk} and the levels $D = hp + ks + lq$, it will be possible to construct the two-dimensional rectilinear texture network or grid. If, then, a system of circles with experimentally measured radii is constructed, the places where they intersect the two-dimensional grid points will give the theoretically possible hk indices, which are the same as the indices of the corresponding vertical straight line b_{hk}, and the index l, equal to the last index of the horizontal straight line D_{hkl}. Then, by comparing these results with the actual diffraction pattern, the indices of the reflections actually present can be determined.

An example of the construction of such a rectilinear texture grid will be given below in the description of an electron-diffraction study of sepiolite.

Vainshtein (1956) has previously noted that although the curvature of the reflecting sphere is extremely small, it nonetheless has an effect on

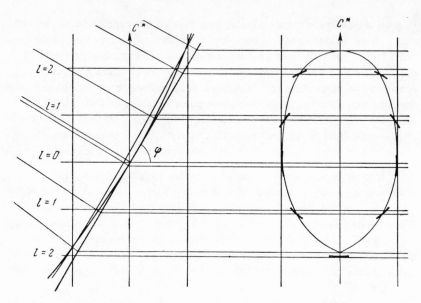

Fig. 58. Distortion of the diffraction pattern due to curvature of the reflecting "plane." The diagram shows the projection on two perpendicular vertical planes of the section where the reflecting "plane" is cut by the reciprocal lattice cylinder.

texture patterns, destroying the mirror symmetry relative to the minor axes of the ellipses. From Fig. 58 it can be seen that the upper part of the ellipse (which is not really a true ellipse, but is somewhat distorted) corresponds, as it were, to a larger value of φ, at which the ellipse is more drawn out and the reflections spread wider, while the lower part of the ellipse has the opposite character. Here it may be added that since the intersection of the reflecting sphere with the cylinder is projected on the plane along the directions of the sphere radii, the ellipse is compressed below, the degree of compression being greater, the greater the radius of the cylinder. So the zero line on the diffraction pattern is not a straight line at all, but bends downward as it moves away from the center. This point must be taken into account when measuring the heights D, and using them in geometrical analysis of the diffraction pattern.

The geometrical analysis of texture patterns described above was worked out in studies on layer silicates, mostly clay minerals, and relates to the special case of a textured specimen with a nonrectangular cell, but with a rectangular base ($\gamma = \pi/2$). It is based on the direct relationship between the diffraction pattern and the reciprocal lattice, and makes use of quantities which are directly tied up with the unit cell constants (p,s represent the inclinations of the a^* and b^* axes to the ab plane; q corresponds to c^*, etc.). Thanks to the graphic nature of the formulas describing the distribution of the reflections, it is possible to see in advance the relationship between the true and minimum c constants, and whether the lattice is triclinic or monoclinic, which is shown by whether the well-known $02l$ reflections are split or not. Extraneous reflections which might cause a structural possibility to be discounted can be detected reliably, and the reflections can be indexed in a most simple manner. Moreover, if it is known where reflections must theoretically lie, it is possible to search for weak reflections and determine whether a particular arc represents one or several overlapping reflections (and the degree of overlap).

It is not difficult to see that the methodical approach and results given above can be used, to a greater or lesser extent, in analysis of electron-diffraction patterns from other types of specimens, and also in X-ray structural analysis when dealing with rotation photographs, and even powder photographs.

In the geometrical analysis of electron-diffraction photographs, Vainshtein's methods (1956) can also be used. They are universal in the sense that they work from measurements of the radii of reflections, and do not depend on the shape of the curves (ellipses, hyperbolas, etc.) on which the reflections lie or on a relationship between this and the unit cell and its orientation relative to the texture axis. In the particular case where reflections grouped on ellipses are used, then, taking into account the fact that these represent lattice points lying on straight lines parallel to the c^* axis, formulas (37) and (38) can be used, these also being applicable in analysis of point-diffraction patterns.

As well as transmission electron-diffraction texture patterns, it is also important to obtain reflection patterns, although this is a much more complicated business.

Reflection patterns must show a direct section of the reciprocal lattice ($\varphi = \pi/2$), and this section must pass through the $00l$ lattice points which lie along the c^* axis. These patterns show the same relationships as oblique texture patterns, except that the ellipses degenerate into parallel lines, and an additional zero "ellipse" comes in, i.e., a straight line passing through the origin, with the $00l$ basal reflections lying on it. Moreover, since this is a reflection pattern, it will show only the upper half of the reciprocal lattice section. In its analysis we can therefore use the rules given above for oblique texture specimens, but with $\varphi = \pi/2$. In reflection studies on powdered specimens, layer silicates may show side effects due to diffraction by crystallites lying in positions in which they are more "permeable" to electrons. Thus, as well as a texture pattern with a $c*$ axis, there may appear a texture pattern with a $b*$ axis (see Fig. 23, a pattern obtained from celadonite by the author and Shakhova, 1957), or a point pattern of a single crystal standing end-on to the reflecting surface (see Fig. 35, obtained from kaolinite by the author and Popov, 1959). There have been very few reflection studies on powdered silicate specimens as yet, and it is possible that new methods of analyzing these will be developed in the future.

3. The Distribution of Reflections on Ellipses on Electron-Diffraction Texture Patterns from Layer Silicates

Above we noted some of the features of electron-diffraction texture patterns from layer silicates, namely, those due to the similarity between flat silicate layer motifs all possessing a centered ab cell face with $b = a\sqrt{3}$. Here the reflections were grouped on certain sets of ellipses, in which the ratios of the lengths b_{hk} of the minor axes were the same for all the layer silicates, while the absolute values of b_{hk} were determined by the cell constants a,b, which in their turn depended on the chemical constitution of the layers.

The ellipses were characterized by the unmixed hk indices given in Table 14, with identical values of $3h^2 + k^2$.

The individual features of the three-dimensional arrangements of layers in silicate structures, their relative positioning, and the ordering, or conversely, the disordering of the structures, are also reflected in their electron-diffraction patterns. They show up in the particular arrangement, qualitative character, and intensity distribution of the reflections in each ellipse.

The constants c and β in the unit cell, and thus the spacing of reflections along each ellipse, depend primarily on the types of layer making up the structure. Two-storied kaolinite-type layers, three-storied mica-type layers, and chlorite packets give c constants which are approximate multiples of 7, 10, and 14 A, respectively.

It was shown above [formulas (12)-(13) in Chapter 1] that the lattices of all layer silicate structural modifications may be described in terms of an orthogonal cell with a repeat unit of one layer (1M'), two layers (2O, $2M_3$, $2M_1'$, $2M_2'$), three layers (1M, 3T), or six layers ($2M_1$, $2M_2$, 6H). It is natural that lattices described in terms of the same cell would have the same reciprocal lattice geometry and corresponding diffraction patterns. However, additional conditions tied up with the presence of monoclinic symmetry, and with the rules for the display of a minimum repeat unit given above, lead to the elimination of certain reciprocal lattice points and diffraction-pattern reflections, thus bringing about certain differences between the modifications. Actually, from condition (12) of Chapter 1, and structural crystallography formulas (Belov, 1951b) (for clarity we will only let $\beta \neq \pi/2$), the following relationship will apply between monoclinic cells extending over one or two layers and orthogonal cells with heights of three or six layers:

$$\mathbf{a}_{3,6} = \pm\, \mathbf{a}_{1,2}, \quad \mathbf{c}_{3,6} = \mathbf{a}_{1,2} + 3\mathbf{c}_{1,2}. \tag{61}$$

In reciprocal space,

$$\mathbf{a}^{*}_{3,6} = \pm\, \mathbf{a}^{*}_{1,2} \mp \frac{\mathbf{c}^{*}_{1,2}}{3}, \quad \mathbf{c}^{*}_{3,6} = \frac{\mathbf{c}^{*}_{1,2}}{3}, \tag{62}$$

and in the diffraction patterns, for the reflections observed,

$$h_{3,6} = \pm\, h_{1,2}, \quad l_{3,6} = h_{1,2} + 3l_{1,2}. \tag{63}$$

Thus, a monoclinic structure with a reciprocal lattice described by the orthogonal cell $\mathbf{a}^{*}_{3,6}$, $\mathbf{c}^{*}_{3,6}$ could contain only those lattice points, and its diffraction pattern only those reflections, which would at the same time apply to the monoclinic cell $\mathbf{a}^{*}_{1,2}$, $\mathbf{b}^{*}_{1,2}$, $\mathbf{c}^{*}_{1,2}$. The hkl indices of the corresponding reflections, calculated for an orthogonal cell in accordance with (63), would be subject to the condition that

$$l \mp h = 3l' \tag{64}$$

for the modifications 1M and $2M_1$, and

$$l \mp k = 3l' \tag{65}$$

for $2M_2$ modifications. Here, of course, the indices h and k may have either sign.

If, in addition, $k = 3k'$ in the case of the $2M_1$ modification, or $h = 3h'$ for the $2M_2$ modification, then the diffraction patterns can only include reflections corresponding to an orthogonal repeat unit of three layers and a nonorthogonal repeat unit of one layer, i.e., reflections with only the l index even. The relationships between the indices of these reflections, calculated for monoclinic and orthogonal cells with different numbers of layers in the cell, have the forms

1) $n = 2$; $h_2 = -h_1$; $l_2 = 2l_1 + h_1$; $l_2 + h_2 = 2l_1$;

2) $n = 3$; $h_3 = \pm h_1$; $l_3 = 3l_1 + h_1 = l_2 + l_1$; $l_3 \mp h_3 = 3l_1$;

3) $n = 6$; $h_6 = \pm h_2 = \mp h_1$; $l_6 = 3l_2 + h_2 = 6l_1 + 2h_1 = 2\,(l_2 + l_1) = 2l_3$; $l_6 \mp h_6 = 3l_2$; $l_6 \mp 2h_6 = 6l_1$.

$$\hspace{10cm} (66)$$

Thus, in modifications 1M and $2M_1$ with h $\neq$ 3h', and in modification $2M_2$ with k $\neq$ 3k', reflections with $l \neq 3l\,'$ are missing. Also, in modification $2M_1$ with k = 3k', and in modification $2M_2$ with h = 3h', reflections with $l = 2l\,' + 1$ are missing. On the other hand, if in modifications 1M and $2M_1$, h = 3h', or in modification $2M_2$, k = 3k', then according to equations (64)-(65), only reflections with $l = 3l\,'$ are possible, which is a reflection of the one-layer orthogonal repeat unit for 1M, and the similar two-layer unit for $2M_1$, $2M_2$. If h = 3h' and k = 3k', then in all cases l must be simultaneously multiplied by two and three, i.e., $l = 6l\,'$. Reflections with other l values will be absent.

In the case of the modification 2O with h $\neq$ 3h', reflections with any value of the index l_2 corresponding to an orthogonal two-layer repeat unit are possible, and when h = 3h', whatever the values of k, only reflections with $l = 2l\,'$ are present.

To sum up, a distribution containing all reflections with h = 3h', k = 3k' is common to all modifications. They determine the minimum one-layer orthogonal repeat unit. For reflections with k = 3k', the structures 1M, 3T, and $2M_1$ (Figs. 59-62) do not differ, and thus not only the positions, but also the intensities of these reflections, are the same. These fix the three-layer orthogonal repeat unit. When k $\neq$ 3k', the reflections are distributed identically for 1M and 3T and twice as frequently for $2M_1$. The individual properties of modifications $2M_2$ and 2O (Fig. 74) are shown up by the reflections with k $\neq$ 3k'. Here, the reflections with h = 3h' fix an orthogonal repeat unit of three layers in $2M_2$, and one layer in 2O.

The distribution of reflections with k = 3k' is identical in these modifications, and when h $\neq$ 3h' they fix an orthogonal two-layer repeat unit.

Since the layers have identical scattering abilities in directions with h = 3h', k = 3k', and in the majority of cases with k = 3k' alone, then for scattering in these directions, the c unit formed by the layers in the lattices will extend over only one layer. If the true c repeat unit extends over n layers, then for these indices only reflections with $l = nl'$ will be possible. The arrangement of these reflections will determine only $1/n$-th of the unit cell, embracing the atoms of one layer. The true c repeat distance, defining the true unit cell, can only be found from the arrangement of reflections with $h \neq 3h', k \neq 3k'$ ($h', k' = 1, 2, \ldots$). If it is not required that $h \neq 3h'$ also, then these may be the reflections $02l, \overset{\pm}{11}l, 04l, \overset{\pm}{22}l, \overset{\pm}{15}l, \overset{\pm}{31}l, \overset{\pm}{24}l$, etc.; otherwise, reflections with $h = 3h'$ are unsuitable.

The distance between two silicate layers along the c axis, i.e., c/n if the c repeat unit extends over n layers, is fixed by the distances, along the normals to the zero line, between reflections having the indices $l = nl'$ [nl', $n(l' + 1)$, $n(l' + 2)$, etc.]. If $k = 3k'$ ($h = 3h'$), there will be no other reflections [with the same k(h) index] between these reflections. If $k = 3k' \pm 1$ ($h = 3h' \pm 1$), then between these reflections on the particular ellipse there will be another $n - 1$ reflections, with indices $l = nl' + 1, nl' + 2, \ldots, nl' + n - 1$. For sufficiently large n (and, in particular, when $n \to \infty$, i.e., there is no overall strict c repeat unit), the reflections of those ellipses for which $k \neq 3k'$ ($h \neq 3h'$) overlap one another, and are not separately distinguishable. In this case, the true c repeat unit becomes indeterminate. Since the loss of strict periodicity is connected with displacements of networks and layers in the direction of the b axis, by amounts which are multiples of $b/3$, then the arrangement of reflections with $k = 3k'$ in this case can be used only to give information on one n-th of the unit cell, pertaining to one layer having a repeat unit of c/n, which can be called the minimum c unit.

The lack of a strict c repeat unit must show itself in the appearance of an additional intervening "background" between the reflections with indices having $k \neq 3k'$ on the particular ellipse, and this background can be considered as made up of reflections arranged as close to one another as is wished.

Thus, the qualitative features of the reflections with indices $k = 3k'$ and $k \neq 3k'$ indicate whether the structure is ordered, or disordered because of the relative displacements of layers along the b-axis direction. If all the reflections appear equally sharp and clear, then the structure is perfect and its lattice is strictly periodic. If the reflections with $k = 3k'$ are sharp, but those with $k \neq 3k'$ are smeared out into continuous ellipses, then the corresponding layer networks are irregularly displaced relative to one another, by multiples of $b/3$, along the direction of the b axis.

If all the reflections are smeared out into continuous ellipses, whatever the value of the k index, this indicates either that networks or complete layers

Fig. 59. Diffraction pattern from kaolinite with a monoclinic cell
(modification $1Tk_1$, $\varphi = 55°$).

are irregularly displaced in both the a- and b-axis directions, or that the
layers are chemically nonuniform. Any other type of structural imperfection
must inevitably affect the properties of the reflections and their arrangement
on the ellipses. When reflections on texture patterns are spread out in two dimensions, this makes it easier to discern all their characteristics.

The geometrical characteristics of oblique texture patterns are given below for the different layer silicate varieties. The values of p, s, and q which
determine the heights D of reflections according to formulas (49) and (52), are
given for modifications with c repeat units of 1, 2, 3, and 6 layers. The corresponding values of c, β, α are also given, and the limits of variation of the
constants a and b. The tables give the sequence of reflections in the most
important ellipses, i.e., the first, second, and fifth ones, with a note of relative heights and hk$l$ indices. The values of p, s, and q correspond to a value
of $2L\lambda = 40.5$, which was the value applying to the EM-4 electron-diffraction
apparatus we used, for the optimum angle of specimen tilt $\varphi = 55°$. The
tables can also be used to construct the reflection layout scheme for any value
of φ and for any scale. The relative heights are given in the tables, and the

ratios of the lengths of ellipse minor axes to heights can be found from the expression

$$\frac{b_{hk}}{p} = \frac{\sqrt{3h^2 + k^2}\, b^* \sin \varphi}{a^* \cos \beta} = \frac{\sqrt{3h^2 + k^2} \sin \varphi}{\sqrt{3} \cos \beta}.$$

In view of the geometric property that the sequence of reflections along a given ellipse is periodic, with a repeat unit in index l and height D of $\triangle l = 1$ and $\triangle D = q$, the sequence can be defined by giving just one or a few such repeat units. The distribution of intensities in these electron-diffraction reflections will be discussed in Chapter 4.

Electron-Diffraction Patterns from Minerals Made Up of Two-Storied Silicate Layers. The constants a and b have values of 5.14 and 8.90 A, respectively, for the Al varieties, and ~5.3 and ~9.2 A for the Mg varieties.

Fig. 60. Diffraction pattern from dickite (modification $2M_1$, structure I, 4; $\varphi = 55°$).

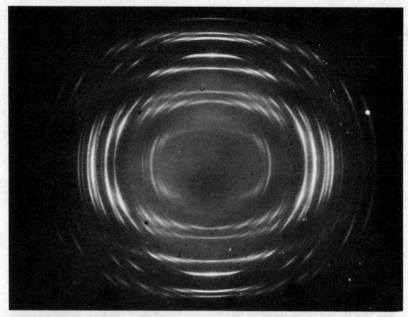

Fig. 61. Diffraction pattern from celadonite (dioctahedral modification 1M, $\varphi = 55°$).

Sequence of Reflections for Kaolinite with a Triclinic
Cell (Fig. 21, Tables 16-18).

Table 16. First Ellipse

hkl	D/s	hkl	D/s
020	2	021	44
1$\bar{1}$0	14	1$\bar{1}$1	56
110	16	111	58
$\bar{1}\bar{1}$1	26	$\bar{1}\bar{1}$2	68
$\bar{1}$11	28	$\bar{1}$12	70
0$\bar{2}$1	40	0$\bar{2}$2	72

Table 17. Second Ellipse

hkl	D/s
1̄30, 2̄01	12
130	18
1̄31	24
1̄31, 200	30
1̄31, 2̄02	54
131	60
1̄32	66
1̄32, 201	72

Table 18. Fifth Ellipse

hkl	D/s
06̄0, 3̄3̄1	—6
3̄31, 33̄1̄	0
060, 33̄1̄	6
06̄1, 3̄32	36
3̄32, 3̄3̄0	42
061, 330	48

Fig. 62. Diffraction pattern from sericite (dioctahedral modification $2T_1$, $\varphi = 55°$).

For a one-layer c repeat unit, $q \sim 7$; $p \sim 2.5$ mm, $c_1 \sim 7.3$ A, $\beta \sim 104°$. In triclinic kaolinite, in addition, $s_1 = \frac{1}{6}$ mm, $\alpha \sim 92°$ ($p = \frac{5}{14}q$, $s \simeq p/15 \simeq q/42$).

In reflections which are common to the monoclinic modifications which have different numbers of layers (n) in the c repeat unit, then from equation (54), $h_1 = -h_2$, or h_3, or $-h_6$; and $l_1 = (l_2 + h_2)/2$, or $(l_3 - h_3)/3$, or $(l_6 - 2h_6)/6$.

Sequence of Reflections in Diffraction Patterns of Varieties (Kaolinites and Serpentines) with Monoclinic Cells (Figs. 59, 60; Tables 19-21).

Table 19. First Ellipse

Number of layers in the c unit (n), and hkl indices				D/p_6 *
1	2	3	6	
020	020	020	020	0
		110	110	1
	110		$\bar{1}11$	6
			021	7
			111	8
		$\bar{1}11$	$\bar{1}12$	13
		021	022	14
110	$\bar{1}11$	111	112	15
			$\bar{1}13$	20
	021		023	21
			113	22
. $\bar{1}11$	111	$\bar{1}12$	$\bar{1}14$	27
	022		024	28
	112		114	29
			$\bar{1}15$	34
			025	35
	$\bar{1}12$		115	36
		$\bar{1}13$	$\bar{1}16$	41
021	022	023	026	42

*It is also convenient to express D in terms of p_6 as it allows direct visual comparison between diffraction patterns of monoclinic and triclinic varieties, since numerically $p_6 = s$.

Table 20. Second Ellipse

n and hk$l$				D/p_6
1	2	3	6	
$\bar{2}01$	200	$\bar{2}01$	$\bar{2}02$	12
130	$\bar{1}31$	131	132	15
$\bar{1}31$	131	$\bar{1}32$	$\bar{1}34$	27
200	$\bar{2}02$	202	204	30
$\bar{2}02$	202	$\bar{2}04$	$\bar{2}08$	54
131	$\bar{1}33$	134	138	57
$\bar{1}32$	133	$\bar{1}35$	$\bar{1}.3.10$	69
201	$\bar{2}04$	205	2.0.10	72

Table 21. Fifth Ellipse

n and hk$l$				D/p_6
1	2	3	6	
$\bar{3}31$	$33\bar{1}$	$\bar{3}30$	$\bar{3}30$	—3
060	060	060	060	0
$33\bar{1}$	$\bar{3}31$	330	330	3
$\bar{3}32$	331	$\bar{3}33$	$\bar{3}36$	39
061	062	063	066	42
330	$\bar{3}33$	333	336	45

With a c repeat unit of two layers, $q_2 = q_1/2 = 3.5$ mm, $p_2 = q_2 - p_1 \simeq 1$ mm, $c_2 \simeq 14.3$ A, $\beta_2 = 96°$.

In reflections common to the different modifications, $h_2 = -h_1$, or $-h_6$; $l_2 = 2l_1 + h$, or $(l_6 + h_6)/3$.

With a c repeat unit of three layers, $q_3 = q_1/3 \simeq 2.3\text{-}2.4$ mm, $p_3 = p_1 - q_3 \simeq 0.3\text{-}0.4$ mm, $c_3 \sim 21.5$ A, $\beta_3 \sim 92°$.

In reflections common to the different modifications, $h_3 = h_1$, or h_6; $l_3 = 3l_1 + h1$, or $l_6/2$.

Fig. 63. Diffraction pattern from montmorillonite ($\varphi = 55°$).

With a c repeat unit of six layers, $q_6 = q_1/6 \simeq 1.2$ mm, $p_6 = -(2q_1/6) + p_1 \sim 0.15$ mm, $c_6 \sim 43$ A, $\beta_6 \sim 90\text{-}91°$.

In reflections common to the different modifications, $h_6 = h_1$, or $-h_2$, or h_3; $l_6 = 6l_1 + 2h_1$, or $3l_2 + h_2$, or $2l_3$.

According to Tables 16-18, the characteristic features of a one-layer kaolinite with a triclinic cell (see Fig. 21) are clearly distinguishable pairs of reflections of indices $0\bar{2}l$ and $02l$ on the first ellipse, separated by an interval of $\Delta D = 4s$ (reflections $1\bar{1}l$ and $11l$ are not separated), sets of four reflections on the second ellipse, separated within by intervals of $\Delta D = 6s$, and separated between sets by $\Delta D = 24s$, and trios of reflections on the fifth ellipse, separated within by $\Delta D = 6s$ and with $\Delta D = 30s$ between the trios.

If the crystal has a low degree of perfection and the diffraction pattern is poorly resolved, it may be found in practice that only the second criterion can be checked.

For nacrite (modification $2M_2$), $p_2 = 0$, $s_2 = 1$ mm. The distribution of its reflections according to their indices is the same as for the micas, and is therefore to be found in Tables 22-24. It will be apparent that the heights D, determined by the relationships $q_2 = 3.5$ mm and $s_2 < q_2/3$, will be different from those in micas.

It should be remarked that the uniform arrangement of reflections on the second and fifth ellipses, noted above for two-storied layer minerals, whatever their true c repeat unit, is not always observed, and in some cases intermediate "forbidden" reflections may be observed. For example, in two-layer dickite the weak reflections 0, 6, 2l and 1.3, 3, 2l may be seen on the fifth ellipse. These are tied up with the presence of distorted layer network motifs, in which the relative displacements of corresponding points in different layers are equal to $\Delta y \neq 0, \pm^1\!/_3$.

A similar thing is also observed in muscovite.

The most obvious feature of the diffraction patterns of these minerals is the interval of alternation of the trios of practically coincident reflections on the fifth ellipse.

Fig. 64. Diffraction pattern from chlorite ($\varphi = 55°$).

Table 22. First Ellipse

1M	2M$_1$	2M$_2$	$D/p_2 = D/s_2$	D/s_2 for nacrite
020	020		0	
	110	110	1	1
110	$\bar{1}11$	$1\bar{1}0$, 020	2	2.5; 2
	021		3	
$\bar{1}12$	111	111, $0\bar{2}2$	4	4.5; 5
	$\bar{1}12$	$1\bar{1}2$	5	6
021	022		6	
	112	112	7	8
111	$\bar{1}13$	$1\bar{1}3$, 022	8	9.5; 9
	023		9	
$\bar{1}12$	113	113, $0\bar{2}4$	10	11.5; 12
	$\bar{1}14$	$1\bar{1}4$	11	13
022	024		12	
	114	114	13	15
112	$\bar{1}15$	$1\bar{1}5$, 024	14	16.5; 16

Table 23. Second Ellipse

1M	2M$_1$	2M$_2$	D/p_1	D/s_2	D/s_2 for nacrite
		$13\bar{1}$, 200, $1\bar{3}1$	0	0	−0.5; 0; 0.5
130, $\bar{2}01$	$\bar{1}31$, 200				
		130, 201, $1\bar{3}2$	1	3	3; 3.5; 4
$\bar{1}31$, 200	131, $\bar{2}02$		2		
		131, 202, $1\bar{3}3$	3	6	6.5; 7; 7.5
131, $\bar{2}02$	$\bar{1}33$, 202		4		
		132, 203, $1\bar{3}4$		9	10; 10.5; 11
$\bar{1}32$, 201	133, $\bar{2}04$		5		
		133, 204, $1\bar{3}5$	6	12	13.5; 14; 14.5
132, $\bar{2}03$	$\bar{1}35$, 204		7		
		134, 205, $1\bar{3}6$		15	17; 17.5; 18
$\bar{1}33$, 202	135, $\bar{2}06$		8		
		135, 206, $1\bar{3}7$	9	18	20.5; 21; 21.5

Table 24. Fifth Ellipse

Modification and hkl indices			$D/p_1 \parallel$ $\parallel D/2s_2$	D/s_2 for nacrite
1M	2M$_1$	2M$_2$		
33$\bar{1}$, 060, $\bar{3}$31	$\bar{3}$31, 060, 33$\bar{1}$	06$\bar{2}$, 33$\bar{1}$, $\bar{3}\bar{3}$1, 0$\bar{6}$2	0	−1; −0.5; 0.5; 1
330, 061, $\bar{3}$32	$\bar{3}$33, 062, 331	060, 331, 3$\bar{3}\bar{3}$, 0$\bar{6}$4	3	6; 6.5; 7.5; 8
331, 062, $\bar{3}$33	$\bar{3}$35, 064, 333	062, 333, 3$\bar{3}\bar{5}$, 0$\bar{6}\bar{6}$	6	13; 13.5; 14.5; 15
332, 063, $\bar{3}$34	$\bar{3}$37, 066, 335	064, 335, 3$\bar{3}\bar{7}$, 0$\bar{6}\bar{8}$	9	20; 20.5; 21.5; 22
333, 064, $\bar{3}$35	$\bar{3}$39, 068, 337	066, 337, 3$\bar{3}$9,0.$\bar{6}$.10	12	27; 27.5; 28.5; 29

In accordance with what has been said above, the varieties with imperfect structures lacking a strict c repeat unit have reflections with k ≠ 3k' smeared out into continuous ellipses. Minerals with nonisometric particles (halloysites) form poorly oriented texture specimens, and, consequently, they give diffraction patterns with reflections in the form of long arcs of circles. It should be noted that under electron-diffraction conditions, the halloysites exhibit a higher degree of structural perfection than under normal X-ray diffraction conditions, showing the spatial reflections hkl which are usually absent on X-ray photographs. The c repeat unit of two layers, which has been found from electron-diffraction texture patterns, has been established only very recently.

Electron-Diffraction Patterns from Minerals Made Up of Three-Storied Silicate Layers: Micas, Hydromicas, Vermiculites, and Montmorillonites (Figs. 61, 62; Tables 22-24):

$$a = 5.16\text{-}5.30 \text{ A}; \qquad b = 8.92\text{-}9.20 \text{ A};$$
$$q_1 = 4.9\text{-}5.5 \text{ mm}; \qquad p_1 = 1.6\text{-}1.8 \text{ mm};$$
$$c_1 = 9.3\text{-}11.0 \text{ A}; \qquad \beta_1 = 99\text{-}101°;$$
$$q_2 = q_1 / 2 \simeq 2.4\text{-}2.7 \text{ mm}; \qquad p_2 = q_2 - p_1 \simeq 0.8\text{-}1.1 \text{ mm}.$$
$$c_2 = 18.5\text{-}20.5 \text{ A};$$

For modification 2M$_2$:

$$s_2 = q_2/3 = 0.8\text{-}0.9 \text{ mm}; \quad \alpha_2 = 99\text{-}100°; \quad \beta_2 = 94\text{-}96°.$$

For the common reflections, $h_2 = -h_1$, $l_2 = 2l_1 + h_1$, $h_1 = -h_2$, $l_1 = (l_2 + h_2)/2$.

The indices for these values of p, q, s, c, and β correspond to repeat units of one and two layers.

For the minerals made up of three-storied layers, the uniform distribution of reflections with k = 3k' only breaks down, and then only to a minor extent, in the case of muscovite,this breakdown being expressed by the appearance on the fifth ellipse of the extremely weak intermediate reflections $0,6,2l$ and $1.3,3,2l$ (see the description of an electron-diffraction study of muscovite given below).

It has already been pointed out that the structural modifications with repeat units of one layer (Fig. 61) and two layers (Fig. 62) do not differ in the geometry of their diffraction patterns from varieties with three and six layers, respectively. A monoclinic one-layer modification may only differ from a trigonal three-layer modification in its reflection intensity distribution, and the same relationship applies between a monoclinic two-layer modification and a hexagonal six-layer modification.

Electron-diffraction conditions take no account of such fundamental differences between montmorillonites and hydromicas as the presence of H_2O molecules between the layers. Nonetheless, the differences between these minerals are revealed quite clearly from a qualitative comparison of their diffraction patterns. Montmorillonites have a much lower degree of structural perfection and a higher degree of dispersion. Their reflections therefore differ in being highly diffuse, and when k ≠ 3k', the reflections are usually run together into continuous ellipses (Figs. 61, 63).

Electron-Diffraction Patterns from Minerals (Chlorites) Made Up of Alternating Three-Storied and One-Storied Silicate Layers (Fig. 64; Tables 25-27):

$$a \simeq 5.3 \text{ A}; b \simeq 9.2 \text{ A}; p \simeq 1.0 \text{ mm};$$
$$q = 3.5 \text{ mm}; c \simeq 14.3 \text{ A}; \beta = 96\text{-}97°.$$

The reflections with indices for which $l + h = 2l'$ indicate that the c ≃ 14 A unit is the minimum one.

Chlorites having ordered structures, with diffraction patterns in which reflections with k ≠ 3k' indices would be expected, are met with extremely rarely, because of the high degree of freedom in the relative arrangement of their layers; chlorite texture patterns show weak reflections with k ≠ 3k' on the continuous background of the corresponding ellipse. On the other hand, the reflections with k = 3k' may be extremely intense and sharp.

Table 25. First Ellipse

hkl	D/p	hkl	D/p
020	0.0	112	8.0
110	1.0	$\bar{1}13$	9.5
$\bar{1}11$	2.5	023	10.5
021	3.5	113	11.5
111	4.5	$\bar{1}14$	13.0
$\bar{1}12$	6.0	024	14.0
022	7.0	114	15.0

Table 26. Second Ellipse

hkl	D/p	hkl	D/p
130	1.0	201	5.5
$\bar{2}01$	1.5	$\bar{1}32$	6.0
200	2.0	132	8.0
$\bar{1}31$	2.5	$\bar{2}03$	8.5
131	4.5	202	9.0
$\bar{2}02$	5.0	133	9.5

Table 27. Fifth Ellipse

hkl	D/p	hkl	D/p
$33\bar{1}$	—0.5	$\bar{3}33$	7.5
060	0.0	332	10.0
$\bar{3}31$	0.5	063	10.5
330	3.0	$\bar{3}34$	11.0
061	3.5	333	13.5
$\bar{3}32$	4.0	064	14.0
331	6.5	$\bar{3}35$	14.5
062	7.0		

Elucidation of the fine details of such a distribution needs a high degree of resolution in the diffraction pattern. In practice, it is necessary to work from a smaller number of reflections than those given in Tables 25-27 for the chlorites.

Electron-Diffraction Patterns from Mixtures. Electron-diffraction patterns from mixtures naturally appear as the sum of the patterns of their individual components. Mixtures can be distinguished fairly easily, particularly if the components have appreciably different a and b constants, which leads to the appearance on the diffraction pattern of separate groups of ellipses for each pair of hk indices, corresponding to the individual components. Splitting of ellipses because several minerals are present in the specimen is particularly apparent in the fifth and more distant ellipses.

The diffraction patterns give a visual demonstration of the presence in the mixture of minerals with different degrees of perfection. Thus, an ellipse containing reflections with the same indices, but corresponding to different mineral components, may allow these reflections to be distinguished from their qualitative character (some may be sharp and others diffuse).

Analysis of the diffraction patterns of polymineralic specimens will yield several unit cells or other structural characteristics for the minerals present in the mixture. It will be obvious that minerals not belonging to the layer silicates will give reflections which do not fall within the ellipse systems described, but have a different qualitative nature, so that they may be detected without any effort.

Diffraction patterns similar to those of mixtures may be obtained from intergrown structures, which may appear as the intermediate product of the transformation of one mineral into another.

Intermediate forms such as those between kaolinite and montmorillonite, or kaolinite and hydromica, are of particular interest. Their crystals must have nonuniform structures, since here their surfaces and inner parts must be different. It appears that many of the specimens which were identified from their general properties as beidellite or monothermite, in the not too distant past (Zvyagin, 1958a), were actually of this nature.

In diffraction patterns from mixtures of these minerals, the patterns due to the clay components may be picked out from the presence or absence of an extremely insignificant splitting of the ellipses, which is an indication of the closeness of the a and b constants of the intergrowth components. Moreover, the kaolinite component in these cases has a monoclinic cell and a low degree of structural perfection, so that the qualitative attributes of its reflections are practically identical to those of a montmorillonite component.

It is obvious that the diffraction patterns from such specimens can only be dealt with using several values of p and q, corresponding to each of the components present.

CHAPTER 4

THE DETERMINATION OF INTENSITIES IN LAYER SILICATE DIFFRACTION PATTERNS

1. General Comments

We have already noted that the general expression for the structure amplitude [formulas (4)-(6) in Chapter 3], i.e., for the scattering by the atoms of one cell, is given by an integral taken over the volume of the unit cell:

$$\Phi_{hkl} = \int_{\Omega} \varphi(\mathbf{r}) e^{2\pi i (\mathbf{r} \mathbf{H})} \, d\tau. \tag{1}$$

The potential $\varphi(\mathbf{r})$ of the cell can be represented by

$$\varphi(\mathbf{r}) = \sum_j \varphi_j(\mathbf{r} - \mathbf{r}_j), \tag{2}$$

where φ_j is the potential of the j-th atom in the cell, and the summation is carried out over all the atoms in the unit cell. If (2) is substituted in the integral (1), the latter is broken up into a sum of integrals each of which involves an integration over the potential of a single atom having an atomic scattering amplitude f_j, so that

$$\Phi_{hkl} = \sum_j f_j e^{2\pi i (\mathbf{r}_j \mathbf{H})} = \sum_j f_j e^{2\pi i (hx_j + ky_j + lz_j)}, \tag{3}$$

where

$$f_j = \int_{\Omega} \varphi_j(\mathbf{r} - \mathbf{r}_j) e^{2\pi i (\mathbf{r} - \mathbf{r}_j)} \, d\tau.$$

It follows from the theory of atomic scattering (Mott, 1930) that the atomic scattering amplitudes of electrons, f_{el}, and X rays, f_X, are related in terms of $\sin \vartheta / \lambda$ by the equation

$$f_{el} = \frac{me^2}{2h^2} \cdot \frac{Z - f_X(\sin \vartheta / \lambda)}{(\sin \vartheta / \lambda)^2}.$$

159

Tables and curves of $f_{el}(\sin \vartheta/\lambda)$ have been constructed by Vainshtein (1956).

In the general case of triclinic symmetry, Φ_{hkl} is a complex quantity, i.e.,

$$\Phi_{hkl} = A_{hkl} + iB_{hkl},$$

where

$$A_{hkl} = \sum_j f_j \cos 2\pi \,(hx_j + ky_j + lz_j),$$

$$B_{hkl} = \sum_j f_j \sin 2\pi \,(hx_j + ky_j + lz_j). \tag{4}$$

If any symmetry elements are present in the structure, relating the coordinates of the symmetrical atoms by certain relationships, then the general expression for (3) can be simplified. In any case, to obtain the simplest expression for Φ_{hkl} it is best to choose the origin of coordinates at the most symmetrical point in the structure. Thus, for example, if the origin is put at a center of symmetry, each atom with coordinates x_j, y_j, z_j will correspond with a similar atom at the point $-x_j$, $-y_j$, $-z_j$, and

$$\Phi_{hkl} = A_{hkl} = 2\sum_j f_j \cos 2\pi \,(hx_j + ky_j + lz_j), \tag{5}$$

where the summation is carried out only over the symmetrically independent atoms. With the space group $c_{2h}^3 = C\ 2/m$, which describes the symmetry of a base-centered monoclinic unit cell with twofold rotation and screw axes passing through it, mirror and glide reflection planes perpendicular to the axes, and centers of symmetry at the points where the axes intersect the planes (see the International Tables, 1952), each atom with coordinates xyz corresponds to three further atoms with coordinates x, $-y$, z; x $+\,{}^1\!/_2$, ${}^1\!/_2 - y$, z; x $+ {}^1\!/_2$, y $- {}^1\!/_2$, z, and to another four atoms which are centrosymmetrically related to the first four. The sum of each such set of eight components reduces to the single expression

$$8f_j \cos^2 2\pi \frac{h+k}{4} \cos 2\pi \,(hx + lz) \cos 2\pi ky. \tag{6}$$

The experimental scattering ability of a crystalline object is found from the intensities I of its diffracted rays. It depends on two basic factors: firstly the composition, structure, dimensions, and shapes of the individual crystals,

Table 28. Values of the L Factor

Specimen type and intensity measurement method	L_{rel}
Mosaic single crystal; integral spot intensity	d_{hkl}
Texture; integral arc intensity	p/r'
Texture; local arc intensity	$d_{hk0}d_{hkl}p$
Polycrystalline; integral ring intensity	$d_{hkl}p$
Polycrystalline; local ring intensity	$d^2_{hkl}p$

and secondly, the manner in which the specimen is made up from its component crystallites. Where kinematic scattering is concerned, the first factor is expressed by the integral of the square of the modulus of the function S [formulas (4) and (7) of Chapter 3], taken over the plane of the diffraction-pattern screen. Since $|S|^2 = |\Phi|^2 \cdot |D|^2$, and $|\Phi|^2$ is independent of the experimental conditions, the intensity is always proportional to the square of the modulus of the structure amplitude or, which comes to the same thing, to the product of that amplitude and its complex conjugate quantity $|\Phi(H)|^2 = \Phi(H)\Phi^*(H)$.

In contrast to this, the intensity of dynamic scattering is proportional to the first power of $|\Phi|$.

Vainshtein (1956) has shown that the second factor can be represented by a factor L, the form of which depends on the character of the specimen. In practical calculations it is usual to use the relative values I^{rel} and $|\Phi|^2$.

If the scattering of waves takes place in accordance with kinematic theory, then, according to Vainshtein (1957),

$$I_{hkl}^{rel} = |\Phi_{hkl}|^2 L^{rel}.$$

Expressions for L for different types of specimens and methods of measuring intensities have been calculated by Vainshtein (1957); these are given in Table 28.

Thus, the squares of the moduli of the structure amplitudes $|\Phi_{hkl}|^2$ are calculated from the intensities I_{hkl} of texture pattern intensities using one of two formulas derived by Vainshtein (1956, 1957):

$$|\Phi_1(hkl)|^2 \approx \frac{I_{hkl}}{pr'_{hk}}, \qquad (7)$$

$$|\Phi_2(hkl)|^2 \approx \frac{I_{hkl}}{pd_{hk0}d_{hkl}}, \qquad (8)$$

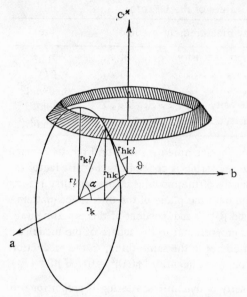

Fig. 65. Formation of the reciprocal lattice ring-node for a textured specimen made up of elongated crystals.

where r' is the horizontal co-ordinate of the reflection in the equation of the corresponding ellipse, and p is the multiplicity factor. In electron-diffraction texture patterns of triclinic specimens, $p = 1$, while for monoclinic specimens $p = 2$ if $k \neq 0$, and $p = 1$ if $k = 0$. Formula (7) corresponds to the assumption that the intensity I_{hkl} is determined by the density of a reciprocal lattice node which has only been "smeared out" into a circle through rotation about the c^* axis, so that this intensity is integral. Formula (8) assumes that the intensity I_{hkl} is determined by the density of a lattice node hkl which has been "smeared out" into a spherical band through rotation about the c^* axis and also through tilting of this axis by a certain solid angle, so that here the intensity is local. The first case is more probable with thin crystallites and well-oriented textures, and the second more probable with thicker crystallites and poorly oriented textures. It is possible to check which of these formulas is the more appropriate, as recommended by Vainshtein (1956), by comparing the fall in the mean values of $|\Phi_{hkl}|^2$ as $|H|$ increases with the fall in the mean values of the squares of the atomic factors $|f_{el}|^2$ for the object under study, and give preference to the $|\Phi_1|^2$ or $|\Phi_2|^2$ values which show the best agreement.

If the crystallites forming the specimen have a nonisometric, platy shape, they will lie at various angles, deviating from strict parallelism with the texture plane, and these angles will not be equally probable for each crystal. The longest crystal directions will be likely to have the smallest angles of tilt, and vice versa. This must lead to unequal "smearing out" of the lattice nodes in reciprocal space, since these points differ in their positions relative to the axes of preferential inclination and the texture axis, and thus correspond to different versions of the intensity formulas (Zvyagin, 1963).

The most obvious manifestation of the nonisometric shape of platy crystals is the way they are drawn out in some definite direction. With mosaic single crystals, the acicular habit of the crystallites was taken into account in the intensity formulas used by Lobachev and Vainshtein in electron-diffraction studies of urea (1961).

With electron-diffraction texture patterns, elongated shapes in the crystals affect not only the distribution of intensities, but also the geometry of the diffraction pattern; this can be seen from the electron-diffraction studies of sepiolite and palygorskite described below.

The appropriate relationships can be derived as previously, from an examination of the reciprocal lattice corresponding to the real specimen.

Let the direction of elongation correspond to the a axis, and the texture axis to the c^* axis. Then the tilting of the crystals with respect to the texture plane will take place mainly through rotations about the a axis. Actually, if one end of a crystal lies an amount t higher than another end, then, depending on the distance l between these extreme points, for an angle of inclination of $\Delta\alpha$, $\tan(\Delta\alpha) = t/l$. In sepiolite and palygorskite, the value of l measured along a is tens or hundreds of times greater than the corresponding distance across the crystal.

It may therefore be assumed that each reciprocal lattice node or ring is formed through a rotation by $\Delta\alpha$ about the a axis and a rotation by 2π about c^* (Fig. 65).

Since the value of $\Delta\alpha$ for crystals in the form of narrow ribbons has an appreciable size, the experimentally observed intensity is essentially a local one. According to Vainshtein (1956), it is determined by the local density of the reciprocal lattice ring-node, which itself is inversely proportional to the area of the surface over which the lattice node is "smeared."

From Fig. 65 it can be seen that the surface area of the spherical band formed by "smearing out" an arbitrary lattice point hkl is

$$M = 2\pi r_{hkl}^2 \Delta(\cos\vartheta) = -2\pi r_{hkl}^2 \sin\vartheta\Delta\vartheta = -2\pi r_{hkl} r_{hk}\Delta\vartheta. \quad (9)$$

Since

$$\cos\vartheta = \frac{r_{kl}\sin\alpha}{r_{hkl}}, \quad (10)$$

then

$$\Delta\vartheta \approx \frac{r_{kl}\cos\alpha}{r_{hkl}\sin\vartheta}\Delta\alpha = -\frac{r_k}{r_{hk}}\Delta\alpha, \quad (11)$$

and

$$M = 2\pi r_{hkl} r_k \Delta\alpha, \quad (12)$$

where r_{hk}, r_{kl}, and r_k are the components of r_{hkl} on the ab and $bc*$ planes and along the b axis.[†]

So to an accuracy of up to the constant factor, and for kinematic scattering,

$$I_{hkl} \approx \frac{p\,|\,\Phi_{hkl}\,|^2}{r_{hkl}\,r_k}. \tag{13}$$

For orthogonal and monoclinic lattices,

$$I_{hkl} \approx p\,|\,\Phi_{hkl}\,|^2\,d_{hkl}\,d_{0k0}. \tag{14}$$

In the opposite case, $1/r_k \approx d'_{0k0}$, where d'_{0k0} is some fictitious interplanar distance.

From these relationships and the value of $\Delta\vartheta$, we can find the angular scatter $\Delta\alpha$ of the crystals in the texture.

In fact,

$$\Delta\alpha = -\frac{r_{hk}}{r_k}\,\Delta\vartheta = -\frac{d_{0k0}}{d_{hk0}}\,\Delta\vartheta. \tag{15}$$

If the specimen is tilted at an angle $\varphi = \pi/2$, the interval $\Delta\vartheta$ is equal to the observed angular length of the reflections. If $\varphi \neq \pi/2$, Vainshtein (1956) has shown that the angular length of a reflection arc is

$$\Delta\vartheta' = \frac{r_{hk}}{r'\sin\varphi}\,\Delta\vartheta, \tag{16}$$

where r' is the coordinate of the corresponding arc along the minor axis of the ellipse in an oblique plane section of the reciprocal lattice.

Therefore,

$$\Delta\alpha = -\frac{r'\sin\varphi}{r_k}\,\Delta\vartheta' = -\frac{d_{0k0}}{d'}\,\Delta\vartheta'. \tag{17}$$

Close to $\alpha = \pi/2$, i.e., when $r_k = 0$, formulas (13) and (14) are not applicable, since they do not allow for the fact that the areas of the bands and the local intensities must be finite. In this case, moreover,

$$\pi/2 - \Delta\alpha' \leqslant \alpha \leqslant \pi/2 + \Delta\alpha - \Delta\alpha'$$

[†]In spite of the fact that $\Delta\alpha$ is not a small quantity, the approximate equations used are acceptable. In fact, in the finite differences $\Delta(\cos\vartheta)$, $\Delta(\sin\vartheta)$ the arguments of the trigonometric functions are either the average values of ϑ and α, or $\Delta\vartheta/2$, $\Delta\alpha/2$. If even these half values are not small enough to be justified, they are nonetheless comparable with the approximations permitted in intensity measurements of normal accuracy.

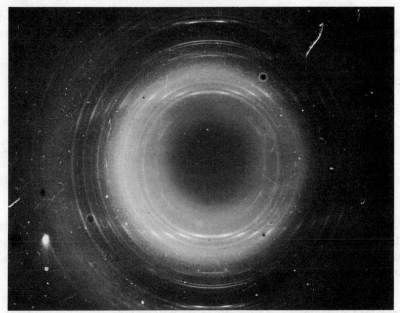

Fig. 66. Diffraction pattern given by sepiolite from Ampandandrava (Malagasy Republic); specimen of A. Preisinger (Brauner, Preisinger, 1956).

($\Delta\alpha'$ is the range of α from $\pi/2$ to one of its limiting values) and the interval $\Delta\vartheta$ corresponds to the change in the angle α, not between its extreme values, but between $\pi/2$ and that one of its limiting values which differs by the largest amount from $\pi/2$. For orthogonal and monoclinic unit cells, these bands lie in the range

$$\frac{\pi - \Delta\alpha}{2} \leqslant \alpha \leqslant \frac{\pi + \Delta\alpha}{2},$$

and the branches of the lattice arc which lie on either side of $\alpha = \pi/2$ make equal contributions to the density of the spherical band on rotation about the c^* axis. In this case,

$$\Delta(\cos\vartheta) \approx \frac{r_l}{r_{h0l}}\left(1 - \cos\frac{\Delta\alpha}{2}\right),$$

$$\tag{18}$$

$$M \approx 2\pi r_{h0l} r_l \frac{(\Delta\alpha)^2}{8},$$

where r_l is the component of r_{h0l} along the c^* axis.

Compared to formulas (13) and (14), the expression for the intensities of reflections with k = 0 has an additional term which depends on $\Delta\alpha$. A similar thing was noted by Vainshtein (1956) for the 00l lattice nodes of ordinary platy textures. Allowing for the doubled lattice-node density because of contributions from both branches of the $\Delta\alpha$ arc,

$$I_{h0l} \approx \frac{4p \mid \Phi_{h0l} \mid^2}{r_{h0l} r_l \Delta\alpha},$$ (19)

or

$$I_{h0l} \approx \frac{4p \mid \Phi_{h0l} \mid^2 d_{h0l} d_{00l}}{\Delta\alpha}.$$ (20)

Because of the particular way in which the reciprocal lattice nodes of an elongated-crystal texture are formed, the h0l reflections are characterized by a sharp fall-off in intensity at the end of the arc most distant from the minor axis of the ellipse, and a gradual tapering off at the closer end. This feature can be used to distinguish the h0l reflections on the diffraction pattern.

As an example, this type of appearance is shown by the 20l reflections on the 18th ellipse of a sepiolite diffraction pattern (Fig. 66). The positions of these reflections on the pattern are not determined from their centers, but from their upper ends.

When k $\neq$ 0 also, the intensity maximum may be shifted from the center of an arc to its most distant end, but without a sharp fall-off, which makes it more difficult to fix the true position of such a reflection, quite apart from the fact that the job is also made difficult by the considerable length of the arc reflection. In such a case it is best to find the heights of reflections from the radii and minor axes of the corresponding ellipses, ignoring the apparent positions of the intensity maxima.

2. Analysis of Intensities in the Investigation of Layer Silicate Structures

An investigation of the crystal structures of layer silicates and clay minerals can be carried out on the plan of finding a structure model, by trial-and-error methods, which gives calculated theoretical intensities corresponding as closely as possible to the experimental data. However, a method which is more likely to prove successful is that of constructing the three-dimensional distribution of material density in the structure, using Fourier series syntheses which have as their coefficients the structure amplitudes Φ_{hkl}.

In order to carry out syntheses of Φ-series (Fourier series), it is necessary to know the signs or phases of the structure amplitudes (depending on whether they are real or complex).

A characteristic feature of structural studies, particularly electron-diffraction structural studies of clay minerals, is that to quite a large extent they may work from predetermined general structural representations, working on the plan of refining a particular ideal model, which can be done with comparatively meager and indefinite diffraction data. This type of approach will be the more successful, the closer the initial model is to the real structure. The possibility is all the more important because it has not as yet been possible to obtain, in electron-diffraction patterns of finely dispersed layer silicates, a set of reflection intensities which has been numerous and perfect enough, without large numbers of gaps and superimpositions, for the direct statistical methods of sign determination (Kitaigorodskii, 1957; Porai-Koshits, 1960) to be usable. So in the case we are examining, we will first use the signs or phases of the theoretical structure amplitudes, calculated from the preliminary structure models, and then refine these from the results of subsequent Fourier syntheses.

Using the general concepts developed in Chapter 1, we can construct a model of a structure, detailed to the point where it includes data on unit cell dimensions and type of layer sequence. The structure model can be additionally refined to some extent using syntheses of $|\Phi|^2$-series (Patterson syntheses), which represent vector diagrams of interatomic distances brought to a single point, the origin of coordinates.

A complete $|\Phi|^2$-series, in which the coefficients are the squares of the structure amplitudes $|\Phi_{hkl}|^2$, obtained directly from the intensities of the hk$l$ reflections, has the form

$$P(x, y, z) = \sum_h \sum_k^{+\infty} \sum_l |\Phi_{hkl}|^2 \cos 2\pi (hx + ky + lz); \quad (21)$$

its projection on the ac (x0z) plane is

$$P'(x, z) = \sum_{-\infty}^{+\infty} h \sum_l l |\Phi_{hol}|^2 \cos 2\pi (hx + lz). \quad (22)$$

The functions $P'(z,y)$ and $P'(x,y)$, which represent projections on the bc (0yz) and ab (xy0) planes, have a similar form.

The derivation of these series and their theoretical foundation are dealt with in handbooks of structural analysis (Vainshtein, 1956; Kitaigorodskii, 1950; Porai-Koshits, 1960), and their summation techniques are the same as for Φ-series.

Of the three projections on the coordinate planes of the lattice, the projection on the x0z plane has the virtue that the ordinary departures from ordering in layer silicates, connected with displacements of networks and layers in the b-axis direction, do not show up on this plane. They naturally do not affect the h0l reflections, and therefore the corresponding projections reflect the order which is left in minerals with imperfect structures. This projection does not usually show all the interatomic distances, but in spite of this it can be used to elucidate some of the structural details of a silicate layer. As far as can be judged from projections obtained previously, the peak representing the interatomic vector from an inner cation to the oxygen atom at the upper corner of an octahedron decides which of the two possible mutually opposite orientations of the octahedral networks applies, depending on whether its normal projection on the a axis (i.e., along the perpendicular to this axis) is equal to $a/6$ or $a/3$. In the first case, the upper triangular octahedron faces are placed with their corners pointing away from the side perpendicular to the x0z plane, to the left, i.e., along the direction of the negative a axis, while in the second case they are to the right, along the positive a axis. For brevity we will call these orientations of the octahedral network left-handed and right-handed, respectively. Although in the general case the face-heights of the tetrahedra are not parallel to the x0z plane, we can nevertheless similarly distinguish between left-handed and right-handed orientations. Thus, in Figs. 5 and 8, the octahedra have right-handed orientations, in Fig. 8a,b the tetrahedra have left-handed, and in Fig. 8d,e they have right-handed orientations. The arrangement shown in Fig. 8c is transitional in this respect between left-handed and right-handed tetrahedral networks. The peaks corresponding to the interatomic vectors from an octahedral cation or O_{oct} to Si and O_{tetr} determine the height of the tetrahedron, and thus the quantity $k = l/l'$, and, moreover, whether the tetrahedral network is left-handed, right-handed, or in between.

It will be appreciated that a projection on the 0yz plane can only be obtained when the diffraction patterns contain separately distinguishable 0kl reflections, which will depend on the degree of ordering where the relative displacements of the polyhedron networks making up the layers, and the complete layers themselves, are concerned (these displacements are multiples of $b/3$). Even in the cases where the diffraction pattern contains separate $0kl$ reflections, however, their intensities may be distorted through partial breakdown of strict ordering in the arrangement of the layers and networks in the structure. This appears to be the reason why projections of a $|\Phi|^2$-series on the $0yz$ plane may sometimes contain less information than one on the $x0z$ plane. Nonetheless, they may result in improvements to the structural information which was derived from an analysis of the unit cell dimensions. This information includes the approximate values of $k = l/l'$ and, for the more perfect structures, a definite scheme of combination (one of the three possible) for each pair of adjacent networks in the silicate layer.

Limitations on the information to be gained from Patterson syntheses are not imposed solely by any loss of ordering in the crystal lattice. Thus, Cowley (1961a,b) has shown that if a clay—mineral crystal is bent around an axis lying in its ab plane, coherent scattering is only given by atoms for the reflections observed when the crystal has this plane perpendicular to the electron beam when these atoms have z coordinates which do not differ by more than a certain amount, which depends on the size of the bending. Thus, a Patterson projection on the $xy0$ plane may only show up those features which relate to crystal fragments which are not too distant from one another along the direction of the z axis. In this way, bending of layers decreases the effective thickness of a crystal, and permits kinematic scattering conditions to be extended to thicker crystals.

In montmorillonites, for example, the above limiting difference Δz is of the order of 3 A, and each network of tetrahedra or octahedra will make its own individual contribution to the reflection intensities. It is for this reason that the hexagonal grids of reflections found on Cowley's diffraction patterns corresponded to the hexagonal symmetry of each individual layer, and not to the symmetry of the whole set of layers, which was at most monoclinic.

For other minerals with crystals less subject to deformation, and for $|\Phi|^2$ projections on other planes, this effect is apparently of lesser significance.

By taking into account all the structural information obtained from analysis of the mineral lattice and synthesis of its $|\Phi|^2$- series, and using its structural chemical formula (which expresses the distribution of its chemical elements among the polyhedra in the structure), it will be possible to construct a model of the structure by combining regular polyhedra into networks,

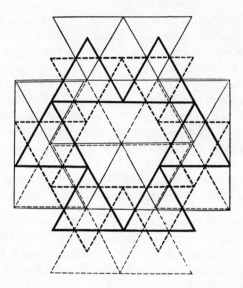

Fig. 67. Combined pattern formed by the
parallel faces in a three-storied silicate layer
(thick continuous line—upper octahedron face;
thick dashed line—lower octahedron face; thin
continuous line—upper tetrahedron face; thin
dashed line—lower tetrahedron face).

networks into layers, and layers into a structure. We have already noted in
Chapter 1 that a structure can be graphically represented using Belov's model
(1949, 1950, 1951), in which all the faces parallel to the ab plane of occupied
tetrahedra and octahedra are depicted simultaneously in the same plane, with
an indication of the levels at which they lie and the type of occupying cation
(Fig. 67). The picture obtained is essentially a normal projection of the structure on the xy0 plane (in distinction to a projection along a direction parallel
to the c axis). Bearing in mind that the anions are located at the corners of
the polyhedra, and the cations at their centers, it is easy to find the atomic
coordinates x' and y' in this a, b coordinate system.

In the particular case of the structural motif for a dioctahedral layer
silicate of three-storied layers, with anions having a cubic close-packed arrangement and with a space group of $C_{2h}^3 = C2/m$ (Fig. 67), the spatially independent atoms have the following x' and y' coordinates (the origin is most

Table 29. Levels at Which Atoms Lie in a Silicate Layer

Level No.	Atoms	$zc \sin \beta$
0	M_{oct}	0
1	OH, O_{oct}	$h/2$
2	Si, Al	$h/2 + 3h'/4$
3	O_{tetr}	$h/2 + h'$
4	K	$c \sin \beta/2$

conveniently chosen at the most symmetrical point, in this case at a center of symmetry lying at the center of a vacant octahedron):

$$M_{oct} - (0, 1/3); \quad OH - (1/3, 0); \quad O_{oct} - (1/3, 1/3);$$

$$(Si, Al) - (1/3, 1/3); \quad O_{tetr} - (0, 1/3), (0, 0); \quad K - (1/3, 0).$$

The other atoms in the unit cell are related to those listed by the symmetry elements in the space group C2/m, and their coordinates are fixed by those already given.

The z coordinates of the atoms are proportional to the heights of the levels on which the atoms are arranged above the zero plane passing through the centers of the octahedra. These heights are therefore fixed by half the height of an octahedron, the distance from the center to the corner of a tetrahedron, and the height of a tetrahedron.

These heights therefore have the values given in Table 29. The z coordinates can be derived from these heights by dividing them by $c \sin \beta$.

If a structure does not have an orthogonal cell, its atomic coordinates x',y' in the normal projection on the ab plane should be converted to its coordinates x,y as determined by an inclined projection of the points (parallel to the axis c) onto the ab plane.

The relationships existing between these coordinates have been derived from structural crystallography formulas (Belov, 1951b), using the relationships between the corresponding coordinate vectors.

In the general case of a nonorthogonal coordinate system a, b, c the normal projection of a structure on the ab plane will be determined by the

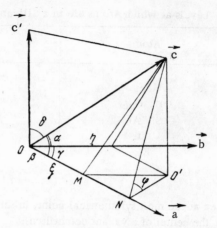

Fig. 68. Derivation of the relationships between
x',y' coordinates (in a normal projection ab) and
x,y coordinates (in an inclined cell).

x',y' coordinates from a coordinate system which differs only in that it includes the vector c', equal to the projection of the vector c on the normal to the ab plane. According to Fig. 68,

$$a = a', \quad b = b',$$

$$c = \frac{\xi}{a}\, a' + \frac{\eta}{b}\, b' + c', \tag{23}$$

where ξ, η are the absolute lengths of the components of the vector c along
the axes a', b' in the a', b', c' coordinate system.

Since the coordinates are transformed using a matrix derived from the
coordinate-vector inverse transformation matrix by interchanging rows and
columns, then

$$x' = x + \frac{\xi}{a} z, \qquad y' = y + \frac{\eta}{b} z, \qquad z' = z. \tag{24}$$

Therefore,

$$x = x' + \Delta x = x' + g_1 z,$$

$$y = y' + g_2 z,$$

where

$$g_1 = -\frac{\xi}{a}, \quad g_2 = -\frac{\eta}{b}. \tag{25}$$

With regard to the quantities ξ, η, these can be expressed in terms of the inclined-cell constants.

We know that the volume of a unit cell is given by the equation

$$\Omega = ([ab]\,c) = ab \sin \varphi \cdot c \cos \delta =$$

$$= abc \sqrt{1 - \cos^2 \alpha - \cos^2 \beta - \cos^2 \gamma + 2 \cos \alpha \cos \beta \cos \gamma},$$

because, for the angle δ between the axes c and c',

$$\cos \delta = \frac{\sqrt{1 - \cos^2 \alpha - \cos^2 \beta - \cos^2 \gamma + 2 \cos \alpha \cos \beta \cos \gamma}}{\sin \gamma}. \tag{26}$$

For the dihedral angle φ between the ab and ac planes,

$$\sin \varphi = \frac{c'}{c \sin \beta} = \frac{\cos \delta}{\sin \beta},$$

and for the angle ψ between the ab and bc planes,

$$\sin \psi = \frac{\cos \delta}{\sin \alpha}.$$

It is now easy to see (Fig. 68) that

$$\xi = \frac{c \sqrt{\sin^2 \alpha - \cos^2 \delta}}{\sin \gamma}, \quad \eta = \frac{c \sqrt{\sin^2 \beta - \cos^2 \delta}}{\sin \gamma}, \tag{27}$$

or, in another form, that

$$\xi = c \left[\cos \beta - \frac{\sqrt{\sin^2 3 - \cos^2 \delta}}{\gamma} \right]; \; \eta = c \left[\cos \alpha - \frac{\sqrt{\sin^2 \alpha - \cos^2 \delta}}{\gamma} \right]. \tag{28}$$

In particular, when $\delta = 0$, $\xi = \eta = 0$; when $\gamma = \pi/2$, $\delta \neq 0$, $\xi = c \cos \beta$, $\eta = c \cos \alpha$. In the monoclinic case,

$$\alpha = \pi/2, \quad \eta = 0. \tag{29}$$

If the structure model is to take account of rotations of polyhedron faces, then to the x,y coordinates, Δx_i, Δy_i should be added as indicated in equation (5) of Chapter 1.

The resulting atomic coordinates will fix the preliminary model of the structure.

After the atomic coordinates in a structure model of this type have been determined, the corresponding theoretical structure amplitudes are calculated from formula (3), for atoms distributed at these coordinates in accordance with the space symmetry group and the structural chemical formula of the mineral. If we take as an example the three-storied structure model of symmetry C_{2h}^3 = C2/m mentioned above, and we work in relative units and assume values of hk are unmixed, i.e., both even or both odd, then on the basis of the formula

$$K_t (M_{p_1}^1 M_{p_2}^2 \ldots) [Si_{4-q}Al_qO_{10}] (OH)_2,$$

the structure amplitudes can be written in the form:

$$\Phi_{hkl} = (p_1 f_1 + p_2 f_2 + \ldots,) \cos 2\pi k / 3 + [qf_{Al} + (4-q)f_{Si}] \times$$
$$\times \cos 2\pi [z_2 l + (^1/_3 + g z_2) h] \cos 2\pi k / 3 + 2f_0 \{(1 + 2\cos 2\pi k / 3) \times$$
$$\times \cos 2\pi [(^1/_3 + g z_1) h + z_1 l] + \cos 2\pi [(g z_3 + \Delta x_2) h + z_3 l] +$$
$$+ 2\cos 2\pi [(g z_3 + \Delta x_1) h + z_3 l] \cos 2\pi k (1/3 + \Delta y_1)\} + tf_k \cos 2\pi \frac{h+l}{2} .$$

Other forms of structure amplitude expressions, for various layer combinations and ordered and disordered structures, will be given below, in section 3 of this chapter.

For convenience in calculation, it is best to use graphs of the atomic factors f as functions of d or sin ϑ/λ, constructed from their values in Vainshtein's tables (1956), and graphs of the trigonometric functions appearing in the structure amplitude according to (4). It is best to put the values of the atomic factors and trigonometric functions for each reflection into a tabular form, and calculate each component of the structure amplitude separately. The list will be valuable in the construction of difference syntheses, in which the contributions to Φ_{hkl} of certain atoms are subtracted, or in the change from the structure amplitudes corresponding to one model to the amplitudes of a different model. In calculation of the structure amplitudes, Belov's nomograms (1954) may also be used with success.

The harmonic synthesis method is based on the fact that the amplitudes of scattering of electrons by a crystal lattice, in diffraction directions which are in accordance with formulas (4)-(7) of Chapter 3 and formula (1) of this chapter, are the expansion coefficients of the Fourier series of the electric potential. Therefore the Fourier series with coefficients Φ_{hkl} gives the structure of the substance in the form of $\varphi(x,y,z)$, the distribution of electric potential in it:

$$\varphi(x, y, z) = \frac{1}{\Omega} \sum_{h,k,l=-\infty}^{+\infty} \sum \sum \Phi_{hkl} e^{-2\pi i(hx+ky+lz)} \ . \qquad (31)$$

Since the potential in the lattice is created by the nuclei of its atoms, these being screened by the electron clouds, and the potential reaches its maximum values at the centers of the atoms, its distribution reproduces the arrangement of the atoms in the structure.

In practice, the series in (31) will not include all terms from $-\infty$ to $+\infty$, but only those corresponding to the set of hkl reflections recorded. This sum of a finite number of terms will give a more accurate representation of the structure, the greater the number of terms used and the smaller the contributions of the terms omitted from the experimental sum or, another way of saying the same thing, the smaller the effect of the Fourier break-of-series. Vainshtein has shown (1956) that in electron diffraction the atomic amplitudes, and with these the mean absolute values of the structure amplitudes, fall more rapidly than the corresponding quantities in X-ray crystallography, and the intensities of the reflections reach minimum practicably recordable values at larger values of d or $\sin\vartheta/\lambda$. For a given value of d or $\sin\vartheta/\lambda$, the fraction of $\varphi(x,y,z)$ due to terms omitted from the series (or the role of the residual terms) is less in electron diffraction than in X-ray diffraction, due to the fact that the Φ-series used in electron diffraction have better convergence than the F-series used in X-ray work [the latter give the structure of a substance in terms of $\rho(x,y,z)$, the three-dimensional distribution of electron density in it]. As they express $\varphi(x,y,z)$, the three-dimensional distribution in the substance, in ideal cases the Φ-series represent the distribution of material not only within the atoms, but also in the spaces between them, and they may show the ionization states of atoms and the nature of the chemical bonds in a substance.

The function (31) is a general expression for the three-dimensional distribution of material in a substance. From this it follows that projections of a structure on the x0z, 0yz, and xy0 planes, respectively (e.g., Vainshtein,1956; Cowley, Rees, 1958), are given by the following two-dimensional series:

$$\varphi'(x, z) = \frac{1}{S_{xz}} \sum_{h,l=-\infty}^{+\infty} \sum \Phi_{h0l} e^{-2\pi i(hx+lz)}, \qquad (32a)$$

$$\varphi'\left(y,\,z\right) = \frac{1}{S_{yz}} \sum_{k,l=-\infty}^{+\infty} \Phi_{0kl} e^{-2\pi i(ky+lz)} , \qquad (32b)$$

$$\varphi'\left(x,\,y\right) = \frac{1}{S_{xy}} \sum_{h,k=-\infty}^{+\infty} \Phi_{hk0} e^{-2\pi i(hx+ky)} , \qquad (32c)$$

where S_{xy}, S_{yz}, S_{xz} are the areas of the faces of the unit cell.

All the syntheses below will be given with an accuracy equal to a constant factor, which will therefore be omitted from the formulas.

By inserting expressions in (31) which relate the coordinates of a surface or line, we can obtain series which express the distribution of material on any surface or line in the crystal lattice. In particular, this method can be used to derive planar or linear sections of the structure, which can be used to refine the fine details of the positioning of the atoms in the substance. As an example, the section through the x0z coordinate plane (y = 0) is given by the series

$$\varphi\left(x,\,0,\,z\right) = \sum_{h,k,l=-\infty}^{+\infty} \Phi_{hkl} e^{-2\pi i(hx+lz)} = \sum_{h,l=-\infty}^{+\infty} C_{hl} e^{-2\pi i(hx+lz)} , \quad (33)$$

where

$$C_{hl} = \sum_{k=-\infty}^{+\infty} \Phi_{hkl}.$$

With monoclinic symmetry, where $\Phi_{h\bar{k}l} = \Phi_{hkl}$,

$$C_{hl} = \Phi_{hol} + 2\sum_{1}^{\infty} {}^{k}\Phi_{hkl}.$$

Detailed accounts of the different types of Φ-series will be found in the structure analysis handbooks mentioned above.

The Φ-series are calculated by reducing them to the simplest possible one-dimensional series. This can be demonstrated using as an example the two-dimensional series (32a) (for simplicity, the index k = 0 will be omitted here):

$$\varphi'(x, z) = \sum\nolimits_h \sum\nolimits_l^{+\infty} \Phi_{hl}\, e^{-2\pi i\,(hx+lz)} =$$

$$= \sum\nolimits_h \sum\nolimits_l^{+\infty} (A_{hl} + iB_{hl})\,[\cos 2\pi\,(hx + lz) - i\sin 2\pi\,(hx + lz)] =$$

$$= \sum\nolimits_h \sum\nolimits_l^{+\infty} [A_{hl}\cos 2\pi\,(hx+lz) + B_{hl}\sin 2\pi\,(hx + lz)] +$$

$$+ i \sum\nolimits_h \sum\nolimits_l^{+\infty} [B_{hl}\cos 2\pi\,(hx + lz) - A_{hl}\sin 2\pi\,(hx + lz)]. \quad (34)$$

Since, according to (4),

$$A_{hl} = A_{\bar{h}\bar{l}}, \; B_{hl} = -B_{\bar{h}\bar{l}}, \quad\quad\quad (35)$$

the imaginary part of $\varphi'(x,z)$ is equal to zero. Conversely, and independently of equation (4), the condition that the value of $\varphi'(x,z)$ must be real leads to equation (35). Therefore,

$$\varphi'(x, z) = \sum\nolimits_h \sum\nolimits_l^{+\infty} [A_{hl}\cos 2\pi\,(hx + lz) + B_{hl}\sin 2\pi\,(hx+ lz)]. \quad (36)$$

After eliminating terms with zero h, l indices,

$$\varphi'(x, z) = A_{00} + \sum\nolimits_{h\neq0}^{+\infty} A_{h0}\cos 2\pi hx + \sum\nolimits_{l\neq0}^{+\infty} A_{0l}\cos 2\pi lz +$$

$$+ \sum\nolimits_h^{+\infty} B_{h0}\sin 2\pi hx + \sum\nolimits_l^{+\infty} B_{0l}\sin 2\pi lz +$$

$$+ \sum\nolimits_{h\neq0}^{+\infty}\sum\nolimits_{l\neq0} [A_{hl}\cos 2\pi\,(hx + lz) + B_{hl}\sin 2\pi\,(hx + lz)]. \quad (37)$$

The general term after the double summation sign transforms into the expression

$$A_{hl}\,(\cos 2\pi hx \cos 2\pi lz - \sin 2\pi hx \sin 2\pi lz) +$$
$$+ B_{hl}\,(\sin 2\pi hx \cos 2\pi lz + \cos 2\pi hx \sin 2\pi lz).$$

The summation over the whole of the reciprocal space (all h,k,l between $-\infty$ and $+\infty$) can be replaced by a summation over one quadrant (h,k,l between 0 or 1 and ∞) if each term with indices hl is replaced by four terms with indices hl , $\bar{h}\bar{l}$, $\bar{h}l$, h$\bar{l}$, and at the same time allowance is made for the way the signs of the trigonometric functions vary with the signs of the arguments. If this is done, the double sum in (37) takes the form

$$\sum_{1}^{\infty}{}^{h}\sum{}^{l}\ [(A_{hl} + A_{\bar{h}\bar{l}} + A_{\bar{h}l} + A_{h\bar{l}})\cos 2\pi hx \cos 2\pi lz -$$

$$- (A_{hl} + A_{\bar{h}\bar{l}} - A_{\bar{h}l} - A_{h\bar{l}})\sin 2\pi hx \sin 2\pi lz +$$

$$+ (B_{hl} - B_{\bar{h}\bar{l}} - B_{\bar{h}l} + B_{h\bar{l}}) \sin 2\pi hx \cos 2\pi lz +$$

$$+ (B_{hl} - B_{\bar{h}\bar{l}} + B_{\bar{h}l} - B_{h\bar{l}}) \cos 2\pi hx \sin 2\pi lz.$$

If we make use of formula (35),

$$\varphi'(x, z) = A_{00} + 2 \sum_{h=1}^{\infty} [A_{h0} \cos 2\pi hx + B_{h0} \sin 2\pi hx] +$$

$$+ 2 \sum_{l=1}^{\infty} [A_{0l} \cos 2\pi lz + B_{0l} \sin 2\pi lz] +$$

$$+ 2 \sum_{h=1}^{\infty} \left\{ \sum_{l=1}^{\infty} [(A_{hl} + A_{\bar{h}l}) \cos 2\pi lz + (B_{hl} + B_{\bar{h}l}) \sin 2\pi lz] \right\} \times$$

$$\times \cos 2\pi hx + 2 \sum_{h=1}^{\infty} \left\{ \sum_{l=1}^{\infty} [(B_{hl} - B_{\bar{h}l}) \cos 2\pi lz -\right.$$

$$\left.- (A_{hl} - A_{\bar{h}l}) \sin 2\pi lz] \right\} \sin 2\pi hx. \qquad (38)$$

After grouping terms with common factors, we have

$$\varphi'(x, z) = A_{00} + \sum_{h=0}^{\infty} [A'_{h}(z) + A''_{h}(z)] \cos 2\pi hx +$$

$$+ \sum_{h=1}^{\infty} [B''_{h}(z) - B'_{h}(z)] \sin 2\pi hx, \qquad (39)$$

where

$$A'_{0}(z) = 2 \sum_{l=1}^{\infty} A_{0l} \cos 2\pi lz, \quad A''_{0}(z) = \sum_{l=1}^{\infty} B_{0l} \sin 2\pi lz,$$

$$A'_{h}(z)\big|_{h\neq0} = 2A_{h0} + 2 \sum_{l=1}^{\infty} (A_{hl} + A_{\bar{h}l}) \cos 2\pi lz, \qquad (40)$$

$$A_h^{''}(z)\big|_{h\neq0} = 2\sum_{l=1}^{\infty} (B_{hl} + B_{\bar{h}l}) \sin 2\pi lz,$$

$$B_h^{'}(z) = 2\sum_{l=1}^{\infty} (A_{hl} - A_{\bar{h}l}) \sin 2\pi lz,$$

$$B_h^{''}(z) = 2B_{h0} + 2\sum_{l=1}^{\infty} (B_{hl} - B_{\bar{h}l}) \cos 2\pi lz. \qquad (40)$$

Thus, the summation of the two-dimensional series (34) can be broken down into two stages. The first stage is to calculate the one-dimensional series (40), which gives numerical values for the quantities $A_h^{'}(z_m)$, $B_h^{'}(z_m)$, $A_h^{''}(z_m)$, $B_h^{''}(z_m)$ for the coordinate series $z_m = m/n$, where n is a certain number of intervals into which the cell is arbitrarily divided, and m takes on all values from 0 to n.

In the second stage of the summation, the one-dimensional series (39) are calculated using the coefficients from (40) for each value of z_m. This yields numerical values for $\varphi'(x,z)$, for the coordinates $x = m_1/n$, $z = m_2/n$. For a graphical picture of this function, a drawing of the ac face on a given scale should be divided up into n^2 cells by straight lines drawn parallel to the a and c directions and passing through the points mc/n, ma/n, and each cell should be labeled with the corresponding value of $\varphi'(m_1/n, m_2/n)$. If isolines, i.e., lines connecting points with equal values of $\varphi'(x,z)$, are drawn on this face, a contour map will be formed which will show a number of peaks; the heights and shapes of these peaks will reflect the distribution of the electric potential $\varphi(x,y,z)$ projected on the ac face of the structure, which, as we have already noted, defines the spatial distribution of material in the substance, that is to say, its structure. If some of the peaks on this projection overlap one another, which happens when atoms are projected onto points lying close together on the plane, then a difference synthesis can be constructed. For this we need to subtract fractions from Φ_{exp} which are proportional to the contributions which the terms relating to the subtracted atoms make to Φ_{theor}. If we construct a projection of the structure using these modified Φ coefficients, the subtracted atoms should not appear, so that the coordinates of the atoms which they overlapped in the original projection can be found more exactly. It will be seen that if this can be done purely through choice of reflections, it would be even better to construct sections through planes in which the sought-after atoms lie but which do not include the atoms which were superimposed on them in the projection, or to carry out other types of three-dimensional synthesis.

Summation of one-dimensional Fourier series of the type described above can be done very rapidly and simply by using a computer. Where there is a comparatively small number of reflections on the diffraction pattern, the very simple and elegant strip method can be applied without complications.

The coordinates of the atoms are fixed by the points at which the Fourier syntheses reach their maxima. To locate these maxima we need to construct curves of the synthesis values close to the maximum, at m/n points lying along the direction of the sought-after coordinate, and find, by interpolation, the point at which the maximum is reached.

In layer silicates the coordinates of all the atoms may be found from just two projections, on the x0z and 0yz planes. Here, also, geometrical considerations stemming from the shapes and relative dimensions of the polyhedra may be taken into account to derive the position of one atom from the positions of others. After the coordinates have been determined, it is necessary to again calculate Φ and construct a new synthesis, and to continue repeating this process as long as it improves the agreement between Φ_{exp} and Φ_{calc}; the extent of this agreement is measured in terms of R, a quantity usually called the reliability factor.

When the atomic coordinates are known, the application of ordinary analytical geometry formulas to the interatomic vector coordinates x,y,z will give the interatomic distances

$$r = \sqrt{x^2a^2 + y^2b^2 + z^2c^2 + 2xz \cos \beta + 2yz \cos \alpha} \,. \tag{41}$$

Here it is convenient to go over to the orthogonal system of axes

$$\boldsymbol{a}, \ \boldsymbol{b}, \ \boldsymbol{c}' \ (c' = c \sqrt{1 - \cos^2 \alpha - \cos^2\beta}),$$

in which

$$r = \sqrt{x'^2a^2 + y'^2b^2 + z^2c'^2} \,, \tag{42}$$

where x',y' are expressed in terms of x,y by formula (24).

The criterion for the reliability of a given structure is the extent of the agreement between the experimental and calculated reflection intensities for the structure. The usual measure of this agreement is the reliability factor

$$R = \frac{\Sigma \, | \, | \Phi_{exp} | - | \Phi_{calc} | \, |}{\Sigma \, | \Phi_{exp} |} \,. \tag{43}$$

To compare Φ_{exp} and Φ_{calc}, and in particular so that R may be calculated, the experimental and calculated (i.e., theoretical) values of $|\Phi|^2$ must be normalized, i.e., brought to the same scale. This scaling can be achieved most naturally by putting $\Sigma|\Phi_{hkl}|$ for both Φ_{exp} and Φ_{calc} equal to any particular single value.

It can be shown that on the basis of the relationship between their values of $|\Phi_{exp}|^2$ and $|\Phi_{calc}|^2$, reflections can be divided up into several groups, depending on their hkl indices. Thus, in layer silicates, as a rule, a distinction exists between reflections close to, and those distant from, the minor axes of the ellipses. In particular, it can be shown that for the former all $|\Phi_{exp}|^2 > |\Phi_{calc}|^2$, and for the latter, conversely, all $|\Phi_{exp}|^2 < |\Phi_{calc}|^2$. This behavior is a consequence of the fact that the two-dimensional ordering and periodicity of the structure within the individual layers and networks is better than its three-dimensional ordering and periodicity over the structure as a whole, and these differences show up in the different projections. This characteristic feature of layer silicates is justified by the very fact of their layered nature, and is in harmony with the fact that layers can be formed from networks, and structures from layers, in more than one way. Other important factors here are the roles which may be played by dislocations, as considered by Drits (1961), and crystal bending, mentioned by Cowley (1961a).

In the same way, the difference in the degree of close-range and long-range order may express itself in additional weakening of the more distant reflections in comparison to the closer ones. The introduction of the temperature factor (Vainshtein, 1956) only reduces this anomaly, not removes it, so that R is improved by 2-3%.

Under these circumstances, several normalizations of $|\Phi_{exp}|^2$ and $|\Phi_{calc}|^2$ should be carried out, the number depending on the number of clearly distinguishable groups of reflections with significantly different ratios of $|\Phi_{exp}|^2$ to $|\Phi_{calc}|^2$.

The accuracy of the coordinates and interatomic distances found can be determined using the theory and formulas specially devised for this purpose by Vainshtein (1956).

3.   Wave-Scattering by a Silicate Layer Sequence

As a consequence of the fact that layer silicate structures are formulated from complete "blocks" in the shape of the individual two-dimensional networks of polyhedra, and because the ways these may be connected together are

governed by certain restrictions, the diffraction properties of all these structures have certain common features.

To explain these, we must start from the general expression for scattering amplitude,

$$S\left(\mathbf{H}\right) = K \int \varphi\left(\mathbf{r}\right) e^{2\pi i \mathbf{H}\mathbf{r}}\, d\tau,$$

(44)

and particularize this for the case we are considering.

If the rules given above for the relative arrangements of layers are followed, then, whatever their type or particular structure, any of the axes a_i, b_i
of each network or pair of attached networks will be parallel to the a_j, b_j axes
of the other networks and combinations. Therefore, a sequence of silicate
layers, as described in terms of the displacements σ, τ, will always be characterized by the two constants a and b ($b = a\sqrt{3}$) in the plane parallel to the
layers used as the basis of the coordinate system.

Nonperiodic Sequences. If a sequence of layers is nonperiodic,
and the sequence possesses variations in layer type, azimuthal orientation, and
relative displacements, then the third vector in the coordinate system may be
assigned an arbitrary absolute magnitude, while its direction is conveniently
chosen to be normal to the layers. The whole sequence of layers may be
broken up, depending on its particular form, into portions made up of one or
more layers, each with its own particular origin of coordinates $\mathbf{r}_{0m}$. The positions of points of the m-th such "zone" will be given by the radius vectors

$$\mathbf{r}_m = \mathbf{r}_{0m} + \mathbf{r}_m',$$

and S($\mathbf{H}$) can be split up into a sum of integrals over the separate zones:

$$S\left(\mathbf{H}\right) = K \sum_m e^{2\pi i \mathbf{H}\mathbf{r}_{0m}} \int \varphi\left(\mathbf{r}_m'\right) e^{2\pi i \mathbf{H}\mathbf{r}_m'}\, d\tau.$$

(45)

From the same reasoning as that used above [formulas (4) and (7) of Chapter 3],
this sum is equivalent to the expression

$$S\left(\mathbf{H}\right) = K D'\left(\mathbf{H}\right) \sum_m e^{2\pi i \mathbf{H}\mathbf{r}_{0m}} F_m\left(\mathbf{H}\right),$$

(46)

where D'($\mathbf{H}$) is the two-dimensional interference function of a layer, a function which can be taken as identical for all the layers, and

$$F_m(\mathbf{H}) = \sum_j f_{jm} e^{2\pi i(\mathbf{Hr}'_{mj})} \tag{47}$$

is the scattering amplitude of one unit cell of the two-dimensional lattice of the m-th zone.*

For a finite number of layers, calculation of S(**H**) is comparatively simple, and the results can be presented in the form of graphs of the distribution of S(**H**) values along straight lines perpendicular to the ab plane and with the two-dimensional coordinates hk. Along each straight line, of course, the reciprocal space coordinate z^* changes continuously, taking on both integral values, $z^* = l$, and other intermediate values.

For an infinite sequence, without a strict repeat unit, a statistical expression which gives the probable distribution of each layer must be settled on. The calculations must take the appropriate distribution functions into account, and this has been done in practice so far only for the simplest $00z^*$ reflections, in analysis of the so-called mixed-layer products (Brown, 1961).

Sequences of Packets Made Up of Randomly Displaced Networks or Layers. These are sequences which can be split up into groups of layers, where each group of layers is of the same type, but may have different azimuthal orientations and relative displacements parallel to the ab plane. These groups of layers may be conveniently called packets. Thus, the chlorite packet consists of one $(0:1)$ layer and one $(2:1)$ layer, these following one after the other in a definite order.

In the packets, corresponding levels will contain identical atoms having identical z coordinates measured from the assigned zero plane of each packet, but the x,y coordinates may differ. However, in accordance with the displacement components σ and τ (Table 3), these coordinates can generally only differ by multiples of $\frac{1}{3}$. Because of this limitation on the differences between the x,y coordinates, this type of sequence cannot be called completely disordered. These features inevitably exert an effect on the scattering ability of this type of sequence. In fact, in an orthogonal system of coordinates,

---

*In this section, and up to the end of Chapter 4, the structure amplitudes will be denoted by F, to distinguish them from Φ, the amplitudes of the individual layers.

$$S\left(\mathbf{H}\right) = KD'\left(\mathbf{H}\right) \sum_{0}^{\infty} F_m\left(\mathbf{H}\right)\ e^{\ 2\pi i\left(hx_{0m} + ky_{0m} + z_m^*\right)}, \tag{48}$$

$$F_m\left(\mathbf{H}\right) = \sum_{j} f_j e^{\ 2\pi i\left(hx'_{mj} + ky'_{mj} + z^* z'_j\right)}, \tag{49}$$

where $F_m(\mathbf{H})$ relates to identical cells of like composition. It will be quite obvious that for reflections with h = 3h', k = 3k', $F_m(\mathbf{H})$ is independent of the individual features of the arrangement of the networks in the m-th packet. Remembering that $x_{0m}, y_{0m} = 0$, $\pm \frac{1}{3}$, for these reflections we have $S(\mathbf{H}) = KD(\mathbf{H})F(\mathbf{H})$, where $D(\mathbf{H}) = D'(\mathbf{H}) \sum e^{2\pi i z^* m}$ is the three-dimensional interference function, which picks out from the continuous F($\mathbf{H}$) distribution a discrete set of values for integral $z^* = l$, along the line in reciprocal space with the given h,k indices. So, for reflections with h,k = 3n, the layer sequence has a repeat unit of one layer along directions normal to the layer. This repeat unit, which applies to only some of the properties of the sequence (in this example to scattering abilities in directions with h,k = 3n), can be called a nonrigorous repeat unit, in contrast to the strict repeat unit, which must apply to all the properties of the sequence, including the scattering abilities in directions with any h,k indices.

Sequences with rotationally identical packets may also possess properties like those just described, particularly sequences with layers of a single type related by relative rotations through multiples of $2\pi/6$, where these rotations lead to differences in the x and y coordinates of corresponding points in different packets, like those described above.

Sequences of Packets with Randomly Displaced Networks or Layers Along Only One of the Axes a or b. These are sequences of packets or layers, all of the same type, in which a c-axis direction can be chosen such that not only do the z coordinates of corresponding points coincide, but also either the x or y coordinates, while the third coordinates (y or x) will differ by multiples of $\frac{1}{3}$. This is the case when the corresponding σ and τ values of the sequence packets have identical components along one of the axes x or y. Thus, in a sequence of identical layers of identical parity (all even or all odd, i.e., only σ_{2n} or only σ_{2n+1} layers), the normal coordinates x_{0m} of the origins of successive layers will vary by the same amount, or remain equal to zero.

If the c axis (inclined or perpendicular to the a axis) is directed through the normal projections of the layer origins onto a plane lying perpendicular to the b axis, then in the new system of coordinates, corresponding points of different layers will have identical x coordinates and different y coordinates, in this setting. In another special case, where the sequence contains both odd and even layers related by reflection in a plane perpendicular to the a axis (e.g., σ_1 and σ_2, σ_3 and σ_6, σ_4, and σ_5), the c axis can be drawn (inclined or perpendicular to the b axis) along a plane perpendicular to the a axis. In this case, the y coordinates of corresponding points will be the same, while the x coordinates will differ by multiples of $\frac{1}{3}$.

These sequences have a nonrigorous orthogonal repeat unit of three layers, determined by the reflections with h,k = 3n. In addition, however, according to formula (48), they have a nonrigorous repeat unit of one layer (inclined or orthogonal), fixed by reflections in which only one index (k or h) satisfies the conditions k = 3k' or h = 3h', while the other index (h or k, respectively) can be arbitrary. In this case, an inclined repeat unit of one layer is equivalent to an orthogonal repeat unit of three layers.

In the expressions for F(hkz*), we can pick out the terms relating to rotationally identical zones in the sequence. These zones may be either complete packets or separate layers of a particular type present in the packets. The corresponding terms are their scattering amplitudes Φ(hkz*). The Φ amplitudes differ from the F amplitudes only in the summation limits, so long as the packets in the sequence are not rotationally identical. In its own system of coordinates, each of the rotationally identical packets or layers has the z coordinates of corresponding atoms the same, while the x,y coordinates fall into six groups of values, depending on the azimuthal orientations of the packets or layers. If these coordinates are expressed in a system in which the a axis has a particular orientation, for example along a_3, then the coordinates will be the same for all packets or layers. Here, the amplitude of a packet oriented with its a_i axis parallel to the a axis of the whole sequence can be transformed using the equation

$$\Phi_i\left(hkz^*\right) = \Phi_3\left(h_ik_iz^*\right). \tag{50}$$

The values h_i, k_i are obtained by replacing x_i, y_i in the power factor by their linear and homogeneous expressions in terms of x_3, y_3.

Thus, if we have a set of values of Φ(hkz*) for a layer or packet in one particular orientation, we can use this set for layers or packets in any of the other five orientations which may exist in the sequence.

Sequences of Identical and Identically Oriented
Packets, Randomly Displaced Relative to One Another
Parallel to the ab Plane. In this case, it is convenient to choose the
system of coordinates given by the vectors a, b, c_1', where c_1' is the distance along the normal between two successive layers.

For a completely defined layer sequence, with given coordinates x_{0m},
y_{0m} for the origin of any arbitrary m-th layer, the scattering amplitude expressed in terms of the amplitude of a separate packet, $\Phi(H)$, has the form

$$S(H) = KD'(H)\,\Phi(H)\sum_{m=1}^{N} e^{2\pi i\,[hx_{0m}+ky_{0m}+z^*\,(m-1)]}. \tag{51}$$

In this formula, z^* is a continuous variable.

If x_{0m}, y_{0m} are not known beforehand, and $W_m(x_0, y_0)$ is the probability
that the origin of the m-th layer has the coordinates x_0, y_0 (here x_0, y_0 should
be considered as continuous variables), then

$$S(H) = KD'(H)\,\Phi(H)\sum_{m=1}^{N}\int_{ab} W_m(x_0, y_0)\,e^{2\pi i\,[hx_0+ky_0+z^*\,(m-1)]}\cdot dx_0 dy_0 \tag{52}$$

where the integration is taken over the ab face of the unit cell. The equation
obtained from (52),

$$C(H) = \frac{S(H)}{KD'(H)\,\Phi(H)} = \int_{ab}\left[\sum_{m=1}^{N} W_m(x_0, y_0)\,e^{2\pi i z^*\,(m-1)}\right] e^{2\pi i\,(hx_0+ky_0)}\,dx_0 dy_0, \tag{53}$$

is the expression for the expansion coefficients of the Fourier series for the
function

$$W(x_0, y_0) = \sum_{1}^{N} W_m(x_0, y_0)\,e^{2\pi i z^*\,(m-1)}.$$

Therefore,

$$W(x_0, y_0) = \sum_{hk} C(H)\,e^{-2\pi i\,(hx_0+ky_0)}. \tag{54}$$

The function $W(x_0, y_0)$ defines the combined distribution of layer origin positions, in projection on the ab plane, for the whole layer sequence. It is worth

noting that this expression is given by reflections which have any z^* index value common to all.

For integral $z^* = l$, $W = \Sigma W_m$ is the projection of the x_0, y_0 distribution on the ab plane. For nonintegral z^*, the projection represents a dummy x_0, y_0 distribution, which as m varies favors the appearance of some layers and suppresses others. When constructed from experimental data, a distribution of this type becomes more effective as the structure of an individual layer or packet becomes more accurately known (this can be judged from reflections with $k = 3k'$). In analogy with an F^2-series, it is also possible to construct a C^2-series which represents the relative displacement function of layers projected on the ab plane:

$$\sum_{h,k} \left| C\left(H\right) \right|^2 \cos 2\pi \left[h\left(x_0' - x_0\right) + k\left(y_0' - y_0\right)\right]. \tag{55}$$

Syntheses of this type have been used in practice by Cowley (1953, 1957) in electron-diffraction studies of disordered boric acid structures.

Sequences of Identically Displaced Identical Packets. If a sequence of layers can be divided into groups of layers or layer packets, all of which are completely identical in type, internal structure, and relative arrangement of component layers, then the structure will have a strict repeat unit. The displacement undergone by any identical points over the repeat unit of the layers fixes the c axis of the structure. If this axis is inclined, it may be more convenient to triple the height of the unit cell and go over to an orthogonal unit cell, to make it easier to compare different ordered structures made up of the same layers or layer packets.

Analytically, the existence of a strict repeat unit is shown by the use of a repeating group of σ, τ symbols in the layer sequence symbol. Scattering is expressed by the formula $S(H) = KD(H)F(H)$, which in accordance with D(H), has nonzero values for the whole-number values $x^* = h, y^* = k, z^* = l$, and which describes a reciprocal lattice made up of discrete straight lines parallel to the c^* axis. However, whatever the number of layers of packets present in the strict repeat unit, the reflections with both hk indices multiples of three fix an orthogonal repeat unit of one layer. In the same way, if an ordered sequence is characterized by network or layer displacement along only one of the axes a or b, the repeat unit will be one-layer inclined and three-layer orthogonal, or one-layer orthogonal, as shown by the reflections with one of the indices k or h a multiple of three. These units, being equal to nonrigorous repeats units in a disordered sequence, are the minimum repeat units in the present case, provided they do not coincide with the strict repeat unit.

The expression for F can be obtained in a rather more specific form for the particular case where the strict repeat unit embraces n packets or layers of the same type.

If we divide the unit cell into n subcells, each relating to one of the packets or layers in the repeat unit, we can represent the structure amplitude as the sum of n terms, in the form

$$F\left(\mathbf{H}\right) = \sum_{p=1}^{n} \sum_{j} f_{j} e^{2\pi i \left(hx_{pj} + ky_{pj} + l\frac{z'_{j}+p-1}{n}\right)}, \qquad (56)$$

where z' represents coordinates calculated for the cell with a c' constant of c/n. After taking the common factor out of the first summation sign,

$$F\left(\mathbf{H}\right) = \sum_{p=1}^{n} e^{\frac{2\pi i l(p-1)}{n}} \Phi_{p}, \qquad (57)$$

where

$$\Phi_{p} = \sum_{j} f_{j} e^{2\pi i \left(hx_{pj} + ky_{pj} + lz'_{j}/n\right)} \qquad (58)$$

is the expression for the structure amplitude of the hypothetical sequence which would be obtained from the original sequence if all the layers were removed except the p-th layer of each repeat unit, and the absolute value of c remained unaltered.

If $l = nl'$, then

$$F\left(hkl\right) = \sum_{p=1}^{k} \Phi_{p}\left(hkl\right),$$

where

$$\Phi_{p}\left(hkl\right) = \sum_{j} f_{j} e^{2\pi i \left(hx_{pj} + ky_{pj} + l'z_{j}\right)} = \frac{F_{p}\left(hkl'\right)}{n}, \qquad (59)$$

where $F_p(hkl')$ is the structure amplitude for the reflection hkl', corresponding to a sequence of identical layers of the p-th sort, arranged without gaps with a repeat unit of one layer. The $1/n$ factor is introduced because Φ_p refers to an n-times more sparse distribution of layers. Generally speaking, it would have been more correct to include it in $D(\boldsymbol{H})$, but formula (59) is equally correct if it is assumed that F_p and Φ_p are scaled to the same number of unit cells in the crystal.

Thus, when $l = nl'$,

$$F(hkl) = \frac{\sum_{p=1}^{n} F_p(hkl')}{n}.$$

(60)

If $l \neq nl'$, then $S(\boldsymbol{H})$ essentially depends on the hkl indices. For reflections with h = 3h', k = 3k', or only with k = 3k', depending on the particular example, then $\Phi_p(\boldsymbol{H}) = \Phi(\boldsymbol{H}) = \text{const}$ for the same reasons as those noted above, and

$$F(\mathbf{H}) = \Phi(\mathbf{H}) \sum_{p=1}^{n} e^{\frac{2\pi i l(p-1)}{n}} = 0.$$

(61)

This means that for scattering in directions determined by these hk indices, only reflections with $l = nl'$ are possible, and these correspond to a repeat unit one n-th the size of the true one, all intervening reflections being missing, in spite of the fact that they are allowed by the interference function $D(\boldsymbol{H})$, corresponding to a repeat unit of n layers. This is only natural, since the packets are identical in respect to scattering in directions with these hk, and their sequence has a repeat unit of one packet. This unit, as already pointed out, is the minimum repeat unit.

The true repeat unit of n layers is given by the reflections with h $\neq$ 3h', k $\neq$ 3k', or k $\neq$ 3k' alone, depending on whether or not different subcells contain antiparallel corresponding networks and displacements τ with different components along the a axis. At these hk, between the reflections with $l = nl'$, there are a further n − 1 intermediate reflections with $l = nl' + 1, \ldots, nl' + n - 1$.

4. The Distribution of Intensities in Layer Silicate Structure Modifications

When comparing the diffraction properties of the different possible layer silicate structure modifications, it is expedient to use a single system of coordinates for all modifications. The axes a and b of the whole layer sequence can serve as the axes of any particular layer, for example layer σ_3, since a_3 lies in its plane of symmetry. The c axis must pass along a normal to the layers, and the corresponding c unit must extend over a certain number of layers, where this number is an integral multiple of the number of layers in an orthogonal unit cell and an integral multiple of three times the number of layers in the inclined unit cells of the different structure modifications. In accordance with Tables 6 and 10, these conditions are satisfied by an orthogonal c repeat unit of six layers.

If a structure is formed of layers of a single type, and is described in terms of the axes a_3, b_3 and a cell containing n layers, the coordinates of atoms in the m-th layer will be

$$x_m = x_{0m} + x'_m, \quad y_m = y_{0m} + y'_m, \quad z_m = \frac{m}{n} + z', \qquad (62)$$

where x_{0m}, y_{0m} are the coordinates of the origin of the m-th layer. Here, then,

$$F_{hkl} = \sum_{m=0}^{n-1} e^{2\pi i \left(h x_{0m} + k y_{0m} + l \frac{m}{n} \right)} \sum_j f_j e^{2\pi i \left(h x'_{mj} + k y'_{mj} + l z'_j \right)} =$$

$$\sum_{m=0}^{n-1} \Phi_m (hkl) \, e^{2\pi i \left(h x_{0m} + k y_{0m} + l \frac{m}{n} \right)}; \qquad (63)$$

$$\Phi_m (hkl) = \sum_j f_j e^{2\pi i \left(h x'_{mj} + k y'_{mj} + l z'_j \right)}. \qquad (64)$$

In expression (64), the z'_j coordinates do not depend on the kind of layer concerned; the coordinates x'_m, y'_m can fall into six groups of values, according to which one of the six possible kinds of layer the m-th layer. If the coordinates x'_m, y'_m are expressed in a system of coordinates with the x axis lying in a plane of symmetry of the m-th layer, which can be brought about by rotation through the angle indicated by the layer transformation scheme (Fig. 69), then these coordinates will be the same for all layers and will be equal to the coordinates of atoms in a layer of the σ_3 (or σ_6) kind, i.e., to x_3, y_3 (or

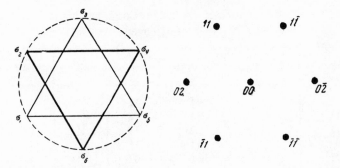

Fig. 69. Transformation scheme for rotationally identical layers and the corresponding transformation of reciprocal lattice point indices and the reflections $0\bar{2}l$, $1\bar{1}l$, $11l$, $02l$, $\bar{1}1l$, $\bar{1}\bar{1}l$.

x_6, y_6). Under these conditions, expression (64) takes the form

$$\Phi_m\,(hkl) = \Phi_3\,(h_m,\,k_m,\,l) = \sum_j f_j e^{2\pi i\,(h_m x_{3j} + k_m y_{3j} + lz'_j)}. \qquad (65)$$

Here, as above, in considering the phase relationships between the reciprocal lattice points, the equality $\boldsymbol{H}\boldsymbol{r}_m = \boldsymbol{H}_m\boldsymbol{r}_3$. applies. If the vector $\boldsymbol{r}_m$ can be derived from the vector $\boldsymbol{r}_3$ by rotating it, together with the σ_3 layer, through an angle φ (where φ is a multiple of $2\pi/6$), so that σ_3 transforms into σ_m, then the vector $\boldsymbol{H}_m$ is derived from the vector $\boldsymbol{H}$ by rotating the latter through the angle $-\varphi$. Thus, the transformation of the hk indices to h_m, k_m corresponds to the transformation of reciprocal lattice point indices for a structure of σ_3 layers when it is rotated through the angle necessary to convert the σ_3 layers to the kind of layer present at the m-th level. The numerical values of h_m, k_m can be obtained directly by seeing which reciprocal lattice point h'k' stands in place of the given hk point on rotation through the appropriate angle (see Table 30). In Fig. 69, the h_m, k_m points corresponding to the different σ_m form the same cyclic sequence as the coordinate vectors $\boldsymbol{a}_m$.

Table 30 shows the h_m, k_m values which figure in equation (65).

The values of x_{0m}, y_{0m} in (63) are made up of the σ and τ displacements which preceded the m-th layer, and the structure amplitude of any regular structure can be wholly expressed in terms of the layer amplitude of a single particular layer, in this case the layer σ_3.

DETERMINATION OF INTENSITIES

Table 30. Transformation of h,k Indices Corresponding to Transformation of σ_i Layers

hk	σ_3	σ_4	σ_5	σ_6	σ_1	σ_2
$0\bar{2}$	$0\bar{2}$	$\bar{1}\bar{1}$	$\bar{1}1$	02	11	$1\bar{1}$
$\bar{1}\bar{1}$	$\bar{1}\bar{1}$	$\bar{1}1$	02	11	$1\bar{1}$	$0\bar{2}$
$\bar{1}1$	$\bar{1}1$	02	11	$1\bar{1}$	$0\bar{2}$	$\bar{1}\bar{1}$
02	02	11	$1\bar{1}$	$0\bar{2}$	$\bar{1}\bar{1}$	$\bar{1}1$
11	11	$1\bar{1}$	$0\bar{2}$	$\bar{1}\bar{1}$	$\bar{1}1$	02
$1\bar{1}$	$1\bar{1}$	$0\bar{2}$	$\bar{1}\bar{1}$	$\bar{1}1$	02	11

The general expressions obtained for F may be transformed and simplified in different specific cases using the same relationships as those applied to $\Phi_3(hkl)$.

In a σ_3 layer there is a symmetry plane perpendicular to the b axis, and

$$\Phi_3(h\bar{k}l) = \Phi_3(hkl). \qquad (66)$$

The trioctahedral two-storied σ_3 layers have the symmetry p31m, and their threefold rotation axes are displaced by $a/3$ away from the chosen origins. In three-storied layers or chlorite packets, the upper and lower halves have separate threefold axes which pass the origin (in the σ_3 setting) at distances of $+a/3$ and $-a/3$, respectively (Figs. 11 and 12).

The structure amplitudes of two-storied layers, or the parts of the structure amplitudes relating to half a three-storied layer or chlorite packet, are calculated with respect to the symmetry axes indicated, which are invariant for rotations of $\pm120°$.

If the amplitude of a trioctahedral σ_3 layer is denoted by Φ', and the invariant amplitudes are denoted by Φ_{inv} and $\Phi_{\pm\frac{1}{2}}$ for a two-storied layer and half a three-storied layer or chlorite packet, respectively (with invariant amplitudes calculated for origins on the three-fold axes), then for a two-storied layer

$$\Phi'_3(hkl) = \Phi_{\text{inv}}e^{2\pi i\frac{h}{3}}; \quad \Phi_{\text{inv}} = \Phi'_3(0kl), \qquad (67)$$

and for a three-storied layer or chlorite packet,

$$A_3'(hkl) = \Phi_{1/2}e^{2\pi i \frac{h}{3}} + \Phi_{1/2}^* e^{-2\pi i \frac{h}{3}}. \tag{68}$$

In the latter expression, A_{hkl}' is real, since the corresponding origin lies at a center of symmetry.

Therefore,

$$\Phi_{-1/2} = \Phi_{1/2}^*. \tag{69}$$

Assuming that $\Phi_{\frac{1}{2}} = A_{\frac{1}{2}} + iB_{\frac{1}{2}}$, we can transform (68) into the form

$$A_3'(hkl) = 2A_{1/2} \cos 2\pi h/3 + 2B_{1/2} \sin 2\pi h/3. \tag{70}$$

From the expression for the amplitude of the σ_3 layer [see formula (73) below, and also (68)], it follows that when $l = 0$, $B_{\frac{1}{2}} = 0$. When $h = 0$,

$$A_{hkl}' = 2A_{1/2}. \tag{71}$$

If the layers are dioctahedral, then their corresponding layer amplitudes are

$$\Phi_3 = \Phi_3' - f_M, \tag{72}$$

where f_M is the atomic amplitude of the cation occupying an octahedron.

With these formulas it is possible to calculate either general analytical expressions, or the numerical values of $|\Phi_{hkl}|^2$, which characterize the distribution of reflection intensities on single-crystal diffraction patterns, or of $\Sigma|\Phi_{hkl}|^2$, for the intensities of reflections which coincide on diffraction patterns from true or textured polycrystalline specimens.

Previously, numerical values of these quantities could only be found for a model of the layer, which would only approximately represent its true structure and composition. Therefore, the theoretical relationships between $|\Phi_{hkl}|^2$ and $\Sigma|\Phi|^2$ will show a better agreement with the experimental intensity relationships, the closer the reflections used lie to the central spot, since any deviations of the models from the real structures will have less effect on these.

It will be quite obvious that in their diffraction patterns, the different structure modifications can only be distinguished rigorously and reliably using their reflections which, in the general case, have both indices $h,k \neq 3n$, or if they have σ,τ values with identical components along one of the axes a or b,

from reflections with k ≠ 3k' or h ≠ 3h', respectively. Modifications consisting of layers or packets all of the same type will have their reflections with
h,k = 3n identical in arrangement and intensities, but can be divided up into
groups, each having identical reflections with either k = 3k' or h = 3h', depending on the particular example. Thus, for example, the mica modifications
1M, $2M_1$, and 3T have identical reflections with k = 3k', while modifications
$2M_2$, 2O, and 6H have identical reflections with h = 3h'. All the mica modifications have identical reflections for which both h and k = 3n. *

The conditions given above are best satisfied by the reflections $02\overset{\pm}{l}, 11\overset{\pm\pm}{l}$.
In electron-diffraction texture patterns these lie on the first ellipse, while in
X-ray powder photographs the majority of them occupy a self-contained area
in the low interplanar distance region, with d from ~4.5 to ~2.6 A, which may
also include one or two easily distinguishable basal reflections.

In this range of d, the fall-off in the atomic scattering curves is the
same, in broad outline, for both electrons and X rays, so the general diagnos-
tic features of the $|\Phi|^2$ and $\Sigma|\Phi|^2$ distributions ("larger−smaller" and
"stronger−weaker") are equally true for the use, in equations (63)-(65), of
either the X-ray or the electron scattering amplitudes for f.

As has already been noted (Chapter 1), the sequence of layers characteristic of a strictly ordered modification may be deviated from, and in practice
these deviations involve displacements of networks or layers in the b-axis direction by multiples of b/3. This type of loss of ordering cannot affect reflections with k = 3k', but it can lead to a smearing out, to a greater or lesser extent, of reflections with k ≠ 3k', which in some cases may appear as a continuous diffuse band. Because these structures contain some element of order,
as shown by the reflections with k = 3k', Brown and Bailey (1962) have suggested that they should be called semirandom structures. It will be appreciated
that the repeat unit given by the reflections with k = 3k' must be considered
as a nonrigorous unit or a pseudoconstant. It is equal to the strict repeat unit
of the corresponding ordered structures in which this unit is fixed by reflections with h = 3h', for example, the $2M_2$ and 2O micas, and also equal to the
partial repeat unit of corresponding ordered structures in which the strict re-

---

*Because of distortion of the regular shapes of their polyhedra, certain min-
  erals (e.g., muscovite, dickite) have intermediate reflections on the diffraction patterns, although these are admittedly weaker than the main reflections
  with h,k = 3n, which correspond to the strict c repeat unit.

peat unit is fixed by reflections with k = 3k', like the 1M, $2M_1$, and 3T micas. Semirandom structures can also be classed into various modifications, but there are much fewer of these than in the case of ordered structures. These semirandom modifications are characterized by the properties of their reflections with k = 3k'. In these cases, also, it is necessary to use the reflections closest to the center of the diffraction pattern in diagnostic work, these here being the $13l$,$20l$ reflections on the second ellipse.

The Scattering Properties of Two-Storied Layer Structures. For a two-storied dioctahedral layer, as shown in Fig. 11, with the origin at the center of a vacant octahedron,

$$\Phi_3 = A_3 + iB_3,$$

$$A_3 = 2f_{Al} \cos 2\pi k/3 + 2f_{Si} \cos 2\pi \, (h/3 + lz_2/6) \cos 2\pi k/3 +$$

$$+ \, 2f_0 \, (1 + 2\cos 2\pi k/3) \cos 2\pi \, (h/3 + lz_1/6) +$$

$$+ \, f_0 \Big[\cos 2\pi \Big(\frac{-h + lz_3}{6} \Big) + 2 \cos 2\pi \, (h/12 + lz_3/6) \cos 2\pi k/4 \Big], \qquad (73)$$

$$B_3 = 2f_{Si} \sin 2\pi \, (h/3 + lz_2/6) \cos 2\pi k/3 +$$

$$+ \, f_0 \Big[\sin 2\pi \Big(\frac{-h + lz_3}{6} \Big) + 2 \sin 2\pi \, (h/12 + lz_3/6) \cos 2\pi k/4 \Big].$$

For the $02\overset{\pm}{l}$, $11\overset{\pm\pm}{l}$ reflections, if we take it that $\Phi_3 = \Phi_3' - f_{Al}$ (in what follows the suffixes 3 and Al will be omitted for brevity),

$$\Phi_{11l}' = \Phi_{02l}' \, e^{\frac{2\pi i}{3}}, \qquad \Phi_{\bar{1}1l}' = \Phi_{02l}' \, e^{-\frac{2\pi i}{3}}. \qquad (74)$$

These relationships follow directly from (73), and also from formula (67), in accordance with Fig. 11, from which it can be seen that if the layer were trioctahedral, it would have a threefold axis at a distance $\Delta x = {}^1/_3$ away from the chosen origin.

From (74) we obtain the relationships

$$\Phi_{11l}' + \Phi_{\bar{1}1l}' = - \, \Phi_{02l}',$$
$$\Phi_{11l}' - \Phi_{\bar{1}1l}' = i \, \sqrt{3} \, \Phi_{02l}', \qquad (75)$$

which are useful in the simplification of expressions for the structure amplitudes of different modifications.

The method of calculation can be illustrated by a particular example of the calculation of F, for example, for structure II,6 (Table 6).

From the analytical symbol for this structure, $\sigma_3 \tau_6 \sigma_4 \tau_1 \sigma_5 \tau_2 \sigma_6 \tau_3 \sigma_1 \tau_4 \sigma_2 \tau_5 \sigma_3$, it follows that the origins of successive layers have the coordinates $0,0; -\frac{1}{3},0;$ $0, \frac{1}{3}; -\frac{1}{3}, -\frac{1}{3}; 0, -\frac{1}{3}; -\frac{1}{3}, \frac{1}{3}; 0,0;$ etc. Because the structure is base-centered, the projections of these six-layer origins on the xy0 plane are equivalent to the set of six points $0,0; -\frac{1}{3}, 0; -\frac{1}{2}, -\frac{1}{6}; -\frac{1}{3}, -\frac{1}{3}; 0, -\frac{1}{3}; \frac{1}{6}, -\frac{1}{6}$, lying at the corners of a hexagon with center at $-\frac{1}{6}, -\frac{1}{6}$. Through this point, of course, a 6_1 axis passes, and the origin of the whole structure should be taken at this point, or at its equivalent axis, which has the coordinates $\frac{1}{3}, \frac{1}{3}$.

Therefore,

$$
F_{hkl} = [\Phi_3 + \Phi_4 e^{2\pi i \left(-\frac{h}{3} + \frac{l}{6}\right)} + \Phi_5 e^{2\pi i \frac{l+k}{3}} + \Phi_6 e^{2\pi i \left(\frac{-h-k}{3} + \frac{l}{2}\right)} +
$$
$$
+ \Phi_1 e^{-2\pi i \frac{l+k}{3}} + \Phi_2 e^{2\pi i \left(\frac{-h+k}{3} - \frac{l}{6}\right)}\Big] e^{-2\pi i \frac{h+k}{3}} = \Phi_3 e^{-2\pi i \frac{h+k}{3}} + \Phi_5 e^{2\pi i \frac{l-h}{3}} +
$$
$$
+ \Phi_1 e^{-2\pi i \frac{l+h-k}{3}} + (-1)^l \left(\Phi_6 e^{2\pi i \frac{h+k}{3}} + \Phi_2 e^{2\pi i \frac{l+h}{3}} + \Phi_4 e^{2\pi i \frac{h-k-l}{3}}\right).
$$

When $l = 6l'$, then from equations (63)-(65), (74), (75), and Table 30,

$$
F_{02l} = \Phi_{02l} e^{-\frac{2\pi i}{3}} + \Phi_{\bar{1}1l} + \Phi_{11l} e^{\frac{2\pi i}{3}} + \Phi_{02l} e^{\frac{2\pi i}{3}} + \Phi_{11l} +
$$
$$
+ \Phi_{\bar{1}1l} e^{-\frac{2\pi i}{3}} = \Phi'_{\bar{1}1l} - fe^{-\frac{2\pi i}{3}} + \Phi'_{\bar{1}1l} - f + \Phi'_{\bar{1}1l} - fe^{\frac{2\pi i}{3}} +
$$
$$
+ \Phi'_{11l} - fe^{\frac{2\pi i}{3}} + \Phi'_{11l} - f + \Phi'_{11l} - fe^{-\frac{2\pi i}{3}} =
$$
$$
= 3\,(\Phi'_{11l} + \Phi'_{\bar{1}1l}) - 2f - 2f\left(e^{\frac{2\pi i}{3}} + e^{-\frac{2\pi i}{3}}\right) =
$$
$$
= -3\Phi'_{02l} - 2f + 4f \cos\frac{2\pi}{3} = -3\Phi'_{02l};
$$

$$
F_{\bar{1}1l} = \Phi_{11l} + \Phi_{\bar{1}1l} e^{-\frac{2\pi i}{3}} + \Phi_{02l} e^{\frac{2\pi i}{3}} + \Phi_{\bar{1}1l} + \Phi_{11l} e^{\frac{2\pi i}{3}} + \Phi_{02l} e^{-\frac{2\pi i}{3}} =
$$
$$
= 3\,(\Phi'_{11l} + \Phi'_{\bar{1}1l}) = -3\Phi'_{02l}.
$$

Similarly,

$$
F_{11l} = F_{02l} = F_{\bar{1}1l} = \Phi_{\bar{1}\bar{1}l} = -3\Phi'_{02l}.
$$

Thus, all six reciprocal lattice points, related by a sixfold axis and lying in a single plane with $l = 6l'$, are characterized by identical values of F.

If $l = 6l' + 1$,

$$F_{0\bar{2}l} = \Phi_{02l}e^{-\frac{2\pi i}{3}} + \Phi_{\bar{1}1l}e^{\frac{2\pi i}{3}} + \Phi_{11l} - \Phi_{02l}e^{\frac{2\pi i}{3}} - \Phi_{11l}e^{\frac{2\pi i}{3}} - \Phi_{\bar{1}1l}e^{\frac{2\pi i}{3}} =$$

$$= 3fe^{\frac{2\pi i}{3}}.$$

The next reflection, $1\bar{1}l$ in the cyclic sequence of reflections related by the sixfold axis, which lies counterclockwise of $0\bar{2}l$ separated by an angle of $2\pi/6$, has the structure amplitude

$$F_{1\bar{1}l} = -F_{0\bar{2}l}\,e^{-\frac{2\pi i}{3}} = F_{0\bar{2}l}\,e^{\frac{2\pi i}{6}}.$$

Similarly, the rest of the reflections in the cyclic sequence have the amplitudes

$$F_{11l} = F_{0\bar{2}l}e^{\frac{2\pi i}{3}}, \quad F_{02l} = -F_{0\bar{2}l} = F_{0\bar{2}l}e^{\frac{2\pi i}{2}},$$

$$F_{\bar{1}1l} = F_{0\bar{2}l}\,e^{-\frac{2\pi i}{3}} = F_{0\bar{2}l}\,e^{2\pi i\frac{2}{3}}; \quad F_{\bar{1}\bar{1}l} = F_{0\bar{2}l}\,e^{-\frac{2\pi i}{6}} = F_{0\bar{2}l}\,e^{2\pi i\frac{5}{6}}.$$

When $l = 6l' + 2$,

$$F_{0\bar{2}l} = 3fe^{\frac{2\pi i}{6}},$$

and the structure amplitudes of the other five reflections, $1\bar{1}l$, $11l$, $02l$, $\bar{1}1l$, $\bar{1}\bar{1}l$, differ successively by a phase factor of $\exp(2\pi i/3)$.

When $l = 6l' + 3$,

$$F_{0\bar{2}l} = 3\,(\Phi'_{\bar{1}1l} - \Phi'_{11l}) = 3\sqrt{3}i\Phi'_{02l}.$$

The other reflections have amplitudes differing successively by the phase $(2\pi i/2) = -1$, so that the amplitudes of reflections related by a rotation through $2\pi/6$ have opposite signs.

When $l = 6l' + 4$,

$$F_{0\bar{2}l} = 3fe^{-\frac{2\pi i}{6}};$$

and the successive reflections differ by the amplitude phase factor $\exp(-2\pi i/3) = \exp[2\pi i(2/3)]$.

When $l = 6l' + 5$,

$$F_{0\bar{2}l} = 3fe^{\frac{2\pi i}{6}}.$$

The remaining reflections differ successively by the amplitude phase factor

$$e^{-\frac{2\pi i}{6}} = e^{2\pi i \frac{5}{6}}.$$

In both this example and the others, these phase relationships can be justified in terms of the reciprocal lattice representation described above of the three-dimensional symmetry differences in the crystal lattices concerned. These phase relationships correspond to a 6_1 screw axis, which appears in the reciprocal lattice as a unique colored sixfold symmetry axis, and the relationships hold for any hkl reflections related by such an axis.

Table 31 gives general expressions for $|F_{hkl}|^2$ for the $0\overset{\pm}{2}l$, $\overset{\pm\pm}{11}l$ reflections. Their distribution defines the symmetry of the reciprocal lattice-point density distribution, which must also appear on single-crystal diffraction patterns.

Rounded-off numerical values of $\Sigma |F|^2$, normalized for scattering by a three-layer cell, for those $0\overset{\pm}{2}l$, $\overset{\pm\pm}{11}l$ reflections which coincide in electron-diffraction texture patterns and X-ray powder photographs, are given in Table 32. The distribution can be used for the diagnosis of modifications from their diffraction patterns. Values given in the same row of the table correspond to reflections which should have identical heights on texture patterns. The number of the row is equal to the l index for those modifications in which the orthogonal cell contains six layers.

For modifications in which the orthogonal repeat unit includes three or two layers or one layer, the reflections appear 2, 3, or 6 times less frequently. They will have l indices the same number of times smaller than the row number in the table.

Table 31. Distribution of Values of $|F_{hkl}|^2$ for the $0\overset{\pm}{2}l$, $\overset{\pm\pm}{11}l$ Reflections in Two-Storied Layer Minerals

Modification	Structure No.	l \ hk	$3l'$	$3l'+1$	$3l'+2$
				$|F_{hkl}|^2$	
1Tk₁	I, 1	02	$9\,|\Phi_{\bar{1}1l}|^2$	0	0
		$0\bar{2}$	$9\,|\Phi_{11l}|^2$	0	0
		11	0	$9\,|\Phi_{02l}|^2$	0
		$1\bar{1}$	0	$9\,|\Phi_{11l}|^2$	0
1Tk₁ ($k=k'$)	I, 2	$\bar{1}1$	0	0	$9\,|\Phi_{\bar{1}1l}|^2$
		$\bar{1}\bar{1}$	0	0	$9\,|\Phi_{02l}|^2$
		Σ	Reflections not fully coincident		
1Tk₂ (3T)	IV, 1 (II', 1)	$0\bar{2}$	0	$9\,|\Phi_{02l}|^2$	0
		11	0	$9\,|\Phi_{\bar{1}1l}|^2$	0
		$\bar{1}1$	0	$9\,|\Phi_{11l}|^2$	0
		02	0	0	$9\,|\Phi_{02l}|^2$
	IV, 2 (II', 2)	$1\bar{1}$	0	0	$9\,|\Phi_{\bar{1}1l}|^2$
		$\bar{1}\bar{1}$	0	0	$9\,|\Phi_{11l}|^2$
		Σ	0	$27\,(|\Phi'|^2 + f^2)$	$27\,(|\Phi'|^2 + f^2)$
1M	VI, 1 (I', 1)	$0\overset{+}{\bar{2}}$	$9\,|\Phi_{02l}|^2$	0	0
		$1\overset{+}{\bar{1}}$	0	$9\,|\Phi_{\bar{1}1l}|^2$	0
		$\bar{1}\overset{+}{\bar{1}}$	0	0	$9\,|\Phi_{11l}|^2$
		Σ	$18\,|\Phi_{02l}|^2$	$18\,|\Phi_{\bar{1}1l}|^2$	$18\,|\Phi_{11l}|^2$
	VI, 3, VI, 4	hk	$3\,|\Phi_{02l}|^2$	$3\,|\Phi_{\bar{1}1l}|^2$	$3\,|\Phi_{11l}|^2$
		Σ	$18\,|\Phi_{02l}|^2$	$18\,|\Phi_{\bar{1}1l}|^2$	$18\,|\Phi_{11l}|^2$
3T	I, 6	$0\bar{2},\ 1\overset{+}{\bar{1}}$	$3\,|\Phi_{\bar{1}1l}|^2$	$3\,|\Phi_{11l}|^2$	$3\,|\Phi_{02l}|^2$
	I, 8	$02,\ \bar{1}\overset{+}{\bar{1}}$	$3\,|\Phi_{11l}|^2$	$3\,|\Phi_{02l}|^2$	$3\,|\Phi_{\bar{1}1l}|^2$
	I, 7	$0\bar{2},\ 1\overset{+}{\bar{1}}$	$3\,|\Phi_{11l}|^2$	$3\,|\Phi_{02l}|^2$	$3\,|\Phi_{\bar{1}1l}|^2$
	I, 9	$02,\ \bar{1}\overset{+}{1}$	$3\,|\Phi_{\bar{1}1l}|^2$	$3\,|\Phi_{11l}|^2$	$3\,|\Phi_{02l}|^2$
		Σ	$9\,\{|\Phi_{11l}|^2 + |\Phi_{\bar{1}1l}|^2\}$	$9\,\{|\Phi_{02l}|^2 + |\Phi_{11l}|^2\}$	$9\,\{|\Phi_{02l}|^2 + |\Phi_{\bar{1}1l}|^2\}$

Table 31 (continued)

Modifica-tion	Structure No.	l \\ hk	$3l'$	$3l'+1$	$3l'+2$
3 T	IV, 5	$0\bar{2},\ 1\bar{1}\ (+)$	0	$9f^2$	$9\,|\Phi'_{11l}|^2$
	IV, 6	$02,\ \bar{1}\bar{1}\ (+)$	0	$9\,|\Phi'^{-}_{\bar{1}1l}|^2$	$9f^2$
		Σ	0	$27\,(|\Phi'^{-}_{\bar{1}1l}|^2+f^2)$	$27\,(|\Phi'_{11l}|^2+f^2)$
	VIII, 2	$0\bar{2},\ 1\bar{1}\ (+)$	$9f^2$	$9\,|\Phi'_{02l}|^2$	0
	VIII, 3	$02,\ \bar{1}\bar{1}\ (+)$	$9f^2$	0	$9\,|\Phi'_{02l}|^2$
		Σ	$54f^2$	$27\,|\Phi'_{02l}|^2$	$27\,|\Phi'_{02l}|^2$
	IX, 2	$0\bar{2},\ 1\bar{1}\ (+)$	$9\,|\Phi'_{11l}|^2$	0	$9f^2$
	IX, 3	$02,\ \bar{1}\bar{1}\ (+)$	$9\,|\Phi'^{-}_{\bar{1}1l}|^2$	$9f^2$	0
		Σ	$54\,|\Phi'|^2$	$27f^2$	$27f^2$
	I', 3 \\ I', 4	hk	$3\,|\Phi'|^2$	$3\,|\Phi'|^2$	$3\,|\Phi'|^2$
		Σ	$18\,|\Phi'|^2$	$18\,|\Phi'|^2$	$18\,|\Phi'|^2$

Modifica-tion	Structure No.	l \\ hk	$6l'$	$6l'+1$	$6l'+2$	$6l'+3$	$6l'+4$	$6l'+5$
2M₁	I, 3 (I', 2)	$0\bar{2}\ (+)$	$9\,|\Phi_{02l}|^2$	0	0	$27\,|\Phi_{02l}|^2$	0	0
		$1\bar{1}\ (+)$	0	$27\,|\Phi_{11l}|^2$	0	0	$9\,|\Phi^{-}_{\bar{1}1l}|^2$	0
		$\bar{1}\bar{1}\ (+)$	0	0	$9\,|\Phi^{-}_{\bar{1}1l}|^2$	0	0	$27\,|\Phi^{-}_{\bar{1}1l}|^2$
		Σ	$18\,|\Phi_{02l}|^2$	$54\,|\Phi_{11l}|^2$	$18\,|\Phi^{-}_{\bar{1}1l}|^2$	$54\,|\Phi_{02l}|^2$	$18\,|\Phi_{11l}|^2$	$54\,|\Phi^{-}_{\bar{1}1l}|^2$
	I, 4	$0\bar{2}\ (+)$	$9\,|2\Phi'_{02l}+f|^2$	0	0	$27f^2$	0	0
		$1\bar{1}\ (+)$	0	$27f^2$	0	0	$9\,|2\Phi'_{11l}+f|^2$	0
		$\bar{1}1\ (+)$	0	0	$9\,|2\Phi'^{-}_{\bar{1}1l}+f|^2$	0	0	$27f^2$
		Σ	$18\,|2\Phi'_{02l}+f|^2$	$54f^2$	$18\,|2\Phi'^{-}_{\bar{1}1l}+f|^2$	$54f^2$	$18\,|2\Phi'_{11l}+f|^2$	$54f^2$

Table 31 (continued)

Modification	Structure No.	l \ hk	$6l'$	$6l'+1$	$6l'+2$	$6l'+3$	$6l'+4$	$6l'+5$
2M$_1$	I, 5	$0\bar{2}$ (+)	$9\,\lvert\Phi'_{02l}+f\rvert^2$	0	0	$27\,\lvert\Phi'_{02l}\rvert^2$	0	0
		$1\bar{1}$ (+)	0	$27\,\lvert\Phi'_{\bar{1}1l}\rvert^2$	0	0	$9\,\lvert\Phi'_{11l}+2f\rvert^2$	0
		$\bar{1}\bar{1}$ (+)	0	0	$9\,\lvert\Phi'_{\bar{1}1l}\rvert^2+2f\rvert^2$	0	0	$27\,\lvert\Phi'_{11l}\rvert^2$
		Σ	$18\,\lvert\Phi'_{02l}+2f\rvert^2$	$54\,\lvert\Phi'_{\bar{1}1l}\rvert^2$	$18\,\lvert\Phi'_{\bar{1}1l}+2f\rvert^2$	$54\,\lvert\Phi'_{02l}\rvert^2$	$18\,\lvert\Phi'_{11l}+2f\rvert^2$	$54\,\lvert\Phi_{11l}\rvert^2$
	VI, 2	$0\bar{2}$ (+)	$9\,\lvert\Phi_{02l}\rvert^2$	0	0	$27\,\lvert\Phi'_{02l}+f\rvert^2$	0	0
		$1\bar{1}$ (+)	0	$27\,\lvert\Phi'_{11l}+f\rvert^2$	0	0	$9\,\lvert\Phi_{11l}\rvert^2$	0
		$\bar{1}\bar{1}$ (+)	0	0	$9\,\lvert\Phi_{\bar{1}1l}\rvert^2$	0	0	$27\,\lvert\Phi'_{\bar{1}1l}+f\rvert^2$
		Σ	$18\,\lvert\Phi_{02l}\rvert^2$	$54\,\lvert\Phi'_{11l}+f\rvert^2$	$18\,\lvert\Phi_{\bar{1}1l}\rvert^2$	$54\,\lvert\Phi'_{02l}+f\rvert^2$	$18\,\lvert\Phi_{11l}\rvert^2$	$54\,\lvert\Phi'_{\bar{1}1l}+f\rvert^2$
2M$_2$	II, 1	$0\bar{2}$	0	0	0	0	$36\,\lvert\Phi_{\bar{1}1l}\rvert^2$	0
		11	0	$27f^2$	0	0	$9\lvert2\Phi_{\bar{1}1l}+f\rvert^2$	0
		$\bar{1}1$	0	$27f^2$	0	0	$9\lvert2\Phi'_{\bar{1}1l}+f\rvert^2$	0
		02	0	0	$36\,\lvert\Phi_{11l}\rvert^2$	0	0	0
		$1\bar{1}$	0	0	$9\,\lvert2\Phi'_{11l}+f\rvert^2$	0	0	$27f^2$
		$\bar{1}\bar{1}$	0	0	$9\,\lvert2\Phi'_{11l}+f\rvert^2$	0	0	$27f^2$
	II, 2	$0\bar{2}$	0	0	0	0	$36\,\lvert\Phi_{11l}\rvert^2$	0
		11	0	$27f^2$	0	0	$9\,\lvert2\Phi'_{11l}+f\rvert^2$	0
		$\bar{1}1$	0	$27f^2$	0	0	$9\,\lvert2\Phi_{11l}+f\rvert^2$	0
		02	0	0	$36\,\lvert\Phi_{\bar{1}1l}\rvert^2$	0	0	0
		$1\bar{1}$	0	0	$9\,\lvert2\Phi'_{\bar{1}1l}+f\rvert^2$	0	0	$27f^2$
		$\bar{1}\bar{1}$	0	0	$9\,\lvert2\Phi_{\bar{1}1l}+f\rvert^2$	0	0	$27f^2$

Table 31 (continued)

Modification	Structure No.	hk \\ l	$6l'$	$6l'+1$	$6l'+2$	$6l'+3$	$6l'+4$	$6l'+$
2M₂ (6T)	II, 3· (V', 1)	$0\bar{2}$	0	0	0	0	$36\,|\Phi_{02l}|^2$	0
		11	0	$27f^2$	0	0	$9\,|2\Phi'_{02l}+f|^2$	0
		$\bar{1}1$	0	$27f^2$	0	0	$9\,|2\Phi'_{02l}+f|^2$	0
		02	0	0	$36\,|\Phi_{02l}|^2$	0	0	0
		$1\bar{1}$	0	0	$9\,|2\Phi'_{02l}+f|^2$	0	0	27
		$\bar{1}\bar{1}$	0	0	$9\,|2\Phi'_{02l}+f|^2$	0	0	27
		Σ	0	$54f^2$	$54\,(2|\Phi'|^2+f^2)$	0	$54\,(2|\Phi'|^2+f^2)$	54
2M₂	V, 4 (III', 4)	$0\bar{2}$	0	0	0	0	$36\,|\Phi_{02l}|^2$	0
		11	0	$27\,|\Phi'_{02l}+f|^2$	0	0	$9\,|\Phi_{02l}|^2$	0
		$\bar{1}1$	0	$27\,|\Phi'_{02l}+f|^2$	0	0	$9\,|\Phi_{02l}|^2$	0
		02	0	0	$36\,|\Phi_{02l}|^2$	0	0	0
		$1\bar{1}$	0	0	$9\,|\Phi_{02l}|^2$	0	0	$27\,|\Phi'{}+f$
		$\bar{1}\bar{1}$	0	0	$9\,|\Phi_{02l}|^2$	0	0	$27\,|\Phi'{}+f$
6H	V, 5, 6	hk	0	$9\,|\Phi'_{02l}+f|^2$	$9\,|\Phi_{02l}|^2$	0	$9\,|\Phi_{02l}|^2$	$9\,|\Phi'{}+f$
		Σ	0	$54\,|\Phi'_{02l}+f|^2$	$54\,|\Phi_{02l}|^2$	0	$54\,|\Phi_{02l}|^2$	$54\,|\Phi'{}+f$
2M₂	VII, 1	$0\bar{2}$	0	0	0	0	$36\,|\Phi_{\bar{1}1l}|^2$	0
		11	0	$27\,|\Phi'_{\bar{1}1l}+f|^2$	0	0	$9\,|\Phi_{\bar{1}1l}|^2$	0
		$\bar{1}1$	0	$27\,|\Phi'_{\bar{1}1l}+f|^2$	0	0	$9\,|\Phi_{\bar{1}1l}|^2$	0
		02	0	0	$36\,|\Phi_{11l}|^2$	0	0	0
		$1\bar{1}$	0	0	$9\,|\Phi_{11l}|^2$	0	0	$27\,|\Phi'_{1}+f$
		$\bar{1}\bar{1}$	0	0	$9\,|\Phi_{11l}|^2$	0	0	$27\,|\Phi'_{1}+f$

Table 31 (continued)

tion	Structure No.	l / hk	$6l'$	$6l'+1$	$6l'+2$	$6l'+3$	$6l'+4$	$6l'+5$
					$\|F_{hkl}\|^2$			
3H	VII, 2, 3	hk	0	$9\ \|\Phi'_{\bar{1}1l}+f\|^2$	$9\ \|\Phi_{11l}\|^2$	0	$9\ \|\Phi_{\bar{1}1l}\|^2$	$9\ \|\Phi'_{11l}+f\|^2$
		Σ	0	$54\ \|\Phi'_{\bar{1}1l}+f\|^2$	$54\ \|\Phi_{11l}\|^2$	0	$54\ \|\Phi_{\bar{1}1l}\|^2$	$54\ \|\Phi'_{11l}+f\|^2$
2M₂	X, 1	$0\bar{2}$	0	0	0	0	$36\ \|\Phi_{11l}\|^2$	0
		11	0	$27\ \|\Phi'_{11l}+f\|^2$	0	0	$9\ \|\Phi_{11l}\|^2$	0
		$\bar{1}1$	0	$27\ \|\Phi'_{11l}+f\|^2$	0	0	$9\ \|\Phi_{11l}\|^2$	0
		02	0	0	$36\ \|\Phi_{\bar{1}1l}\|^2$	0	0	0
		$1\bar{1}$	0	0	$9\ \|\Phi_{\bar{1}1l}\|^2$	0	0	$27\ \|\Phi'_{\bar{1}1l}+f\|^2$
		$\bar{1}\bar{1}$	0	0	$9\ \|\Phi_{\bar{1}1l}\|^2$	0	0	$27\ \|\Phi'_{\bar{1}1l}+f\|^2$
6H	X, 2; X, 3	hk	0	$9\ \|\Phi'_{11l}+f\|^2$	$9\ \|\Phi_{\bar{1}1l}\|^2$	0	$9\ \|\Phi_{11l}\|^2$	$9\ \|\Phi'_{\bar{1}1l}+f\|^2$
		Σ	0	$54\ \|\Phi'_{11l}+f\|^2$	$54\ \|\Phi_{\bar{1}1l}\|^2$	0	$54\ \|\Phi_{11l}\|^2$	$54\ \|\Phi'_{\bar{1}1l}+f\|^2$
6H	II, 6, 7, 8, 9	hk	$9\ \|\Phi'_{02l}\|^2$	$9f^2$	$9f^2$	$27\ \|\Phi'_{02l}\|^2$	$9f^2$	$9f^2$
		Σ	$54\ \|\Phi'_{02l}\|^2$	$54f^2$	$54f^2$	$168\ \|\Phi'_{02l}\|^2$	$54f^2$	$54f^2$
	III, 3, 4	hk	$36\ \|\Phi'_{02l}\|^2$	$9f^2$	$9f^2$	0	$9f^2$	$9f^2$
		Σ	$216\|\Phi'_{02l}\|^2$	$54f^2$	$54f^2$	0	$54f^2$	$54f^2$
	III', 5, 6	hk	0	$9\ \|\Phi'\|^2$	$9\ \|\Phi'\|^2$	0	$9\ \|\Phi'\|^2$	$9\ \|\Phi'\|^2$
		Σ	0	$54\|\Phi'\|^2$	$54\|\Phi'\|^2$	0	$54\|\Phi'\|^2$	$54\|\Phi'\|^2$

tion	Structure No.	l / hk	$2l'$			$2l'+1$		
2M'₁ (2T)	IV, 3 (II', 3)	$0\overset{+}{\bar{2}}$	$\|\Phi_{02l}\|^2$			$3\ \|\Phi_{02l}\|^2$		
		$1\overset{+}{\bar{1}}$	$\|\Phi_{11l}\|^2$			$3\ \|\Phi_{11l}\|^2$		
		$\bar{1}\overset{+}{1}$	$\|\Phi_{\bar{1}1l}\|^2$			$3\ \|\Phi_{\bar{1}1l}\|^2$		
		Σ	$6(\|\Phi'\|^2+f^2)$			$18(\|\Phi'\|^2+f)^2$		

Table 31 (continued)

Modification	Structure No.	l / hk	$2l'$	$2l'+1$
			$\lvert F_{hkl} \rvert^2$	
	IV, 4	$0\overset{+}{\bar{2}}$	$\lvert \Phi_{02l} \rvert^2$	$3\,\lvert \Phi'_{02l} + f \rvert^2$
		$1\overset{+}{\bar{1}}$	$\lvert 2\Phi'_{\bar{1}1l} + f \rvert^2$	$3f^2$
		$\overset{+}{\bar{1}}\overset{}{\bar{1}}$	$\lvert 2\Phi'_{11l} + f \rvert^2$	$3f^2$
		Σ	$6\,(3\,\lvert \Phi' \rvert^2 - 2A'_{02l}f + f^2)$	$6\,(\lvert \Phi' \rvert^2 + 2A'_{02l}\cdot f + 3f^2)$
$2M_1'$	VIII, 1	$0\overset{+}{\bar{2}}$	$\lvert \Phi_{02l} \rvert^2$	$3\,\lvert \Phi'_{02l} \rvert^2$
		$1\overset{+}{\bar{1}}$	$\lvert \Phi_{11l} \rvert^2$	$3\,\lvert \Phi'_{11l} \rvert^2$
		$\overset{+}{\bar{1}}\overset{}{\bar{1}}$	$\lvert \Phi_{\bar{1}1l} \rvert^2$	$3\,\lvert \Phi'_{\bar{1}1l} \rvert^2$
		Σ	$6\,(\lvert \Phi' \rvert^2 + f^2)$	$18\,\lvert \Phi' \rvert^2$
	IX, 1	$0\overset{+}{\bar{2}}$	$\lvert 2\Phi'_{02l} + f \rvert^2$	$3f^2$
		$1\overset{+}{\bar{1}}$	$\lvert 2\Phi'_{11l} + f \rvert^2$	$3f^2$
		$\overset{+}{\bar{1}}\overset{}{\bar{1}}$	$\lvert 2\Phi'_{\bar{1}1l} + f \rvert^2$	$3f^2$
		Σ	$6\,(4\lvert \Phi' \rvert^2 + f^2)$	$18f^2$
$2M_2'$	III, 1	$0\bar{2}$	$4\,\lvert \Phi_{11l} \rvert^2$	0
		02	$4\,\lvert \Phi_{\bar{1}1l} \rvert^2$	0
		$\overset{+}{\bar{1}}1$	$\lvert 2\Phi'_{11l} + f \rvert^2$	$3f^2$
		$\overset{+}{1}\bar{1}$	$\lvert 2\Phi'_{\bar{1}1l} + f \rvert^2$	$3f^2$
		Σ	$24\,(\lvert \Phi' \rvert^2 + f^2)$	$12f^2$
	V, 1	$0\bar{2}$	$4\,\lvert \Phi_{11l} \rvert^2$	0
		02	$4\,\lvert \Phi_{\bar{1}1l} \rvert^2$	0
		$\overset{+}{\bar{1}}1$	$\lvert \Phi_{11l} \rvert^2$	$3\,\lvert \Phi'_{11l} + f \rvert^2$
		$\overset{+}{1}\bar{1}$	$\lvert \Phi_{\bar{1}1l} \rvert^2$	$3\,\lvert \Phi'_{\bar{1}1l} + f \rvert^2$
		Σ	$12\,(\lvert \Phi' \rvert^2 + A'_{02l}\cdot f + f^2)$	$12\,(\lvert \Phi' \rvert^2 - A'_{02l}\cdot f + f^2)$

Table 31 (conclusion)

Modification	Structure No.	hk \\ l	$|F_{hkl}|^2$	
			$2l'$	$2l'+1$
2M$_3$ (2H)	V, 2 (III', 2)	$0\overset{+}{2}$	$|\Phi_{02l}|^2$	$3\,|\Phi_{02l}|^2$
		$\overset{++}{1\bar{1}}$	$|2\Phi'_{02l}+f|^2$	$3f^2$
		$\overset{+-}{1\bar{1}}$	$|\Phi'_{02l}+f|^2$	$3\,|\Phi'_{02l}|^2$
	V, 3 (III', 3)	$0\overset{+}{2}$	$|\Phi_{02l}|^2$	$3\,|\Phi_{02l}|^2$
		$\overset{++}{1\bar{1}}$	$|\Phi'_{02l}+f|^2$	$3\,|\Phi'_{02l}|^2$
		$\overset{+-}{1\bar{1}}$	$|2\Phi'_{02l}+f|^2$	$3f^2$
		Σ	$12\,(|\Phi'|^2 + A'_{02l}\cdot f + f^2)$	$12\,(|\Phi'|^2 - A'_{02l}\cdot f + f^2)$
	II, 4 (V', 2)	$0\overset{+}{2}$	$|\Phi'_{02l}+2f|^2$	$3\,|\Phi'_{02l}|^2$
		$\overset{++}{1\bar{1}}$	$|\Phi_{02l}|^2$	$3\,|\Phi'_{02l}+f|^2$
		$\overset{+-}{1\bar{1}}$	$|\Phi_{02l}|^2$	$3\,|\Phi_{02l}|^2$
(2H)	II, 5 (V', 3)	$0\overset{+}{2}$	$|\Phi'_{02l}+2f|^2$	$3\,|\Phi'_{02l}|^2$
		$\overset{++}{1\bar{1}}$	$|\Phi_{02l}|^2$	$3\,|\Phi_{02l}|^2$
		$\overset{+-}{1\bar{1}}$	$|\Phi_{02l}|^2$	$3\,|\Phi'_{02l}+f|^2$
		Σ	$6\,(|\Phi'|^2+2f^2)$	$3\,(3|\Phi'|^2+2f^2)$
(2H)	III, 2 (VI', 1)	$0\overset{+}{2}$	$4\,|\Phi_{02l}|^2$	0
		$\overset{++}{1\bar{1}},\ \overset{+-}{1\bar{1}}$	$12\,\Phi'_{02l}+f|^2$	$3f^2$
20		Σ	$12\,(2|\Phi'|^2+f^2)$	$12f^2$
	XII, 1 (III', 1)	$0\overset{+}{2}$	$4\,|\Phi_{02l}|^2$	0
		$\overset{++}{1\bar{1}},\ \overset{+-}{1\bar{1}}$	$|\Phi_{02l}|^2$	$3\,|\Phi_{02l}|^2$
		Σ	$12\,|\Phi_{02l}|^2$	$12\,|\Phi_{02l}|^2$
1M' 1T)	XI, 1 (IV', 1)	$0\overset{+}{2}$		$|\Phi_{02l}|^2$
		$\overset{+}{1\bar{1}}$		$|\Phi_{11l}|^2$
		$\overset{+}{\bar{1}1}$		$|\Phi_{\bar{1}1l}|^2$
		Σ		$6(|\Phi'|^2+f^2)$

Table 32. Rounded-Off Numerical Values of $\Sigma|F_{hkl}|^2$ for the $02l$, $11l$ Reflections in Two-Storied Layer Minerals

Dioctahedral varieties

l	1TK₁ I,1,2	1TK₂ IV,1,2	1M,3T VI,3,4	3T I,6,7,8,9	3T IV,5,6	3T VIII,2,3	3T IX,2,3	2M₁ I,3	2M₁ I,4	2M₁ I,5	2M₁ VI,2	2M₂ II,2,3	2M₂,6H V,4,5,6	2M₂,6H VII,2,3	2M₂,6H X,2,3	6H II,6,7,8,9	6H III,3,4	2M' IV,3
0	55, 55	0	380	110	0	170	430	95	65	5	95	0	0	0	0	105	420	150
1								40	40	100	240	40	25	160	250	40	40	
2	150, 10	280	250	160	290	450	85	160	120	60	60	240	230	15	180	40	40	
3								170	40	95	95	0	0	0	0	290	40	410
4	150, 75	250	45	230	250	230	75	10	190	140	10	220	110	230	35	35	35	
5								220	35	85	10	35	170	170	10	35	310	
6	130, 65	0	35	200	0	120	300	130	170	120	10	0	0	0	0	75	30	110
7								35	30	65	65	30	190	35	65	30	25	
8	5, 95	170	140	100	170	20	50	20	75	40	35	150	10	140	100	45	0	
9								45	20	55	130	0	0	0	0	170	20	130
10	25, 25	140	170	50	140	75	40	10	30	1	1	110	40	35	130	20	20	
11								25	20	40	45	20	60	10	100	20	145	
12	5, 55	0	95	55	0	65	150	55	40	20	110	0	0	0	0	35	15	55
13								5	15	30	25	15	15	85	35	15	15	
14	50, 20	80	15	75	80	160	25	55	60	45	35	70	80	30	10	15	0	
15								1	10	15	5	0	0	0	0	50	10	95
16	20, 35	60	5	60	60	120	20	40	45	40	1	50	55	30	5	10	10	
17								10	10	15	20	10	10	50	20	15	60	
18	25, 1	0	35	30	0	30	60	10	20	15	10	0	0	0	0	10	5	25
19								1	5	10	35	10	25	1	35	5	5	
20	5, 10	35	45	10	35	10	15	5	5	1	10	30	5	10	35	30	5	
21								10	5	10	30	0	0	0	0	5	5	45
22	10, 5	25	25	10	25	0	10	10	5	5	5	20	0	20	20	5	5	
23								5	5	5	5	5	20	1	10	5	5	
24	5, 10	0	5	15	0	15	20	1	10	10	10	0	0	0	0	5	20	10

Table 32 (continued)

	$2M_1'$			$2M_2'$	$2M_2'$, $2M_3$	$2M_3$	$2O$		$1M'$	$3T$	$1M$, $3T$	$2M_1$	$6T$ (6R)	$2M$, $6H$	$2T$	$2H$	$2M_3$, $2O$	$2H$	$1T$
IV, 4	VIII, 1	IX, 1	III, 1	V', 2, 3	II, 4, 5	III, 2	XII, 1	XI, 1	II', 1, 2	I', 3, 4	I, 2	V', 1	III', 5, 6	II, 3	V', 3	III', 1, 2, 3	VI', 1	IV', 1
460	150	470	510	170	190	510	570	600	0	140	95, 40	0, 0	0, 100, 100	100	100	210	420	420
170	290	110	470	310	180	80	340	—	200, 180	130, 120	60, 170	200, 0	0, 90, 85	280	140	190	0	—
180	110	340	370	300	140	370	55	430	0	100	10, 230	180, 0	0, 70, 60	75	75	160	310	310
180	160	70	240	100	100	45	45	—	120, 90	80, 60	10, 130	0, 0, 120	0, 45, 40	160	70	110	0	—
140	55	160	180	85	70	180	140	210	0	50	35, 20, 45	0, 90	0, 30, 30	35	35	75	145	145
25	65	35	120	100	35	25	110	—	55, 40	35, 25	10, 25, 55	0, 0, 0	0, 20, 20	40	25	35	0	—
70	25	70	90	45	30	75	50	95	0	20	5, 55, 1	55, 0	0, 10, 10	15	15	30	60	60
40	25	20	50	15	20	10	5	—	20, 15	15, 10	40, 10, 5	40, 0	5, 5	25	15	20	0	—
15	10	25	30	25	15	30	5	35	0	5	10, 1, 5, 10, 1	0, 0, 20, 15, 0	0	5	5	10	20	20

Dioctahedral varieties (columns $2M_1'$ through $1M'$); Trioctahedral varieties (columns $3T$ through $1T$).

Fig. 70.   Texture pattern from halloysite ($\varphi \sim 65°$).

For the modification $1Tk_1$, the $|F|^2$ values are for reflections which are split because $\alpha \neq \pi/2$.

In transforming to the indices corresponding to an inclined cell, use can be made of the relationships for reflection heights D in texture patterns, i.e., $D = hp + l'q_k' = lq_0$ or $D = ks + l'q_k^t = lq_0$, according to whether the a or the b axis forms the oblique angle with the c axis (p,s $\sim$ q /3; q_k and q_0 correspond to inclined and orthogonal cells). Equations (63)-(66) of Chapter 3 may also be used.

The distributions of $\Sigma|F|^2$ values for structures I, 1,2; I,4; and II, 2 are in good agreement with the intensity distributions for kaolinite, dickite, and nacrite, respectively. The intensity ratios for the 02l reflections from diffraction patterns of individual halloysite particles correspond most closely to the $|\Phi|^2$ values for these reflections ($l_{orth} = 3l'$) for the I,3 structure (modification $2M_1$). Table 33 is a comparison of relative values of $|F|^2$ for the distribu-

Table 33. Calculated Values of $|F_{hkl}|^2$

$\pm$ hkl	d	Monoclinic halloysite	Triclinic halloysite (hkl)	Triclinic halloysite (hkl)	Dickite	Kaolinite	hkl	d (X-ray)
020	4.45	90	15	15	65	{ 55 / 55	0$\bar{2}$0 / 020	4.46
110	4.43	40	70	140	40			
$\bar{1}$11	4.35	60	40	3	120	{ 150 / 10	1$\bar{1}$0 / 110	4.37 / 4.33
021	4.25	170	110	10	40			
111	4.13	10	40	20	190	{ 150 / 75	$\bar{1}$11 / $\bar{1}$11	4.17 / 4.12
$\bar{1}$12	3.95	230	30	30	35			
022	3.80	10	35	15	170	{ 130 / 65	0$\bar{2}$1 / 021	3.84 / 3.74
112	3.64	125	80	5	30			
$\bar{1}$13	3.43	35	2	25	75	{ 5 / 95	1$\bar{1}$1 / 111	3.42 / 3.37
023	3.28	25	35	70	20			
113	3.14	45	6	6	30	{ 24 / 25	$\bar{1}$12 / $\bar{1}$12	3.14 / 3.10
$\bar{1}$14	2.97	10	30	50	20			
024	2.82	25	1	15	40	{ 5 / 55	0$\bar{2}$2 / 022	2.84 / 2.75
114	2.70	55	3	40	15			
$\bar{1}$15	2.54	5	15	5	60	{ 50 / 20	1$\bar{1}$2 / 112	2.52 / 2.48
025	2.44	55	5	30	10			
115	2.34	2	5	10	45	{ 20 / 35	$\bar{1}$13 / 113	2.34 / 2.30
$\bar{1}$16	2.22	40	20	1	10			
026	2.14	10	10	0	20	{ 25 / 1	0$\bar{2}$3 / 023	2.13 / 2.09
116	2.08	5	20	10	5			

tion of intensities on texture patterns, for the $\sigma_3 \tau_+ \sigma_3 \tau_- \sigma_3$ structure (I,3), the "tubular kaolin" model (Honjo et al., 1954, see Chapter 1), and for dickite and kaolinite. The values of d_{hkl} given in Table 33 were calculated from unit cell constants found from halloysite electron-diffraction patterns, these values being $a = 5.14$, $b = 8.90$, $c = 14.7$ A, $\beta = 96°$. The indices and d-values given separately for kaolinite were calculated from X-ray data.

Table 33 clearly demonstrates the individual diffraction features of all the structures, as shown by their $|F|^2$ distributions for both 02l and 11l reflections. The particular intensity distributions observed for the 02l reflections in electron-microdiffraction patterns [Fig. 24b], where the 020 and 0.2.2l +1 reflections appear separately, in contrast to the 0.2.2l ($l \neq 0$) reflections, would actually satisfy either halloysite model. The monoclinic model is to be preferred from the 020 and 022 reflections, while the triclinic model is supported only by the 0$\overset{\pm}{2}$3 reflections. It is, however, true that these latter re-

Fig. 71. Texture pattern from platy serpentine in which a phase with the 1T structure predominates ($\varphi = 55°$).

Fig. 72. Texture pattern from platy serpentine with a type D structure containing an admixture of type B structure (φ = 55°).

flections are more sensitive to small departures of the accepted model from
the real structure, and are not such a reliable indication of the basic features
of the structure.

The author, together with S. I. Berkhin and A. I. Gorshkov, has carried
out X-ray diffraction and electron-diffraction studies of a collection of halloysite specimens, obtained from F. V. Chukhrov. Many of these specimens
gave electron-diffraction texture patterns (Fig. 70) with $02l$, $11l$ reflections,
of which the most intense were the 021, $\overline{1}12$, and 112 reflections. This behavior can be taken as strong evidence for the monoclinic halloysite model
(structure I, 3) and against the triclinic model (Honjo et al., 1954).

To obtain the $|F|^2$ values for triclinic modifications, we need only put
$\Phi = \Phi'$ and $f = 0$ in Table 31. The corresponding values of $\Sigma|F|^2$ are listed in
Table 32.

Table 34 gives general expressions for $|F|^2$ for the $\overset{\pm}{20l}$, $\overset{\mp\pm}{13l}$ reflections,
which define the symmetry of the intensity distribution in diffraction patterns
from single crystals of trioctahedral varieties.

For an idealized model of a σ_3 serpentine layer of composition
$Mg_3Si_2O_5(OH)_4$, numerical values were calculated for $\Phi_3' = A_3 + iB_3$ on the

Table 34. Distribution of $|F_{hkl}|^2$ Values for the $\overset{\mp\pm}{13}l$, $\overset{\pm}{20}l$ Reflections of Trioctahedral Two-Storied Layer Minerals

Modification	Structure No.	hk \ l	$6l'$	$6l'+1$	$6l'+2$	$6l'+3$	$6l'+4$	$6l'+5$
2M₁	I',2	∓±13,20	0	0	0	0	$36\lvert\Phi'_{13}\rvert^2$	0
		±1̄3,20	0	0	$36\lvert\bar{\Phi}'_{12}\rvert^2$	0	0	0
		Σ	0	0	$108\lvert\Phi'_{13}\rvert^2$	0	$108\lvert\Phi'_{13}\rvert^2$	0
6T	V',1	∓±13,20	$9(C-D+E)$	0	0	$9(C+D-E)$	0	0
		Σ	$54(C-D+E)$	0	0	$54(C+D-E)$	0	0
2M₂	III',4	±±13,20	$9(C-D-E)$	0	0	$9(C+D+E)$	0	0
		±±1̄3,20	$9(C-D-E)$	0	0	$9(C+D+E)$	0	0
6H	III',5,6	±±13,20	$9(C-D-E)$	0	0	$9(C+D+E)$	0	0
		±±1̄3,20	$9(C-D-E)$	0	0	$9(C+D+E)$	0	0
		Σ	$54(C-D-E)$	0	0	$54(C+D+E)$	0	0

Modification	Structure No.	hk \ l	$3l'$	$3l'+1$	$3l'+2$
3T	II',1	∓±13,20	$9\lvert\bar{\Phi}'_{13}\rvert^2$	0	0
		±1̄3,20	$9\lvert\Phi'_{13}\rvert^2$	0	0
		Σ	$27\left(\lvert\bar{\Phi}'_{13}\rvert^2+\lvert\Phi'_{13}\rvert^2\right)$	0	0
	II',2	±13,20	0	$9\lvert\Phi'_{13}\rvert^2$	0
		∓±1̄3,20	0	0	0
1M	I',1	13,2̄0	0	0	$9\lvert\Phi'_{13}\rvert^2$
		∓±13,20	0	0	0

Modification	Structure No.	$l\backslash hk$	$3l'$	$3l'+1$	$3l'+2$
3T	I', 3	$\mp\pm13,\ 20$	0	$9\lvert\Phi'_{13}\rvert^{2}$	0
	I', 4	$\pm\mp13,\ 20$	0	0	$9\lvert\Phi'_{13}\rvert^{2}$
		Σ	0	$27\lvert\overline{\Phi'_{13}}\rvert^{2}$	$27\lvert\overline{\Phi'_{13}}\rvert^{2}$

Modification	Structure No.	$l\backslash hk$	$2l'$	$2l'$	$2l'+1$
2O, 2M₃	III', 1; III', 2; III', 3	$\pm\pm\mp13,\ 20$	0	$C-D-E$	$C+D+E$
		Σ	0	$6(C-D-E)$	$6(C+D+E)$
2H	V', 2	$\pm\pm\mp13,\ 20$		$C-D+E$	$C+D-E$
	V, 3, VI, 1	Σ		$6(C-D+E)$	$6(C+D-E)$
2T	II', 3	$\mp\pm13,\ 20$		$4\lvert\overline{\Phi'_{13}}\rvert^{2}$	0
		$\pm\mp13,\ 20$		$4\lvert\Phi'_{13}\rvert^{2}$	0
		Σ		$12\left(\lvert\overline{\Phi'_{13}}\rvert^{2}+\lvert\Phi'_{13}\rvert^{2}\right)$	0
1T	IV', 1	$\mp\pm13,\ 20$		$\lvert\overline{\Phi'_{13}}\rvert^{2}$	
		$\pm\mp13,\ 20$		$\lvert\Phi'_{13}\rvert^{2}$	
		Σ		$3\left(\lvert\overline{\Phi'_{13}}\rvert^{2}-\lvert\Phi'_{13}\rvert^{2}\right)$	

Notes: $\Phi'_{hk} = A_{hk} + iB_{hk}$; $\Phi'^{\pm}_{\substack{13\\20}} = \Phi'^{\pm}_{\substack{\bar{1}\bar{3}\\20}}$; $D = A_{13}A_{\bar{1}\bar{3}} + B_{13}B_{\bar{1}\bar{3}}$; $C = A^{2}_{13} + A^{2}_{\bar{1}\bar{3}} + B^{2}_{13} + B^{2}_{\bar{1}\bar{3}}$; $E = (A_{13}B_{\bar{1}\bar{3}} - A_{\bar{1}\bar{3}}B_{13})\sqrt{3}$.

basis of hk$l$ indices referred to an orthogonal unit cell with a = 5.32, b = 9.20, c = 7.15 × 6 = 42.9 Å.

Table 35 gives $\Sigma |F|^2$ values for the $20l$, $13l$ reflections, which coincide in texture patterns and X-ray powder photographs. For convenience in comparison, these values have been scaled for scattering by a three-layer cell. Table 32 is of use in diagnosis of ordered structures with strict repeat units, and Table 35 is valuable for diagnosis of both ordered and disordered structures which do not give results which can be compared with the data in Table 32.

The results in Table 32 do not allow certain of the trioctahedral modifications falling in the same column to be distinguished. These modifications are 1M and the two enantiomorphic 3T structures, $2M_2$ and the two enantiomorphic 6T modifications, the two $2M_3$ structures and 2O,1T, and 2H. Table 35, on the other hand, allows modifications 1T and 2H to be distinguished easily, but as a whole allows for fewer structures to be distinguished. On the basis of the diffraction behavior shown in it, Table 35 splits the serpentine structures up into four groups, denoted by A, B, C, and D, respectively. The first group is characterized by an orthogonal one-layer c pseudo repeat unit, and includes four structures, the second has an inclined one-layer pseudo repeat unit, possessed in its pure form by modification 1M, while the third and fourth groups have orthogonal two-layer pseudo-units, and include six and four structures, respectively, easily distinguishable from the intensities of the $20l$,

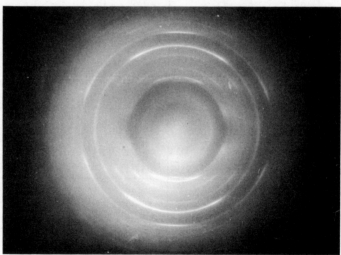

Fig. 73. Texture pattern from clino-chrysotile (φ = 60°).

Table 35. Values of $\Sigma|F|^2$ for the $\overset{\pm}{2}0l$ Reflections of Serpentine Structures $\left(F_{\overset{\pm}{2}0l} = F_{\overset{\mp\pm}{1}3l}\right.$, for Coincident Reflections)

l orth	d	IV',1 II',1,2,3 (A)	I',1,2,3,4 (B)	III',1-6 (C)	V',1,2,3 VI,1 (D)	IV',1 II,1,2,3 (1T)	I',1,2,3,4 (1M)	(2M$_1$)	III',1-6 V',1,2,3 VI',1
0	2.65	375		95	95	200			200
1	2.65								
2	2.64		360				$\bar{2}01$	200	
3	2.61			60	335				$\pm\bar{2}01$
4	2.58		140				200	$\bar{2}02$	
5	2.54								
6	2.49	935		365	795	$\pm\bar{2}01$			$\pm\bar{2}02$
7	2.44								
8	2.38		870				$\bar{2}02$	202	
9	2.32			635	220				±203
10	2.26		55				201	$\bar{2}04$	
11	2.20								
12	2.13	515		145	225	$\pm\bar{2}02$			$\pm\bar{2}04$
13	2.07								
14	2.01		300				$\bar{2}03$	204	
15	1.94			125	330				±205
16	1.89		70				202	$\bar{2}06$	
17	1.83								
18	1.77	460		300	110	$\pm\bar{2}03$			±206
19	1.72								
20	1.67		465				$\bar{2}04$	206	
21	1.62			370	310				±207
22	1.57		135				203	$\bar{2}08$	
23	1.53								
24	1.48	520		40	265	$\pm\bar{2}04$			±208
25	1.44								
26	1.40		62				$\bar{2}05$	208	
27	1.36			400	210				±209
28	1.33		505				204	$\bar{2}.0.10$	
29	1.29								
30	1.26	440		265	310	±205			$\pm2.0.10$

DETERMINATION OF INTENSITIES

Table 36. Relative Displacements of Origins of $\sigma_i \sigma_k$ Layer Pairs

σ_i \ σ_k	σ_1	σ_3	σ_5	σ_2	σ_4	σ_6
σ_1	—1/3, —1,3	—1/3, 1/3	—1/3, 0	0, —1/3	0,0	0,1/3
σ_3		—1/3, 0	—1/3, —1/3	0, 1/3	0, —1/3	0,0
σ_5			—1/3, 1/3	0,0	0, 1/3	0, —1/3
σ_2				1/3, —1/3	1/3, 0	1,3, 1/3
σ_4					1/3, 1/3	1,3, —1/3
σ_6						1/3,0

Table 37. The Sequence of x_0, y_0 Values

Modification	Layer sequence	Layer No.						
		0	1	2	3	4	5	6
1M	$\sigma_3\sigma_3\sigma_3$	0,0	—1/3, 0	1/3, 0	0,0			
3T	$\sigma_3\sigma_5\sigma_1$	0,0	—1/3, —1/3	1/3, —1/3	0,0			
	$\sigma_3\sigma_1\sigma_5$	0,0	—1/3, 1/3	1/3, 1/3	0,0			
6H	$\sigma_3\sigma_4\sigma_5\sigma_6\sigma_1\sigma_2$	0,0	0, —1/3	0,0	0, —1/3	0,0	0, —1/3	0,0
	$\sigma_3\sigma_2\sigma_1\sigma_6\sigma_5\sigma_4$	0,0	0,1/3	0,0	0, 1/3	0,0	0, 1/3	0,0
2M₁	$\sigma_2\sigma_4\sigma_2\sigma_4$	0,0	1/3, 0	—1/3, 0	0,0	1/3, 0	—1/3, 0	0,0
2M₂	$\sigma_5\sigma_4\sigma_5\sigma_4$	0,0	0,1/3	0, —1/3	0,0	0,1/3	0, —1/3	0,0
2O	$\sigma_3\sigma_6\sigma_3\sigma_6$	0,0	0,0	0,0	0,0	0,0	0,0	0,0

$13l$ reflections. The indices of these reflections, and also theoretical values of d calculated for the cell used, are given in the columns of Table 35.

The data given in Tables 32 and 35 allow ordered and disordered (A, B, C, D) modifications to be distinguished, if they are represented in separate columns, from their X-ray photographs or electron-diffraction patterns; combinations of these modifications and structure types can also be recognized. In practice, electron-diffraction studies of serpentines and two-storied nickel minerals have revealed the presence of the following modifications and combinations: 1T (Fig. 71), 1T + 2M₁, 2T + 2H (D), 1T + 1M (A) (particularly the antigorites), B (particularly clino-chrysotiles, Figs. 46, 47), D (in particular, ortho-chrysotiles), A + B, A + D, and D + B (Fig. 72).

In addition to the usual layer silicate pattern, electron-diffraction texture patterns from antigorites also contain extra reflections which correspond to the presence of a superlattice constant, A. From the arrangement and intensities of the distributions, however, it can be deduced that the structural changes which take place in conversion of platy serpentine to antigorite all occur within the limits of structure type A.

It should be noted that the texture patterns given by clino-chrysotile (Fig. 73) have a rather unusual appearance. The reflections lie on deformed ellipses, with the maxima of the $h0l$ reflections lying away from the minor axes and the $0kl$ reflections lying precisely on the minor axes. At present, it is difficult to put forward a complete and rigorous explanation for these features. The one thing clear, which follows from single-crystal point diffraction patterns, is that this behavior is not due to variable monoclinic axes ($\alpha \neq \pi/2$) but to particular morphological properties which cause the most favorable scattering conditions to be ones in which the lattice is oriented with its ab plane not parallel to the texture plane. Under these conditions, the distribution of relative intensities of the successive reflections on the second ellipse, according to their distance from the minor axis, is in excellent agreement with the values listed in the Table 35 column for structure type B; this thus emerges as a reliable and basic piece of evidence in support of the view that clino-chrysotile possesses a monoclinic lattice.

The diffraction characteristics listed in Tables 32 and 35 are worth comparing with the diffraction patterns of individual chrysotile tubes, obtained by the microdiffraction method, since these contain the $20l$ reflections and sometimes $11l$ and $02l$ as well. These establish unambiguously that clino-chrysotile has the structure type B (Figs. 46, 47), and that ortho-chrysotile has a type D structure (Fig. 48). In the absence of separate hkl reflections with $l \neq 0$, $k \neq 3k'$, it is natural to assume that $c \sim 7$ A for clino-chrysotile, and index the reflections accordingly. There is no reason to follow the usual practice of indexing the reflections on the basis of a two-layer cell, so long as there are no additional $13l, 20l$ reflections present, which would mean that the value of c would have to be doubled.

The Scattering Properties of Three-Storied Layer Structures.   The coordinates of the origins of successive layers in structures made up of three-storied layers, which coordinates are used in the phase factors of the individual structure amplitude terms, are derived from the sum of the displacements for the corresponding value of σ_i.

Table 36 gives the relative displacements of origins of coordinates in projection on the xy0 plane for each pair of layers $\sigma_i\sigma_k$, and Table 37 gives the x_0, y_0 values for the different structure modifications.

The following are the F_{hkl} values, calculated from formulas (63)-(65), for the polymorphic mica varieties listed in Table 10:

1) 1M:
$$F = A_3 + A_3e^{2\pi i \frac{l-h}{3}} + A_3e^{-2\pi i \frac{l-h}{3}} =$$
$$= A_3\left(1 + 2\cos 2\pi \frac{l-h}{3}\right). \tag{76}$$

2) 3T:
$$F_{3.5.1} = [A_3 + A_5e^{2\pi i \frac{-h-k-l}{3}} + A_1e^{2\pi i \frac{h-k-l}{3}}]\, e^{-\frac{2\pi ik}{9}},$$
$$F_{3.1.5} = F^*_{3.5.1}. \tag{77}$$

3) 6H:
$$F_{3.4.5.6.1.2} = [A_3 + A_4e^{2\pi i(l/6-k/3)} + A_5e^{2\pi il/3} +$$
$$+ A_6e^{2\pi i(l/2-k/3)} + A_1e^{-2\pi il/3} + A_2e^{2\pi i(-l/6-k/3)}]\, e^{-2\pi i \frac{k}{3}},$$
$$F_{3.2.1.6.5.4} = F^*_{3.4.5.6.1.2}. \tag{78}$$

4) $2M_1$:
$$F = A_2 + A_4e^{2\pi i(l/6+h/3)} + A_2e^{2\pi i(l/3-h/3)} + A_4e^{2\pi il/2} +$$
$$+ A_2e^{2\pi i(-l/3+h/3)} + A_4e^{-2\pi i(l/6-h/3)} = \left(1 + 2\cos 2\pi \frac{l-h}{3}\right)[A_2+(-1)^l A_4]. \tag{79}$$

5) $2M_2$:
$$F = \left(1 + 2\cos 2\pi \frac{l-k}{3}\right)[A_5 + (-1)^l A_4]. \tag{80}$$

6) 2O:
$$F = A_3 + (-1)^l A_6. \tag{81}$$

If all $A_i(hkl)$ in formulas (76)-(81) are expressed in terms of $A_3(h'k'l')$ in accordance with formulas (64), (65), and Table 30, then to calculate F_{hkl} for any sequence of layers it will be possible to limit ourselves to using a single structure only, namely, one made up of identical σ_3 layers. Taking the experimental evidence into account, the amplitude of one layer of such a model (we will assume the layer is trioctahedral to avoid confusion) may be written in a form which is approximate but nonetheless quite accurate enough for explaining the general features of the F distribution, as follows:

$$A_3\,(hkl) = f_{Mg}\,(1 + 2\cos 2\pi k/3) +$$
$$+\, 2f_O\,(1 + 2\cos 2\pi k/3)\cos 2\pi\left(h/3 + l\,\frac{0.105}{n}\right) +$$
$$+\, (3f_{Si} + f_{Al})\cos 2\pi\left(h/3 + l\,\frac{0.270}{n}\right)\cos 2\pi k/3 +$$
$$+\, 2f_O\left[\cos 2\pi\left(-h_i'^r + l\,\frac{0.325}{n}\right) + 2\cos 2\pi\left(h/12 + l\,\frac{0.325}{n}\right)\times\right.$$
$$\left.\times\cos 2\pi k/4 + f_K\cos 2\pi\,(h/3 + l/2n)\right]. \tag{82}$$

According to formula (82), $A_{hkl} = A_{h\overline{k}l}$, and for the $02l$, $\overset{\pm}{1}1l$ reflections the following special condition holds as well [see equations (68)-(71)]:

$$A_{11l} + A_{\overline{1}1l} = -A_{\underset{02l}{\pm}}. \tag{83}$$

For simplicity, the subscript 3 is left off the A here and in what follows. For dioctahedral layers,

$$A_D = A_T - f_{Mg} \tag{84}$$

and equation (83) is not obeyed.

When considering the diffraction properties of the structure modifications listed in Table 10, we need only obtain a set of A_{hkl} values for n = 6, which will correspond directly to modifications 6H, $2M_1$, and $2M_2$, and include as special cases the values of A_{hkl} for the modifications 1M, 3T (n = 3, $A^1_{hkl} = A_{h.k.2l}$) and 2O (n = 2, $A^{''}_{hkl} = A_{h.k.3l}$).

Table 38 shows $|F_{hkl}|^2$ values when h k are equal to 02 and 11, these characterizing the symmetry of the reciprocal lattice of these modifications, and values of $\sum_{h,k}|F|^2$, which determine the intensities of those reflections which are observed on electron-diffraction texture patterns and X-ray powder photographs because of the appropriate coincidences. These values are expressed in terms of A_{hkl} and the dioctahedral coefficient α $(0 \le \alpha \le 1)$, defined by the condition that $(3-\alpha)$ out of 3 octahedra are occupied.

The distribution of $|F_{hk\,l}|^2$ values given in Table 38 is in agreement with the geometrical features noted above for the reciprocal lattice (section 3

Table 38. Distribution of $|F_{hkl}|^2$ Values for the $0\overset{\pm}{2}l$, $1\overset{\pm\pm}{1}l$ Reflections of Three-Storied Layer Minerals

$|F_{hkl}|^2$

1M

hk \ l	$3l'$	$3l'+1$	$3l'+2$
$0\overset{\pm}{2}$	$9(A_{02l} - \alpha f)^2$	0	0
$1\overset{\pm}{1}$	0	$9(A_{11l} - \alpha f)^2$	0
$\overset{\pm}{1}1$	0	0	$9(A_{\bar{1}1l} - \alpha f)^2$
$\sum\limits_{hk} F^2$	$18(A_{02l} - \alpha f)^2$	$18(A_{11l} - \alpha f)^2$	$18(A_{\bar{1}1l} - \alpha f)^2$

3T

hk \ l	$3l'$	$3l'+1$	$3l'+2$
$0\overset{\pm}{2}, 1\overset{\pm\pm}{1}, \overset{\pm}{1}1$ $\ \sum\limits_{hk} F^2$	$3(A^2_{02l} + A_{02l}\cdot\alpha f + \alpha^2 f^2)$	$3(A^2_{11l} + A_{11l}\cdot\alpha f + \alpha^2 f^2)$	$3(A^2_{\bar{1}1l} + A_{\bar{1}1l}\cdot\alpha f + \alpha^2 f^2)$
$\sum\limits_{hk} F^2$	$18[(A_{02l} - \alpha f)^2 + 3\alpha f/A_{02l}]$	$18[(A_{11l} - \alpha f)^2 + 3\alpha f/A_{11l}]$	$18[(A_{\bar{1}1l} - \alpha f)^2 + 3\alpha f/A_{\bar{1}1l}]$

6H

hk \ l	$6l'$	$6l'+1$	$6l'+2$	$6l'+3$	$6l'+4$	$6l'+5$
$0\overset{\pm}{2}, 1\overset{\pm\pm}{1}, \overset{\pm}{1}1$	$(A_{02l}+A_{11l}+A_{\bar{1}1l}-3\alpha f)^2$	$3(A_{02l}-A_{11l})^2$	$9A^2_{11l}$	$3(A_{02l}+A_{11l}+A_{\bar{1}1l}-3\alpha f)^2$	$9A^2_{\bar{1}1l}$	$3(A_{02l}-A_{\bar{1}1l})^2$
$\sum\limits_{hk} F^2$	$6(A_{02l}+A_{11l}+A_{\bar{1}1l}-3\alpha f)^2$	$18(A_{02l}-A_{11l})^2$	$54A^2_{11l}$	$18(A_{02l}+A_{11l}+A_{\bar{1}1l}-3\alpha f)^2$	$54A^2_{\bar{1}1l}$	$18(A_{02l}-A_{\bar{1}1l})^2$

Table 38 (continued)

$|F_{hkl}|^2$

$hk \backslash l$	$6l'$	$6l'+1$	$6l'+2$	$6l'+3$	$6l'+4$	$6l'+5$
2M₁						
$\pm 0\bar{2}$	$9\,(A_{02l}+2\alpha f)^2$	0	0	$8\,(A_{11l}-A_{\bar{1}1l})^2$	0	0
± 11	0	$9\,(A_{02l}-A_{11l})^2$	0	0	$9\,(A_{\bar{1}1l}+2\alpha f)^2$	0
$\bar{1}1$	0	0	$9\,(A_{11l}+2\alpha f)^2$	0	0	$9\,(A_{02l}-A_{\bar{1}1l})^2$
$\sum_{hk} F^2$	$18\,(A_{02l}+2\alpha f)^2$	$18\,(A_{02l}-A_{11l})^2$	$18\,(A_{11l}+2\alpha f)^2$	$18\,(A_{11l}-A_{\bar{1}1l})^2$	$18\,(A_{\bar{1}1l}\ 2\alpha f)^2$	$18\,(A_{02l}-A_{\bar{1}1l})^2$
2M₂						
$0\bar{2}$	0	0	0	0	$36\,(A_{\bar{1}1l}-\alpha f)^2$	0
± 11	0	$9\,(A_{02l}-A_{11l})^2$	0	0	$9\,(A_{\bar{1}1l}+2\alpha f)^2$	0
02	0	0	$36\,(A_{11l}-\alpha f)^2$	0	0	0
$\pm 1\bar{1}$	0	0	$9\,(A_{11l}+2\alpha f)^2$	0	0	$9\,(A_{02l}-A_{\bar{1}1l})^2$
$\sum_{hk} F^2$	0	$18\,(A_{02l}-A_{11l})^2$	$54\,(A_{11l}^2+2\alpha^2 f^2)$	0	$54\,(A_{\bar{1}1l}+2\alpha^2 f^2)$	$18\,(A_{02l}-A_{\bar{1}1l})^2$

$hk \backslash l$	$2l'$	$2l'+1$
20		
$\pm 0\bar{2}$	$4\,(A_{02l}-\alpha f)^2$	0
$\pm 11,\ 1\bar{1}$	$(A_{02l}+2\alpha f)^2$	$(A_{11l}-A_{\bar{1}1l})^2$
$\sum_{hk} F^2$	$12\,(A_{02l}+2\alpha^2 f^2)$	$4\,(A_{11l}-A_{\bar{1}1l})^2$

Table 39. Rounded-Off Numerical Values of $\Sigma |F_{hkl}|^2$ for the $02l^{\pm}$, $11l^{\pm\pm}$ Reflections of Three-Storied Layer Minerals

Level No.	ΣF^2 (trioctahedral)				ΣF^2 (dioctahedral)						hkl					
	1M,3T	2M$_1$	2M$_2$,6H	2O	1M	3T	6H	2M$_1$	2M$_2$	2O	1M	3T	6H	2M$_1$	2M$_2$	2O
0	160	40	0	250	380	130	30	0.01	0	380	020	$020,0\bar{2}0$	020	020		$020,110$
1		115	115				115	115	115				021	110	110	
2	110	30	85		20	220	85	135	145		110	$021,0\bar{2}1$	022	$\bar{1}11$	$1\bar{1}1,020$	
3		45	0	85			90	45	0	85			023	021		111
4	15	3	10		95	30	10	20	70		$\bar{1}11$	$022,0\bar{2}2$	024	111	$111,0\bar{2}2$	
5		0.2	0.2				0.2	0.2	0.2				025	$\bar{1}12$	$1\bar{1}2$	
6	30	5	0	40	125	35	25	10	0	145	021	$023,0\bar{2}3$	026	022		$022,112$
7		75	75				75	75	75				027	112	112	
8	200	50	150		75	315	150	160	200		111	$024,0\bar{2}4$	028	$\bar{1}13$	$1\bar{1}3,022$	
9		240	0	460			65	240	0	460			029	023		113
10	450	110	320		650	325	320	30	350		$\bar{1}12$	$025,0\bar{2}5$	0.2.10	113	$113,0\bar{2}4$	
11		390	390				390	390	390				0.2.11	$\bar{1}14$	$1\bar{1}4$	
12	570	140	0	850	360	705	20	280	0	920	022	$026,0\bar{2}6$	0.2.12	024		$024,114$
13		440	440				440	440	440				0.2.13	114	114	
14	530	130	400		750	445	400	50	430		112	$027,0\bar{2}7$	0.2.14	$\bar{1}15$	$1\bar{1}5,024$	
15		370	0	750			40	370	0	750			0.2.15	025		115
16	420	105	330		265	545	330	200	350		$\bar{1}13$	$028,0\bar{2}8$	0.2.16	115	$115,0\bar{2}6$	
17		240	240				240	240	240				0.2.17	$\bar{1}16$	$1\bar{1}6$	
18	230	60	0	340	360	185	10	15	0	390	023	$029,0\bar{2}9$	0.2.18	026		$026,116$
19		110	110				110	110	110				0.2.19	116	116	
20	80	20	60		30	130	60	65	80		113	$0.2.10,0.2.\overline{10}$	0.2.20	$\bar{1}17$	$1\bar{1}7,026$	
21		30	0	65			30	30	0	65			0.2.21	027		117
22	15	5	10		50	10	10	3	30		$\bar{1}14$	$0.2.11,0.2.\overline{11}$	0.2.22	117	$117,0\bar{2}8$	
23		1	1				1	1	1				0.2.23	$\bar{1}18$	$1\bar{1}8$	
24	0.08	0.02	0	0.4	10	10		10	0	30	024	$0.2.12,0.2.\overline{12}$	0.2.24	028		$028,118$
25		2	2				2	2	2				0.2.25	118	118	
26	5	20	5		0.1	20	5	15	15		114	$0.2.13,0.2.\overline{13}$	0.2.26	$\bar{1}19$	$1\bar{1}9,028$	
27		8	0	15			15	8	0	15			0.2.27	029		119
28	10	2	7		35	10	7	1	20		$\bar{1}15$	$0.2.14,0.2.\overline{14}$	0.2.28	119	$119,0\bar{2}.10$	
29		8	8				8	8	8				0.2.29	$\bar{1}.1.10$, 0.2.10	$1\bar{1}.10$	
30	5	1	0	7	0.1	15	5	10	0	25	025	$0.2.15,0.2.\overline{15}$	0.2.30	0.2.10		$0.2.10,1.1.10$

of Chapter 3). In particular, it reflects the fact that the reciprocal lattice of the monoclinic modifications ($1M$, $2M_1$, $2M_2$) has its lattice points arranged at different levels along the c^* axis, forming inclined lattice point planes, each of which will have the same l index for a monoclinic cell. In $2M_2$ lattice points with $l = 3l'$ are absent for given hk, in accord with the cell having an angle $\alpha \neq \pi/2$, and the points $0.\overset{\pm}{2}.2l + 1$ are also missing, which is due, as for the case of 2O, to the fact that the reflections with $h = 3h'$ correspond to the minimum layer repeat unit.

In the modifications $1M$ and $2M_1$, $F^2_{hk\bar{l}} = F^2_{hkl}$, and in $2M_2$, $F^2_{\bar{h}kl} = F^2_{hkl}$, which is in agreement with the existence of a plane of symmetry parallel to the x axis in the former case, and one parallel to the y axis in the latter, in the system of coordinates used.

In the reciprocal lattices of 3T and 6H, for given hk, all $|F|^2$ with the same l index have identical values, and corresponding lattice points are related by a sixfold axis, but in 6H the lattice planes follow each other along the c^* axis twice as frequently as in the other modification. The arrangement of lattice points in the 2O reciprocal lattice is subject to two mutually perpendicular symmetry planes parallel to the c^* axis, so that

$$F^2_{11l} = F^2_{\bar{1}1l} = F^2_{1\bar{1}l} = F^2_{\bar{1}\bar{1}l} \neq F^2_{02l} = F^2_{0\bar{2}l}.$$

If we allow ourselves to be guided by all the indications, we can determine these polymorphic modifications rigorously and unambiguously from their single-crystal photographs. Possible ways of diagnosing them from diffraction patterns in which the reflections with identical orthogonal l values coincide (X-ray powder photographs and electron-diffraction texture patterns) may be derived, for the majority of the modifications, from the features of their reflection patterns, as shown above. The only ones still left in doubt are $1M$ and 3T, which have an identical three-layer orthogonal cell and thus give geometrically identical texture patterns, so that they can only be distinguished from their reflection intensities. The values of $\Sigma|F|^2$ given in Table 38 show that when $\alpha = 0$, the intensities of these reflections coincide and the trioctahedral varieties $1M$ and 3T are indistinguishable. However, the introduction of dioctahedral properties ($\alpha \neq 0$) destroys the equality of the intensities and may remove some of the difficulties.

An unexpected point which follows from equation (83) is that if $\alpha = 0$, modifications $2M_2$ and 6H are also indistinguishable. In this case, it turns out that 6H is extremely sensitive to the presence of unoccupied octahedra, since these lead not only to differences in the intensities, but also to the appearance of new reflections with $l = 3l'$.

Fig. 74. Diffraction pattern from mica of modification $2M_2$ ($\varphi = 55°$).

Fig. 75. Diffraction pattern from mica combining the modifications
$1M + 2M_1$ ($\varphi = 55°$).

For the two limiting cases, when $\alpha = 0$ and $\alpha = 1$, Table 39 gives rounded-off numerical values of $\sum_{h,k} |F|^2$ for different values of l, calculated for $\overset{\pm}{h}, \overset{\pm}{k}$ values of 02 and 11 using formulas (82)-(84) and Table 38. The values of l used in the calculations in Table 39 are represented by the l indices of modifications 2O, 3T, and 6H.

Table 39 also shows the indices of coinciding reflections relative to the cell possessed by each particular modification, but for brevity it only includes those which, in accordance with the above, have different values of $|F|^2$. Reflections lying at the same height above an ellipse minor axis in texture patterns are placed in the same row of the table and are assigned the number which the level would have in the case of a six-layer cell.

For convenience in comparison, the values of $\Sigma|F|^2$ for different modifications are all scaled to the same volume, that of a cell containing three structure layers.

From these numerical values of $\Sigma|F|^2$ it appears easy to distinguish all modifications from their reflection intensities, except for the trioctahedral varieties 1M and 3T, and $2M_2$ and 6H. The intensity distribution can be used to distinguish between dioctahedral and trioctahedral varieties of the same modification, which is particularly important for the lepidolites, which do not always have sufficiently different values of a and b for a distinction to be drawn in this respect. It is noteworthy that in the trioctahedral variety $2M_1$, the reflections which express a repeat unit of two layers (on odd levels) are considerably more intense than those corresponding to a repeat unit of one layer (on even levels), while in the dioctahedral variety $2M_1$, the distribution of intensities is more equal.

Thus the information shown in Table 39 may be used as a basis for diagnosing the polymorphic varieties of the mica-type minerals from texture patterns and polycrystalline specimen photographs. Moreover, from anomalies in the intensity distribution, which do not fit in with Table 39, it is possible to distinguish combinations of modifications (compare the diffraction patterns in Figs. 61, 62, 74, and 75 with Table 39).

It should be noted that our deductions as to the similarity or dissimilarity of the structure modifications still hold where the diffraction properties of other hkl reflections are concerned, when these express the strict periodicity of the corresponding layer sequences. Although the absolute values of the structure amplitudes may differ, the relationships between them, which are de-

Table 40. Nonzero F^2_{hkl} Factors Expressed in Terms of A_3, the Amplitude of a σ_3 Layer

l	hk	Structure		
		$\sigma_4\tau_2\sigma_4$	$\sigma_2\tau_4\sigma_2$	$\sigma_6\tau_6\sigma_6$
$3l'$	$0\bar{2}$	$3A_{\bar{1}1l}$	$3A_{11l}$	$3A_{02l}$
	02	$3A_{11l}$	$3A_{\bar{1}1l}$	$3A_{02l}$
$3l'+1$	11	$3A_{11l}$	$3A_{02l}$	$3A_{\bar{1}1l}$
	$1\bar{1}$	$3A_{02l}$	$3A_{11l}$	$3A_{\bar{1}1l}$
$3l'+2$	$\bar{1}1$	$3A_{02l}$	$3A_{\bar{1}1l}$	$3A_{11l}$
	$\bar{1}\bar{1}$	$3A_{\bar{1}1l}$	$3A_{02l}$	$3A_{11l}$

Table 41. Rounded-Off Theoretical Numerical Values of ΣF^2_{hkl} for Reflections Coinciding on Chlorite Texture Patterns and Powder Photographs

l_{orth}	Modification		hkl for mono-clinic cell	l_{orth}	Modification		hkl for mono-clinic cell
	1Tk	1M			1TK	1M	
0	140	530	020	12	120	40	024
1	400	5	110	13	50	140	114
2	140	600	$\bar{1}11$	14	100	10	$\bar{1}15$
3	350	40	021	15	30	100	025
4	150	400	111	16	70	1	115
5	250	100	$\bar{1}12$	17	20	80	$\bar{1}16$
6	150	250	022	18	50	0	026
7	200	140	112	19	15	60	116
8	150	150	$\bar{1}13$	20	40	2	$\bar{1}17$
9	130	150	023	21	10	40	027
10	140	80	113	22	20	5	117
11	80	160	$\bar{1}14$				

termined by the arguments of the trigonometric functions contained in F_{hkl}, remain the same, since the important thing for them is what relation their indices bear to a multiple of three. The set of indices $0\overset{\pm}{2}l$, $\overset{\pm\pm}{11}l$, $\overset{\mp\pm}{1}1l$ includes all cases of the type h = 3h' ± 1, 3h' which might be met with in the indices of other ellipses.

The Scattering Properties of Chlorites.   The scattering amplitudes of the chlorite modifications derived above can be expressed in terms of the amplitude of a single given packet, calculated for an orthogonal cell with a repeat unit of three layers according to the same procedure as that applied to the micas and kaolinites.

In accordance with the displacements of the centers of symmetry of successive packets, which can be easily worked out from formula (15) of Chapter 1, for the structures listed in Table 11 we have

$$F_{hkl} = A_i \left(1 + 2 \cos 2\pi \frac{l - h}{3}\right),$$ (85)

where i = 4, 2, or 6 for structures 1, 2, or 3 in Table 11, respectively. If, then, in equation (85) we replace $A_i(hkl)$ by $A_3(h'k'l)$, with the hk indices transformed into the h'k' indices through cyclic permutations corresponding to transformation of σ_i into σ_3, then F_{hkl} is given by the quantities listed in Table 40. The squares of these quantities characterize the distribution of lattice-point densities, and thus also the symmetry of single-crystal reciprocal lattices.

Since the hkl indices used relate to an orthogonal unit cell embracing three packets, the reflections shown with the same values of l in Table 40 coincide in texture patterns and powder photographs. Characteristic rounded-off theoretical numerical values of ΣF_{hkl}^2 are shown for these in Table 41.

The A_3 amplitudes used to express these values were calculated for an idealized chloride model consisting of layers of nominal composition $Mg_{4.6}Al_{1.4}[Si_{2.6}Al_{1.4}O_{10}](OH)_8$ with average z coordinates (or rather z/3 coordinates) taken from Steinfink's results (1958):

$$A_3(hkl) = f_0(1 + 2 \cos 2\pi k/3) + 2f_1(1 + 2 \cos 2\pi k/3) \cos 2\pi (h/3 + lz_1) + 4f_2 \cos 2\pi (h/3 + lz_2) \cos 2\pi k/3 + 2f_3 [\cos 2\pi (-h/6 + lz_3) + 2 \cos 2\pi (h/12 + lz_3) \cos 2\pi k/4] + 2f_4 (1 + 2 \cos 2\pi k/3) \times \cos 2\pi lz_4 + f_5(1 + 2 \cos 2\pi k/3) \cos 2\pi (-h/3 + lz_5).$$ (86)

In formula (86), $f_0, f_1, \ldots, f_5$ are the mean atomic amplitudes for atoms lying on the successive levels of a chlorite packet, starting from the center of a three-storied layer octahedron and ending up at the center of a one-storied layer octahedron (see Fig. 6).

In view of the fact that the model was trioctahedral, we did not need to consider how the octahedra were occupied, since for the reflections used,

Table 42. Values of $\Sigma|F|^2$ for the $\overset{\pm}{20l}$, $\overset{\mp\pm}{13l}$ Reflections Coinciding on Texture Patterns of One-Packet Chlorites, According to the Type of packet

l_{orth}	d	$|\sigma'_{2n+1}|$	σ'_{2n}		$|\sigma_{2n}|$	$|\sigma_{2n+1}|$	σ_{2n}		σ'_{2n}	$\overset{\pm}{hkl}_{mono}$	$\overset{\mp\pm}{hkl}_{orth}$
			$(\bar{2}0l)$	$(20l)$			$(\bar{2}0l)$	$(20l)$			
0	2.66		1	1			20	20			130; 200
1	2.66	60			25	100			1	$\begin{cases}130\\ \bar{2}01\end{cases}$	
2	2.65	140			85	60			20	$\begin{cases}200\\ \bar{1}31\end{cases}$	
3	2.62		35	100			140	80			131; 201
4	2.58	120			175	9			200	$\begin{cases}131\\ \bar{2}02\end{cases}$	
5	2.54	100			80	375			255	$\begin{cases}201\\ \bar{1}32\end{cases}$	
6	2.49		760	55			390	160			132; 202
7	2.44	1			620	140			430	$\begin{cases}132\\ \bar{2}03\end{cases}$	
8	2.38	760			9	210			135	$\begin{cases}202\\ \bar{1}33\end{cases}$	
9	2.32		10	40			180	195			133; 203
10	2.25	20			135	330			90	$\begin{cases}133\\ \bar{2}04\end{cases}$	
11	2.19	10			50	110			20	$\begin{cases}203\\ \bar{1}34\end{cases}$	
12	2.13		360	40			50	120			134; 204
13	2.06	50			245	40			30	$\begin{cases}134\\ \bar{2}05\end{cases}$	
14	2.00	295			20	55			300	$\begin{cases}204\\ \bar{1}35\end{cases}$	
15	1.94		50	5			25	190			135; 205
16	1.88	3			110	145			65	$\begin{cases}135\\ \bar{2}06\end{cases}$	
17	1.82	8			4	40			165	$\begin{cases}205\\ \bar{1}36\end{cases}$	
18	1.76		480	1			90	150			136; 206
19	1.71	30			405	140			80	$\begin{cases}136\\ 207\end{cases}$	

Table 42 (continued)

l_{orth}	d	$|\sigma'_{2n+1}|$	σ'_{2n}		$|\sigma'_{2n}|$	$|\bar{\sigma}'_{2n+1}|$	σ_{2n}		σ_{2n}	$\pm hkl_{mono}$	$\pm\pm hkl_{orth}$
			$(\bar{2}0l)$	$(20l)$			$(\bar{2}0l)$	$(20l)$			
20	1.66	570			25	365			90	$\begin{cases} 206 \\ \bar{1}37 \end{cases}$	
21	1.61		30	10			310	100			137, 207
22	1.56_5	130			130	175			365	$\begin{cases} 137 \\ \bar{2}08 \end{cases}$	
23	1.51_5	15			190	15			4	$\begin{cases} 207 \\ \bar{1}38 \end{cases}$	
24	1.47_5		175	310			50	105			138, 208
25	1.43_5	45			55	90			15	$\begin{cases} 138 \\ \bar{2}09 \end{cases}$	
26	1.39_5	15			175	55			365	$\begin{cases} 208 \\ \bar{1}39 \end{cases}$	
27	1.35_5		40	50			45	285			139, 209
28	1.32	440			45	240			7	$\begin{cases} 139 \\ \bar{2}.0.10 \end{cases}$	
29	1.28	65			275	210			95	$\begin{cases} 209 \\ \bar{1}.3.10 \end{cases}$	
30	1.25		30	260			155	75			1.3.10, 2.0.10

only the terms relating to the tetrahedral network differed from zero in the scattering amplitude.

From Table 41 it can be seen that the intensity relationships are widely different for the triclinic and monoclinic modifications and will serve as a clear indication in diagnosis of the ordered polymorphic varieties under consideration, provided that they are present in sufficient quantity in the polycrystalline specimens. Up to the present we have found only one specimen of triclinic chlorite, which revealed itself by its electron-diffraction texture pattern (Fig. 64).

In connection with the fact that the vast majority of chlorites do not show a strict layer repeat unit in their powder photographs and texture patterns, in their case intensity distribution of the $20l$, $13l$ reflections is of considerable diagnostic importance, since these reflections are observed even in chlorites with very imperfect structures. From the arrangement of these reflections it is possible to distinguish clearly between chlorites with $\beta \neq \pi/2$ and those with $\beta = \pi/2$. Table 42 lists numerical values of $\Sigma |F|^2$, calculated in the usual manner for electrons scattered in the directions of the corresponding reflections, and by comparing intensities of reflections on texture patterns with those in the table it will be possible to determine uniquely the types of packet making up a structure.

5. Geometry of Layer Silicate Structural Features as Revealed Through the Distribution of F or $|F|^2$

The relationship which exists between the structural features of layer silicates and the intensities of their diffraction pattern reflections is most sensibly shown by expressing the structure amplitude in the form used in (63):

$$F (\mathbf{H}) = \sum_{m=0}^{n-1} \Phi_m (hkl) e^{2\pi i \left(hx_{om}+ky_{om}+l \frac{m}{n}\right)},$$

since in this form the factors relating to the structures of the individual layers and to their relative disposition are separately represented.

We must first consider the properties of Φ_m, the amplitudes of the individual layers, and the effects which these properties have on the structure amplitude $F(hkl)$.

We have already pointed out above that because the combination of a trioctahedral network with one tetrahedral network has the symmetry p31m, the amplitude of a σ_3 layer with a dioctahedral coefficient α can be written in the form

$$\Phi_3 (hkl) = \Phi_{inv} e^{2\pi i \frac{h}{3}} - \alpha f$$

for a two-storied layer (1:1), and

$$\Phi_3 (hkl) = \Phi_{1/2} e^{2\pi i \frac{h}{3}} + \Phi_{1/2}^* e^{-2\pi i \frac{h}{3}} - \alpha f$$

for three-storied (1:2) or four-storied (2:2) layers, where Φ_{inv}, $\Phi_{\frac{1}{2}}$ are invariant with respect to rotations about the c* axis through angles of $\pm 2\pi/3$. From this it follows that Φ_3 is also invariant with respect to such rotations, but with the additional condition that these rotations relate points in reciprocal space with indices of

$$h = \pm 1, \ \mp 2, \ \pm 4, \ \mp 5, \ \ldots \pm (3h' + 1), \ \mp (3h' - 1) \quad (87)$$

or of $h = 3h'$. Since, for a σ_6 layer, $\Phi_6(hk l) = \Phi_3(\overline{hk} l)$, the same property of rotational| invariance is also shown by the amplitude Φ_6.

In the case of a lattice with orthogonal unit cells, in each plane perpendicular to the $c*$ axis and with l = const, reflections with equal values of $3h^2 + k^2$ are distributed at equal distances from the $c*$ axis. The absolute values of the corresponding hk indices are given in Table 14. In the reciprocal lattice, the number of these lattice points which are rotationally congruent on rotation about $c*$ is determined by the number of hk combinations with the same value of $3h^2 + k^2$, where the signs of the indices h and k can vary. According to Table 14, each of these lattice point groups contains points with h indices satisfying the conditions in (87). However, they are by no means al-. ways related by rotations of $\pm 2\pi/3$ about $c*$. This turns out to be true only for points with indices in which k = 3k'. For example, when $3h^2 + k^2 = 12$, the points $13l$, $1\overline{3}l$, $\overline{2}0l$, and $\overline{1}3l$, $\overline{1}\overline{3}l$, $20l$ are related by such rotations (Fig.30). At the same time, these points satisfy (87). When $3h^2 + k^2 = 28$, equilateral triangles are formed by the following sets of points: $15l$, $2\overline{4}l$, $\overline{3}1l$; $24l$, $1\overline{5}l$, $\overline{3}1l$; $31l$, $\overline{1}\overline{5}l$, $\overline{2}4l$; $3\overline{1}l$, $\overline{2}\overline{4}l$, $\overline{1}5l$ (Fig. 30). Within each set of three, the indices do not satisfy the conditions in (87).

If the lattice has a monoclinic cell, then the same plane perpendicular to the $c*$ axis contains points lying at the same height D' above the ab plane. In electron-diffraction patterns these correspond to reflections with identical heights of $D = hp + l q$ or $D = ks + l q$, depending on whether it is the angle α or β which differs from $\pi/2$. Because the relationships $c \cos \beta = -a/3$ or $c \cos \alpha = -b/3$ applies here, it follows that $p = q/3$ or $s = q/3$. At the same height D there may be reflections with h indices satisfying the conditions in (87) required for the above-noted rotational invariance of Φ_3. Thus, reciprocal lattice points with hk indices for which Φ_3 is invariant, in the above sense, in the case of an orthogonal cell, will also lie in a single plane perpendicular to $c*$ in the case of a monoclinic cell, and Φ_3 will remain invariant for them. These lattice points will only differ in their l indices. When $3h^2 + k^2 = 12$, these points will be $13l$, $1\overline{3}l$, $\overline{2}.0.l + 1$ and $\overline{1}3l$, $\overline{1}\overline{3}l$, $2.0.l - 1$, if p = q/3, and $1.3.l - 1$, $1.\overline{3}.l + 1$, $\overline{2}0l$ and $\overline{1}.3.l - 1$, $\overline{1}.\overline{3}.l + 1$, $20l$, if s = q/3. As regards the reflections with $3h^2 + k^2 = 28$ when p = q/3, the reflections with the same D, for example $15l$ and $\overline{2}.4.l + 1$, belong to different triangles, and when s = q/3, the reflections in each triangle, for example $1.5.l - 1$, $2.\overline{4}.l + 1$, $\overline{3}1l$, although lying at the same level D, have h indices which do not satisfy (87).

It should be noted that for σ_3, σ_6 layers with strictly hexagonal or trigonal arrangements of tetrahedral layers, i.e., with $s = 0, \frac{1}{4}$, $\phi = 0°, 30°$, the amplitudes Φ_3, Φ_6 are invariant with respect to any rotations about the c^* axis which bring integral hk$l$ lattice points into coincidence, if the h indices satisfy (87); this can be seen from equations (73), (82), and (86). Here, the Φ_3 and Φ_6 amplitudes of the corresponding reflections are also identical even when $k \neq 3k'$, for example, in the reflections $15l$ and $\overline{2}4l$ or $42l$ and $17l$.

For reflections with $k = 3k'$, $\Phi_3 = \Phi_5 = \Phi_1$, $\Phi_6 = \Phi_2 = \Phi_4$. In fact, the layers σ_5, σ_1 and σ_2, σ_4 differ from the layers σ_3 and σ_6 only in that the tetrahedral networks are displaced by $\pm b/3$, so that for the same x coordinates, the y coordinates of the atoms differ by $\Delta y = \pm \frac{1}{3}$, which does not alter the harmonic functions in the expressions for Φ if $k = 3k'$. We can retain the form of the structure amplitude for the σ_3 layers [formulas (73), (82), (86)] on going over to Φ_5, Φ_1 by replacing $2\pi(hx + lz)$, the argument for displaced networks in σ_1, σ_5 layers, by the argument $2\pi(hx + lz \pm k/3)$, which differs in the case where $k = 3k'$ only by the additional term $\pm 2\pi k'$, where k' is a whole number.

Thus rotations about the c^* axis by $\pm 120°$, where the index h fulfills the conditions in (87), and where $k = 3k'$ for the reciprocal lattice points, turn out to be invariant in the case of all amplitudes Φ_i. It is not difficult to see that since $x_{0m}, y_{0m} = 0, \pm \frac{1}{3}$, the phase factors in (63) do not impose any additional limitations on the lattice point indices, and the above property of invariance holds for the structure factor F(hk$l$) of a structure which is randomly ordered or disordered due to the distribution of its x_{0m}, y_{0m} values.

This means that in the case of reflections with h indices satisfying one of the three relationships $h = 3h'$, $h = 3h' \pm 1$, with indices for which $k = 3k'$ and which coincide on electron-diffraction texture patterns, i.e., they have $3h^2 + k^2 = \text{const}$, then they will all have identical structure amplitudes. This allows their intensities to be evaluated even when they are superimposed on the diffraction pattern, so long as their multiplicity factors are taken into account. For example, in a monoclinic structure with a total intensity of I for the reflections $26l$, $\overline{4}.0.l + 2$, two-thirds I belongs to the reflection $26l$ and one-third of I to the reflection $\overline{4}.0.l + 2$. The same relationships also hold for powder photographs, provided no further reflections are superimposed on the reflections in question.

In electron-diffraction patterns from single crystals, this point leads to the appearance of pseudohexagonal symmetry in the intensity distribution, which at times seemed inconsistent, at the very least, with the monoclinic symmetry of the individual layers forming the structures. This factor should therefore always be borne in mind, together with other factors, such as, in particular, the bending of crystals, which gives the effects described by Cowley (1961) and by Cowley and Goswami (1961).

From the relationships between the x,y coordinates of different σ_i layers, where the layers σ_i and σ_{i+3} are related by a rotation about the normal by π, the layers σ_1 and σ_5, σ_2 and σ_4 are related by reflection in the ac^* plane (which is a plane of symmetry for the layers σ_3, σ_6) and the layers σ_1 and σ_2, σ_5 and σ_4, σ_3 and σ_6 are related by reflection in the bc* plane, it follows that the following relationships will apply:

$$\Phi_{i+3}(\bar{h}\,\bar{k}l) = \Phi_i(hkl), \tag{88}$$

$$\Phi_{2,4,6}(\bar{h}kl) = \Phi_{1,5,3}(hkl), \tag{89}$$

$$\Phi_{5,4,3,6}(h\bar{k}l) = \Phi_{1,2,3,6}(hkl). \tag{90}$$

It was noted above that when k = 3k', for given hkl and any n,

$$\Phi_{2n+1}(hkl) = \text{const}, \quad \Phi_{2n}(hkl) = \text{const}. \tag{91}$$

In addition, when h = 3h',

$$\Phi_3(hkl) = \Phi_6(hkl), \quad \Phi_1(hkl) = \Phi_2(hkl), \quad \Phi_4(hkl) = \Phi_5(hkl). \tag{92}$$

In actual fact, the pairs of layers which are related by reflection in the bc* plane, i.e., σ_3 and σ_6, σ_1 and σ_2, σ_4 and σ_5, have identical y coordinates for their atoms. The x coordinates of all atoms except the O atoms on the tetrahedron faces differ from those of the layer pairs quoted by $\Delta x = 0, \pm \frac{1}{3}$. As regards the O atoms mentioned, if in the layer σ_3 they have coordinates of $x_1 = -2s/3$, $x_2 = s/3$, then in the layer σ_6 they have coordinates of $x_1 = 2s/3(\pm\frac{1}{2})$ and $x_2 = -s/3(\pm\frac{1}{2})$. It should, however, be borne in mind that combinations of antiparallel layers, not all even or all odd, are only likely when the tetrahedral network has approximately hexagonal symmetry, and $s \simeq \frac{1}{4}$. For these atoms, in this case, $\Delta x_1 = 4s/3 = \frac{1}{3}$, $\Delta x_2 = -2s/3 + \frac{1}{2} = \frac{1}{3}$. These differences in coordinates naturally do not show up in the expressions for Φ when h = 3h'.

Using the relationships stated between the amplitudes of layers [formulas (88)-(92)], it is possible to explain and compare the diffraction properties of the different modifications from an analysis of the general expressions for the structure amplitudes F.

The position is best explained by considering concrete examples of each of the modifications listed in Table 13.

1. Modification $1Tk_1$. Structure I,1 (Table 6):

$$F = \Phi_2 \left(1 + 2 \cos 2\pi \frac{l-h}{3}\right).$$

In accordance with the expression in brackets, points with identical hk in the chosen orthogonal reciprocal lattice follow in order along the c* direction at intervals of $\Delta l = 3l'$. If h = 3h' ± 1, then $l = 3l'$ ± 1. When h = 3h', $l = 3l'$. Thus, the reciprocal lattice can also be described in terms of a monoclinic cell with axes $c^*_{mono} = 3c^*_{orth}$, and with a β* angle such that $a^* \cos \beta *$ = $c_{orth} = c^*_{mono}/3$. At the same time, $F(h\bar{k}l) \neq F(hkl)$, since this inequality is characteristic of Φ_2, corresponding to triclinic lattice symmetry.

2. Modification $1Tk_2$. Structure IV,2 (Table 6):

$$F = \Phi_3 \left(1 + 2 \cos 2\pi \frac{l-k}{3}\right).$$

In this case, a monoclinic cell can be chosen from the lattice, with b* cos α* $= c^*_{orth} = c^*_{mono}/3$. Since the σ_3 layer does not have a plane of symmetry parallel to the bc* plane, $F(\bar{h}kl) \neq F(hkl)$, and the lattice has triclinic symmetry.

3. Modification 1M. Structure 1 (Table 10):

$$F = A_3 \left(1 + 2 \cos 2\pi \frac{l-h}{3}\right).$$

In contrast to the previous example, here $F(\bar{h}kl) = F(hkl)$. In addition, some further relationships characteristic for A_3 apply in this case, these being allowed by the factor in the brackets, and resulting from the presence of centers of symmetry and twofold axes in the three-storied σ_3 layer:

$$F(\bar{h}\,\bar{k}\,\bar{l}) = F(\bar{h}k\bar{l}) = F(hkl).$$

4. Modification 1M'. Structure XI,1 (Table 6): F = Φ_3. In this case, lattice points with given hk indices have a complete set of all l indices. The unit cell is orthogonal, but since the only relationship holding here is $F(h\bar{k}l) = F(hkl)$, the symmetry of the lattice is only monoclinic.

5. Modification $2M_1$. Structures I,3 (probably halloysite) and I,4 (dick-ite) (Table 6):

a) $$F = \left(1 + 2 \cos 2\pi \frac{l-h}{3}\right)\left[e^{2\pi i \frac{k}{3}} + (-1)^l e^{-2\pi i \frac{k}{3}}\right]\Phi_3,$$

b) $F = \left(1 + 2\cos 2\pi \dfrac{l-h}{3}\right)\left[\Phi_1 e^{-2\pi i \frac{k}{3}} + (-1)^l \Phi_5 e^{2\pi i \frac{k}{3}}\right].$

When h = 3h' ± 1, l = 3l' ± 1, and when h = 3h, l = 3l'. Also, F(hk$l$) = $(-1)^l$F(hk$l$). When k = 3k', l = 2l'. If, in addition, h = 3h', then l = 6l'. A monoclinic cell may be chosen from the reciprocal lattice, with

$$a^* \cos \beta^* = c^*_{orth} = c^*_{mono}/3.$$

6. Modification 2M$_1'$. Structures IV,3; VIII, 1 (Table 6):

    ä) $F = \Phi_3\left[e^{2\pi i \frac{k}{3}} + (-1)^l e^{-2\pi i \frac{k}{3}}\right],$

    b) $F = \Phi_1 + (-1)^l \Phi_5.$

As in the previous example, F($\overline{hkl}$) = $(-1)^l$F(hk$l$), but in contrast to this the structure amplitude of modification 2M$_1'$ lacks the factor in the brackets, which reduces the applicability of the h indices, according to whether they are multiples of three or not, to certain l indices. When k = 3k', l = 2l'. The additional equation h = 3h' does not alter this relationship.

7. Modification 2M$_2$. Structures II,2; II,3 (nacrite) (Table 6):

    a) $F = \left(1 + 2\cos 2\pi \dfrac{l-k}{3}\right)\left[\Phi_1 e^{-2\pi i \frac{h}{3}} + (-1)^l \Phi_2 e^{2\pi i \frac{h}{3}}\right],$

    b) $F = \left(1 + 2\cos 2\pi \dfrac{l-k}{3}\right)\left[\Phi_3 e^{-2\pi i \frac{h}{3}} + (-1)^l \Phi_6 e^{2\pi i \frac{h}{3}}\right].$

When k = 3k' ± 1, l = 3l' ± 1. When k = 3k' l = 3l'. F($\overline{hkl}$) = $(-1)^l$ F(hk$l$). When h = 3h', l = 2l'. If, in addition, k = 3k', then l = 6l'.

8. Modification 2M$_2'$. Structure V,1 (Table 6).

$$F = \Phi_1 e^{2\pi i \frac{h}{3}} + (-1)^l \Phi_2 e^{-2\pi i \frac{h}{3}}.$$

In contrast to the preceding example, no relationship exists between the indices k and l. As with 2M$_2$, F($\overline{hkl}$) = $(-1)^l$F(hk$l$). When h = 3h', l = 2l'.

9. Modification 2M$_3$. Structure II,4; V,2 (Table 6):

    a) $F = \Phi_1 e^{-2\pi i \frac{h}{3}} + (-1)^l \Phi_4 e^{2\pi i \frac{h}{3}},$

b) $\quad F = \Phi_3 e^{2\pi i \frac{h+k}{3}} + (-1)^l \Phi_6 e^{-2\pi i \frac{h+k}{3}}$,

$F(\overline{hk}l) = (-1)^l F(hkl)$. If h = 3h' when k = 3k', then $l = 2l'$.

10. Modification 2O. Structure 5 (Table 10):

$$F = A_3 + (-1)^l A_6.$$
$$F(h\overline{k}l) = F(hkl); \quad F(\overline{h}kl) = (-1)^l F(hkl).$$

In addition, because of the centers of symmetry in σ_3 and σ_6, $F(\overline{hkl}) = F(hkl)$. When h = 3h', $l = 2l'$.

11. Modification 3T. Structure 3 (Table 10), IV,5 (Table 6):

a) $\quad F = [A_3 + A_5 e^{2\pi i \frac{-h-k+l}{3}} + A_1 e^{2\pi i \frac{h-k-l}{3}}] e^{-2\pi i \frac{k}{9}}$.

b) $\quad F = [\Phi_3 + \Phi_5 e^{2\pi i \frac{-k+l}{3}} + \Phi_1 e^{2\pi i \frac{k-l}{3}}] e^{2\pi i \frac{h+k}{3}}$.

In the general form given above, the structure amplitudes do not express the phase relationships which are characteristic of modification 3T. These can be brought out by transforming all the A_i, Φ_i values to those for the σ_3 layer. If k = 3k', then

a) $\quad F = A_3 \left[1 + 2 \cos 2\pi \frac{l-h}{3}\right] e^{-2\pi i \frac{k'}{3}}$,

b) $\quad F = \Phi_3 \left[1 + 2 \cos 2\pi \frac{l}{3}\right] e^{2\pi i \frac{h}{3}}$.

In case (a), when h = 3h', $l = 3l'$, and when h = 3h' ± 1, $l = 3l'$ ± 1. To an accuracy up to the phase factor, the distribution of F is the same as that for modification 1M. In case (b), only reflections with $l = 3l'$ are possible.

12. Modification 6H. Structure II,6 (Table 6):

$$F = \Phi_3 e^{2\pi i \frac{-h-k}{3,}} + \Phi_5 e^{2\pi i \frac{l-h}{3}} + \Phi_1 e^{2\pi i \frac{-h+k-l}{3}} + (-1)^l (\Phi_6 e^{2\pi i \frac{h+k}{3}} +$$

$$+ \Phi_2 e^{2\pi i \frac{l+h}{3}} + \Phi_4 e^{2\pi i \frac{h-k-l}{3}}).$$

The above relationships for sixfold screw axes are developed during the transformation of Φ_i to Φ_3 for particular values of the indices hk$l$. When k = 3k', the structure amplitude

$$F = \left(1 + 2\cos 2\pi\, \frac{l}{3}\right)\left[\Phi_3 e^{-2\pi i\frac{h}{3}} + (-1)^l \Phi_6 e^{2\pi i\frac{h}{3}}\right]$$

has the same properties as the amplitude for modifications 2O and 2M$_2$, i.e.,

$$F\,(hkl) = F\,(h\bar{k}l) = (-1)^l F\,(\bar{h}kl).$$

Also, whatever the value of h, $l = 3l'$, and if h = 3h', then $l = 6l'$ as well.

A curious thing is observed in the case of modifications 3T and 6H. If these are formed with the aid of the displacements τ_+, τ_-, τ_0 ($\tau_X = 0$), then their respective $\Sigma |F|^2$ values for points superimposing upon one another on rotation about the c* axis may coincide with the same values of modifications 1M and 2M$_2$, respectively, these being formed from similar layers and displacements (see Tables 6, 10, 31, and 38). In the case of three-storied layer structures, this feature only applies when the layers are trioctahedral. If these modifications include the displacements τ_1, τ_2, . . . , τ_6 ($\tau_X \neq 0$), then it can be seen from Tables 31 and 38 that they possess independent values of $\Sigma |F|^2$. In this case, modification 3T can only have reflections with k = 3k' when they also have indices of $l = 3l'$.

It has often been pointed out in the literature (e.g., Zvyagin, 1954; Brindley, 1955; Brown, 1961), that strict repeat units can only be characterized by reflections with k $\neq$ 3k', since reflections with k = 3k' only determine the minimum c repeat unit. However, nowhere has it been specified that this is true only for structures with identical σ,τ components along the a axis, in particular for structures with identically oriented octahedra and with $\tau_X = 0$ (for example, in modifications 1M, 2M$_1$, and 3T). This follows directly from the above relationships for layers of the same parity (all even or all odd) [formula (91)], according to which such layers are indistinguishable when k = 3k', and also follows from the amplitude examples analyzed above for the different modifications.

In view of the relationships given above between the amplitudes of layers of different parity and the overall expressions for F for the modifications they form, structures made up of the layers σ_3 and σ_6, σ_1 and σ_2, and σ_5 and σ_4 (modifications 2M$_2$, 2O) will have minimum and strict repeat units which are independent of the index k and which are determined by reflections with h = 3h' and h ≠ 3h, respectively. If any other antiparallel layers alternate in the structure (modifications $2M_3$, 6H), then the minimum repeat unit will be fixed by reflections with h = 3h', k = 3k', and the strict unit by reflections with h ≠ 3h', k ≠ 3k'.

From this it follows that when loss of ordering like that noted above occurs, the reflections with k = 3k' and h ≠ 3h' which are left on the diffraction pattern do not always fix a minimum repeat unit of one layer, even along an inclined c-axis direction. For structures described by σ,τ values which separately have identical components along the y axis, they determine an orthogonal repeat unit which is not necessarily of one layer. In a sequence of layers of different parity, for example σ_{2n} and σ_{2n+1}, this unit includes two layers. In ordered structures this is the full or partial repeat unit, while in disordered structures it is a nonrigorous pseudo-repeat unit.

6. Effect of Variations in Chemical Composition on the Intensity Distribution in Layer Silicates

Apart from the geometrical features of a structure, an important factor affecting the distribution of reflection intensities is the scattering ability of the atoms as expressed by their atomic amplitudes f_j. It is of the utmost importance to work out the part played by this factor, and its relative significance, and to see to what extent it may affect the validity of the diagnostic diffraction features noted above for the different structure modifications.

According to the general expression for a structure amplitude,

$$F(\mathbf{H}) = \sum f_j e^{2\pi i (\mathbf{H} r_j)}.$$

It is immediately obvious that particular values of f_j will have a greater effect on the distribution of F(H) values, and thus on the intensities, the smaller is $|\mathbf{H}|$, i.e., the closer are the reflections to the center of the diffraction pattern. As $|\mathbf{H}|$ increases, the atomic scattering curves come closer together and the differences in f_j for different atoms become less. At the same time, F becomes more sensitive to very minor changes in r_j, which may distort the basic features of the intensity distribution, not only with regard to the chemical composition, but also where gross geometrical differences between different structure modifications are concerned.

Thus, in principle, the most effective pointers to the individual chemical features of layer silicates are the intensities of the close-order 00l basal reflections. It is an essential point here that their distribution is always independent of the layer silicate structure modification, as determined by σ, τ displacements perpendicular to the c* axis.

When using electron-diffraction texture patterns, which do not contain $00l$ reflections, it is necessary to fall back on the $02l$, $11l$ reflections of the first ellipse, which lie closest to the center, but in this case a given diagnostic criterion may only apply within a single structure modification.

Minerals made up of three-storied layers are particularly varied in their chemical compositions and isomorphous replacements. To work out the effects of isomorphous replacements on the intensity distributions, curves of $\Phi_3(hkz*)$ were constructed for hk combinations within the range $3h^2 + k^2 \le 52$ (see Table 14), for the intervals $0 \le z* \le 12$ when $c* = \frac{1}{30} A^{-1}$. From this, the following seven alternative types of chemical composition were derived for three-storied layers.

1. A dioctahedral layer with Al octahedra, and with no monovalent cations in the tetrahedral network hollows (analog of a pyrophyllite layer).

2. As alternative 1, but with Fe octahedra (analog of a nontronite layer).

3. As alternative 1, but with K cations on the surfaces of the layer and with $\frac{1}{4}$ of the Si atoms replaced by Al (analog of a muscovite layer).

4. As alternative 3, but with Fe octahedra (analog of a celadonite layer).

5. A dioctahedral layer with Mg octahedra, Si tetrahedra, and K cations on its surface (analog of a lepidolite layer).

6. A trioctahedral layer with Mg octahedra, and with no monovalent cations on its surface (analog of a talc layer).

7. As alternative 6, but with K cations on its surface and $\frac{1}{4}$ of the Si atoms replaced by Al (analog of a phlogopite layer).

Because of the large number of these curves, and because of their awkward size, they cannot all be reproduced here. We will only note that, as might have been expected, the curves for the different alternatives approach one another more and more closely as the distance from the center of the diffraction pattern is increased.

These curves cannot be compared directly with an experimentally observed intensity distribution, because the reflections are recorded in the form of combinations, with various phase factors which depend on the F formulas of the corresponding modifications. From the amplitude of oscillation of these curves it would only be possible to make a qualitative comparison of the average intensity of reflections on different ellipses.

A comparison of this sort leads to one rather unexpected conclusion. It has been found in practice from diffraction patterns of different layer silicates that in the trioctahedral varieties the reflections in the seventh ellipse ($17l$, $35l$, $42l$) are in general considerably weaker than those of the sixth ellipse ($26l$, $40l$), while in the majority of the dioctahedral varieties the reflection intensities on these ellipses are either roughly equal or show the opposite relationship to the trioctahedral case. Now this behavior is in complete disagreement with the theoretical curves, which give the sixth ellipse a much greater amplitude than the seventh, whatever the chemical composition of the layer.

Since the experimental intensity relationships do not appear to depend on the chemical composition of the layers, it is natural to assume that the cause of the observed discrepancy between the experimental and calculated data is to be found in differences between the ideal layer models used, with hexagonal patterns of tetrahedral networks, and the real structures. And, in actual fact, when the arrangement is modified to give ditrigonal patterns of tetrahedral networks with $\varphi = 19°$ ($\psi = 11°$, s = 0.171), the amplitudes of oscillation of the Φ_3(hkz*) curves increase considerably for the seventh ellipse.

Figure 35 shows three curves for a muscovite layer (alternative 3), for $\varphi = 0$, 19, and 30° (s = 0, 1.171, and 0.250), constructed for indices of hk = $\overline{17}$, $\overline{42}$; these illustrate the type of behavior described.

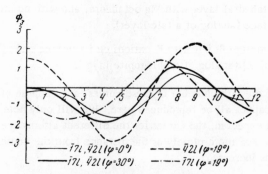

Fig. 76. Values of the Φ_3 amplitudes of the $\overline{17}l$, $\overline{42}l$ reflections, for different arrangements of the tetrahedral network in a three-storied layer.

Table 43. Values of ΣF^2 for the $02l$, $11l$ Reflections of 1M Modifications with Various Chemical Compositions of the Three-Storied Layers

l_{orth}	Alternative No.							hkl
	1	2	3	4	5	6	7	
0	1000	1300	425	625	400	565	165	020
1	185	85	5	2	25	440	110	110
2	470	680	130	250	110	200	15	$\bar{1}11$
3	5	40	140	250	105	25	20	021
4	5	40	40	5	60	20	170	111
5	320	485	675	900	640	140	400	$\bar{1}12$
6	105	40	310	185	360	255	550	022
7	460	640	800	1050	780	250	530	112
8	80	25	240	140	280	190	420	$\bar{1}13$
9	190	300	410	560	370	85	240	023
10	0	15	30	5	40	15	100	113
11	10	40	80	150	65	1	25	$\bar{1}14$
12	70	130	5	35	5	25	1	024

Thus, the experimentally observed relationship between intensities of reflections on the sixth and seventh ellipses may serve as an indication of the ditrigonal nature of the tetrahedral network.

This is shown to be a valid observation by the fact that when an ideal trigonal or hexagonal pattern is lost, to be replaced by a ditrigonal arrangement, the agreement between the Φ_3 values of reflections with h indices satisfying equation (87), but with $k \neq 3k'$, is also lost, and these reflections are described by different curves.

This behavior in itself shows how markedly the intensities of reflections on distant ellipses may be affected by distorting the geometrical structures of the layers.

Among the three-storied layer minerals, the modifications which are most widespread in nature and show the greatest diversity in the chemical composition of their layers are 1M and, to a lesser extent, the modification $2M_1$. Accordingly, for these two modifications, and for the above seven alternative types of chemical compositions of three-storied layers, values of ΣF^2 were calculated for the first ellipse on electron-diffraction texture pat-

Table 44. Values of ΣF^2 for the $02l$, $11l$ Reflections of Modification $2M_1$ for Various Chemical Compositions of the Three-Storied Layers

l_{orth}	Alternative No.							hkl
	1	2	3	4	5	6	7	
0	20	0	1	30	0.5	150	40	020
1	350	380	110	100	125	400	100	110
2	330	500	160	300	140	110	25	$\bar{1}11$
3	220	250	50	50	50	270	50	021
4	0	20	25	90	15	25	5	111
5	50	60	1	0	1	70	1	$\bar{1}12$
6	90	180	20	80	10	5	5	022
7	0	2	55	90	50	0	60	112
8	75	170	170	300	150	5	40	$\bar{1}13$
9	55	50	220	180	260	50	230	023
10	0	20	15	0	25	35	100	113
11	160	170	300	340	300	160	400	$\bar{1}14$
12	200	300	300	450	280	60	140	024
13	200	200	450	360	400	180	400	114
14	5	1	40	5	50	70	130	$\bar{1}15$
15	170	190	350	350	340	160	330	025
16	140	230	230	350	200	50	100	115
17	100	100	230	240	240	100	270	$\bar{1}16$
18	0	10	10	0	15	20	60	026
19	5	30	95	100	100	25	100	116
20	40	90	85	150	75	5	25	$\bar{1}17$
21	2	1	40	40	30	1	30	027
22	20	60	2	20	1	0	5	117
23	5	10	1	2	2	10	1	$\bar{1}18$
24	1	5	15	50	10	5	0	028

terns. Values of Φ_3(hkz*) from the previously prepared curves were used here, together with formulas (76) and (79). Rounded-off numerical values of ΣF^2 for first-ellipse reflections are given in Tables 43 and 44, and for convenience in comparing values corresponding to modifications 1M and $2M_1$, all values are scaled to the same number of layers in the structure.

We must make the reservation that some of the data in these tables may only correspond to the real structures rather approximately. For example, varieties without interlayer cations must differ in unit cell size from varieties possessing these cations, which must lead to certain changes in the intensity relationships. This applies when comparing dioctahedral and trioctahedral varieties. At the same time, it may be assumed that the values given for layers without monovalent surface cations correspond quite closely to ordered dehydrated montmorillonites with various chemical compositions, since their interlayer spaces, although occupied, do not make any appreciable contribution to the reflection intensities, and the cell heights are close to those used in calculation of the F values given in Tables 43 and 44. In a similar way, alternative No. 5, with incompletely filled octahedra, can be applied as an approximation to the case where a proportion of the octahedra are filled by Li atoms, which have small scattering abilities.

Whatever the validity of the above reasoning, in any event the data in Tables 43 and 44 will demonstrate the pure effect of complete replacement of certain atoms by others. It will be obvious that for only partial replacement of these atoms, the resulting changes in the $\Sigma |F|^2$ values will be correspondingly less marked.

Tables 43 and 44 show clearly that replacement of one atom by another has only a very minor effect on the distribution of intensities. A replacement which does have a greater effect than the others is when Fe is substituted for Al or Mg (l_{orth} = 2 and 8 in Table 43, l_{orth} = 2, 3, 9 in Table 44). When this replacement is only partial, the difference in intensities will obviously not be so great.

The intensities are affected much more significantly by the presence of additional atoms in the structure, and this serves to distinguish, for example, trioctahedral varieties from dioctahedral ones, or micas from montmorillonites.

From all this, we can conclude that variations in chemical composition do not significantly alter the relationships given above in Tables 32 and 39 for dioctahedral and trioctahedral varieties of the various layer silicate structure modifications. Among the $\Sigma |F|^2$ distributions derived, none was found to be similar to the distribution which would apply to any combination of other modifications. There is thus a reliable basis for the procedure of determining mixtures of modifications from anomalies in the intensity distributions of the $02l$, $11l$ reflections.

The diagnostic criteria given in Tables 43 and 44 may only be correct within the limits of a given modification with an ordered structure. In the case of structures which are disordered through displacements along the b axis, it is necessary to consider the intensities of the $13l$, $20l$ reflections on the second ellipse, since the $02l$, $11l$ reflections which lie closer to the center

Table 45. Values of ΣF^2 for the 13l, 20l Reflections of Modifications 1M, 3T, and 2M$_1$ for Various Chemical Compositions of the Three-Storied Layers

l orth	Alternative No.							hkl*
	1	2	3	4	5	6	7	
1	300	80	600	275	730	220	475	$\overline{2}$01
2	370	760	700	1210	635	475	845	200
4	220	45	50	1	85	155	20	$\overline{2}$02
5	1160	1750	690	1175	620	1325	815	201
7	110	320	10	115	5	160	30	$\overline{2}$03
8	190	440	420	770	360	260	525	202
10	55	205	180	410	135	90	235	$\overline{2}$04
11	85	250	20	115	5	130	35	203
13	15	8	70	5	85	3	40	$\overline{2}$05
14	570	900	505	1190	795	680	935	204

* The hkl indices correspond to modification 1M.

Table 46. Values of ΣF^2 for the 13l, 20l Reflections of Modifications 2M$_2$, 6H, and 2O for Various Chemical Compositions of the Three-Storied Layers

l orth	Alternative No.							hkl*
	1	2	3	4	5	6	7	
0	102	3	230	10	165	40	130	200,1$\overline{3}$1,13$\overline{1}$
3	565	565	1075	1080	1115	565	1080	201,1$\overline{3}$2,130
6	35	310	125	510	75	90	220	202,1$\overline{3}$3,131
9	1110	1110	575	575	590	1110	575	203,1$\overline{3}$4,132
12	530	1150	315	815	255	700	445	204,1$\overline{3}$5,133
15	20	20	175	175	165	20	175	205,1$\overline{3}$6,134
18	210	602	360	840	300	315	495	206,1$\overline{3}$7,135
21	4	4	25	25	25	4	20	207,1$\overline{3}$8,136
24	70	305	35	195	10	130	60	208,1$\overline{3}$9,137
27	310	310	510	510	545	310	510	209,1.$\overline{3}$.10,138
30	435	830	330	1010	550	575	725	2.0.10 1.$\overline{3}$.11,139

* The hkl indices correspond to modification 2M$_2$.

are smeared out into a continuous ellipse and cannot be separately distinguished.

For this reason, values of ΣF^2 were calculated for the above seven alternative chemical compositions of mica-type minerals, for the second ellipse on texture patterns, and these values are shown in Table 45 (modifications 1M, $2M_1$, 3T) and Table 46 (modifications $2M_2$, 2O, and 6H).

Both Tables 45 and 46 contain data for certain modifications with geometrical differences which are not shown up by the $13l$, $20l$ reflections. The values of l_{orth} and hkl given refer to modifications 1M (Table 45) and $2M_2$ (Table 46).

It follows from general considerations that the further these reflections lie from the center of the diffraction pattern, the less they show the effect of isomorphous replacements. It should, however, be noted that the weakest reflections remain particularly sensitive to such replacements. For example, it is well known that celadonite and glauconite give a 202 reflection of vanishingly small intensity (Fig. 61), which distinguishes them from, say, sericite (Fig. 62). When Table 45 is compared with the diffraction patterns of trioctahedral micas, this reveals the specimens which can be assumed to have incomplete filling of octahedra by Mg cations.

Fig. 77.   Texture pattern from pyrophyllite ($\varphi = 55°$).

Particular attention should be paid to the fact that the intensity distributions on the diffraction patterns of pyrophyllite (Fig. 77) and talc do not correspond with the data in Table 45 (the comparison cannot be made for the first-ellipse reflections, as these are smeared out into a continuous ellipse through loss of ordering in the structures). Particularly remarkable are the diametrically opposite intensity relationships of the reflections 202 and 201 (compare the diffraction patterns of pyrophyllite, Fig. 77, and sericite, Fig. 62). This indicates that the structures of these minerals differ considerably from those of the micas, in ways which have not yet been explained. The data given in columns 1 and 6 of Tables 43-46 cannot be directly applied to pyrophyllite and talc, but only to mica-type minerals with approximately the same compositions as these minerals.

It is necessary to point out that variations in chemical composition may also affect unit cell dimensions, which may sometimes be a more sensitive indication of the presence of particular chemically different varieties of layer silicates. A number of X-ray studies have been devoted to establishing functional relationships between the two types of variable involved (Mikheev,1958; Brown, 1961). This has always meant carrying out special investigations which require carefully selected single-crystal specimens of accurately known chemical compositions, reliably converted into structural formulas. As yet no such investigations have been carried out in the electron-diffraction field.

CHAPTER 5

EXPERIMENTAL ELECTRON-DIFFRACTION
STUDIES OF CLAYS AND RELATED MINERALS

1. The Determination of Crystal Structures

The basic concepts in the field of atomic structures of clay minerals were largely developed from the structural data analyses and brilliant hypotheses made by Pauling (1930a,b), Bragg (1937), and Belov (1947, 1949, 1950, 1951). These proposals have only in recent years been placed on a firmer footing through experimental investigations of large layer silicate single crystals, such as those of Mg vermiculite (Mathieson, Walker, 1954; Mathieson, 1958), dickite (Newnham, Brindley, 1956, Newnham, 1961), the chlorites (Steinfink, 1958), and others.

When structural electron-diffraction methods were developed, it became possible to investigate the structures of finely dispersed layer silicates and clay minerals with perfect structures directly. An account is given below of the results obtained in this way.

Structure determinations are very laborious, and before modern computers were perfected they took a great deal of time. It was therefore desirable to limit the investigations to the smallest number of specimens which would provide a representative sample of all the different types of clay mineral. It will be apparent that to do this it would be best to use varieties of minerals with perfect structures, which could be determined in the greatest detail and with the highest accuracy. It was not essential here that these minerals were in the actual form of clays. On the basis of these ideas, the objects chosen for study were triclinic kaolinite, celadonite, muscovite, and phlogopite—biotite, which had structures containing the characteristic features of minerals made up of two- and three-storied layers, dioctahedral and trioctahedral, with the cations Al, Mg, and Fe, and with various degrees of substitution of Si by Al in the tetrahedra.

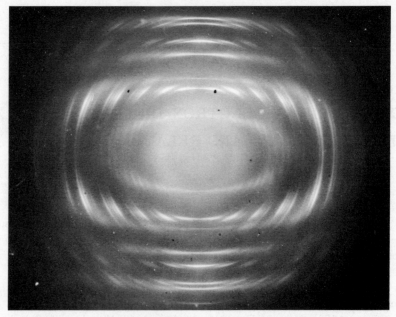

Fig. 78. Texture pattern from triclinic kaolinite ($\varphi = 65°$).

The basic sources of experimental material in all cases were electron-diffraction texture patterns, taken with multiple exposures and at various angles of inclination of the specimen to the electron beam. Where possible, additional photographs were also obtained of single crystal transmission patterns and of texture patterns by reflection.

In this way, a set of diffraction reflections was finally built up sufficient for the construction of a Fourier synthesis. In practice, it was only possible to construct syntheses of structure projections on the x0z and y0z planes, and difference syntheses for these projections. Only in the case of celadonite was it possible to construct a synthesis of a cross section of the structure in the x0z plane. The general methods used in structural studies which are applicable to clay minerals have been described above. Here we will consider particular instances of the investigation of individual specimens, and will note the structural results obtained.

Determination of the Structure of Kaolinite. The specimen investigated here was a sample of Turbov kaolinite (No. 553a in the collection of M. F. Vikulova, described in the "Systematic Manual," 1957), which had a perfect structure and gave a diffraction pattern of high quality.

Electron-diffraction texture patterns from kaolinite (Figs. 21, 78) have a characteristic special arrangement of reflections which gives the impression that the mineral is monoclinic, thus masking its triclinic nature. However, careful analysis of the reflection heights shows that their values can only be expressed in terms of the "triclinic" formula

$$D = hp + ks + lq,$$

although the majority of them can be described by a "monoclinic" formula (s = 0). Within the limits of experimental error, the values of p, s, and q satisfy the equations

$$p = \frac{5}{14}\, q, \quad s = \frac{p}{15} = \frac{q}{42}.$$

On the first ellipse the reflections $\overset{\pm}{11}l$ and $\overset{\pm}{11}\bar{l}$, which are separated by an interval of $\Delta D = 2\underset{\mp}{s}$, overlap one another and give the appearance of single reflections. Of the $02l$ reflections, $0\bar{2}2$ and 023 have extremely low intensities and may be missed if the diffraction pattern quality is not high enough, while the 022 and $0\bar{2}3$ reflections observed may be mistakenly accepted as "monoclinic" $02l$ reflections. It is only because of the pair of reflections $0\bar{2}1$ and 021, which are clearly separated by the interval $\Delta D = 4s$, that it is impossible to ignore the fact that α differs from $\pi/2$.

On the second ellipse the $\overset{\pm}{20}l$ reflections are arranged according to the "monoclinic" law $D = \pm 2p + lq$, with the $\bar{2}.0.(l+1)$ and $20l$ reflections separated by distances, measured along the height, of $\Delta D = 4p - q = 6q/14$. If s had been equal to 0, then between these, at $\Delta D = (p + lq) - [-2p + (l+1)q]$ $= 3p - q = q/14$ from the lower $\bar{2}.0.(l+1)$ reflection, the reflection $13l$ would have been found, and similarly, at $\Delta D = q/14$ from the upper $20l$ reflection, the reflection $\bar{1}.3.(l+1)$. However, since $s \neq 0$ and $3s = q/14$, these intermediate reflections are split, with $13l$ superimposed on $\bar{2}.0.(l+1)$ and $\bar{1}.3.(l+1)$ on $20l$, while between these, separated equally by distances of $\Delta D = 2q/14$, the reflections $13l$ and $\bar{1}.3.(l+1)$ appear (see Table 17). The set of four reflections formed here could be taken as the "monoclinic" group $\bar{2}.0.l+1, 13l, \bar{1}.3.l+1, 20l$. It is, however, easy to check that their heights could not be obtained from the equation $D = hp + lq$ for any single values of p and q. On the other hand, if the system of two similar equations for the outer reflections gives the above values for p and q, the system of equations for the inner reflections

$$lq' + p' = lq + p + 3s \quad \text{and}$$
$$(1 + l) \, q' - p' = (1 + l) \, q - p - 3s$$

gives

$$q' = q, \quad \text{but} \quad p' = p + 3s = \frac{6}{5} \, p.$$

According to the similarity principle, this arrangement of reflections is also repeated on the other 2h.0.l, h.3h.l ellipses. On the fifth ellipse reflection triplets appear which could by analogy be taken as the monoclinic reflections:

$$\bar{3}.3. \, l + 1, \quad 06l, \quad 3.3. \, l - 1.$$

In actual fact they are the six reflections, run together in pairs,

$$0\bar{6}l, \quad \bar{3}.\bar{3}. \, l + 1; \quad \bar{3}.3. \, l + 1, \quad 3.\bar{3}. \, l - 1; \quad 06l, \quad 3.3. \, l - 1,$$

which is shown in similar fashion by the numerical values of their heights.

Other ellipses show both isolated reflections, which like 02$\overset{\pm}{l}$ are directly due to the triclinic cell, and also superimposed reflections.

When all these factors have been cleared up, equations (47)-(50) of Chapter 3 can be applied without difficulty to give the unit cell constants of kaolinite; these were found to have the following values: $a = 5.13$, b = 8.89, c = 7.25 A, $\alpha = 91°40'$, $\beta = 104°40'$, $\gamma = 90°$. In its basic features this result agrees with the cell obtained by X-ray methods (Brindley, 1961). There is a difference for b (in the X-ray results, b = 8.93 A and b /$a \neq \sqrt{3}$).

The intensities were evaluated visually from a series of photographs with multiple exposures. The coincident reflections noted above introduced an unavoidable error into the intensity determinations.

Because of this factor, it was only considered realistic in the past to construct the two-dimensional Φ^2 and Φ series on the coordinate planes.

The superimpositions noted for the reflections used to construct these projections, i.e., 2h.0.l and h.3$\bar{\text{h}}$.l +1, 0.6n.l, and 3n.3n.l ± 1 (n = ±1, ±2, ...), can be taken into account using the relationships given above (in Section 5) of Chapter 4) between the layer silicate structure factors for these reflections.

Values of $|\Phi|$, calculated from I using Vainshtein's formulas (1956), are given in Table 47 for the reflections h0l and 0kl. To obtain the preliminary data necessary for constructing the initial model of the kaolinite structure, the $|\Phi|^2$ series were first calculated in their projections on the x0z and 0yz planes. However, because the I distribution was affected by lack of structural perfection in the kaolinite, it was only possible to work out the orientation of the octahedral network from these projections. Other important features such as the orientation of the tetrahedral network and the way it was joined to the octahedral network could only be established by comparing the experimental intensities with those calculated for the corresponding ideal models, although these models, with reliability factors of R $\sim$ 50%, differed considerably from the real structures. Moreover, the intensity relationships of some reflections turned out to be extremely sensitive to variations in the very structural features being investigated. Thus, the relationship noted above between the intensity relationships of the 0$\underset{+}{2}$2 and 0$\underset{+}{2}$3 reflections is only valid for displacements of the tetrahedral network relative to the octahedral network by Δ = $-b/3$ away from the position where their planes of reflection coincide, which for the coordinate axes chosen ($\alpha > \pi/2$) applies to the layer σ_4. If $\Delta = b/3$, then the opposite position applies, and the reflections 022 and 0$\bar{2}$3 are considerably weaker than 0$\bar{2}$2 and 023. When Δ = 0, the reflection intensities in each pair are not significantly different.

After these points had been established, thus particularizing the general principles of two-storied layer formation for the case of kaolinite, and the sizes of the structural polyhedra as indicated by the unit cell dimensions had been taken into account, the initial model of the kaolinite structure was constructed. In accordance with the triclinic nature of the unit cell, it consisted of polar layers with neither centers nor planes of symmetry. This model was refined by constructing a succession of Fourier projections of the potential on the x0z plane (Fig. 79) and 0yz plane (Fig. 80). The most effective projection here was that on the 0yz plane (Figs. 80, 81), since on this most of the reflections appeared individually. In particular, this projection showed directly the configuration of the ditrigonal pattern formed by the tetrahedron and octahedron bases. The parameter s defining this pattern could be measured from the deviations of the oxygen peak coordinates from the close-packing positions (see Figs. 80, 81), i.e., $\Delta y = s/3$, or from the corresponding change in the y coordinates of the projection of a base edge (e.g., $O_4 - O_6$), i.e., $\Delta y = 2s/3$.

Table 47. Calculated and Experimental (Initial and Scaled) Values of Modulus $|\Phi|$ for $0kl$ and $h0l$ Reflections Recorded on Kaolinite Electron-Diffraction Texture Patterns

hkl	$|\Phi_c|$	$|\Phi_e|$	Scaled $|\Phi_e|$
020	10.7	23.7	11.8
022	11.5	9.2	12.2
024	4.3	2.35	3.15
021	12.6	11.8	15.8
023	3.1	1.1	1.45
$02\bar2$	2.7	1.65	2.2
$02\bar4$	3.1	2.35	3.15
$02\bar1$	18.8	15.0	20.0
$02\bar3$	8.75	5.25	7.0
040	1.4	6.7	3.35
042	5.75	4.2	5.6
044	8.3	4.75	6.35
041	7.95	3.0	4.0
043	5.2	2.4	3.2
$04\bar2$	6.55	5.25	7.0
$04\bar4$	4.75	3.0	4.0
$04\bar1$	1.15	1.85	1.55
$04\bar3$	6.1	5.25	7.0
$04\bar5$	4.85	1.1	1.45
060	19.8	39.5	19.5
062	5.7	4.1	5.5
064	5.75	4.1	5.5
066	5.5	5.0	6.7
068	5.35	3.65	4.9
0.6.10	2.85	1.3	1.75
061	4.5	6.5	5.5
063	3.45	1.3	1.75
065	4.55	4.5	6.0
067	5.0	0.85	1.15

hkl	$|\Phi_c|$	$|\Phi_e|$	Scaled $|\Phi_e|$
$06\bar2$	7.1	6.5	8.7
$06\bar4$	7.0	5.0	6.7
$06\bar6$	5.35	4.1	5.5
$06\bar8$	6.2	3.65	4.9
$0.6.\overline{10}$	3.2	2.95	3.95
$06\bar1$	8.75	9.2	7.75
$06\bar3$	1.9	1.3	3.9
$06\bar5$	3.45	3.65	4.9
$06\bar7$	4.8	1.95	2.6
080	3.35	6.7	3.35
082	2.35	1.5	2.0
084	2.25	1.1	1.45
081	2.15	1.5	2.0
083	2.1	1.5	2.0
085	1.15	1.85	2.45
$08\bar2$	1.75	1.85	2.45
$08\bar4$	1.15	0.85	1.15
$08\bar1$	3.6	4.75	4.0
$08\bar3$	2.1	1.5	2.0
$08\bar5$	1.5	1.1	1.45
0.10.0	2.55	6.5	3.25
0.10.2	2.1	1.2	1.6
0.10.1	2.35	1.65	1.6
0.10.3	2.6	2.35	2.2
$0.10.\bar2$	1.75		3.15
$0.10.\bar1$	0.9	2.35	2.0
$0.10.\bar3$	3.7	2.35	3.15
0.12.0	6.1	6.1	3.1

hkl	$|\Phi_c|$	$|\Phi_e|$	Scaled $|\Phi_e|$
200	14.8	27.8	13.1
202	7.05	8.5	
204	17.5	18.4	
206	5.1	3.15	
201	6.8	11.0	8.0
203	11.6	14.0	
205	2.35	3.15	
$20\bar2$	19.0	27.0	19.4
$20\bar4$	13.8	17.2	
$20\bar6$	8.25	7.0	
$20\bar8$	3.1	2.2	
$20\bar1$	11.9	29.5	14.0
$20\bar3$	16.9	18.0	
$20\bar5$	5.25	2.75	
$20\bar7$	3.7	2.75	
$20\bar9$	6.3	13.9	
400	10.2	8.1	10.1
402	9.5	8.1	
404	2.6	7.65	
401	9.2	0.75	
403	1.5	0.75	
405	1.0		
$40\bar2$	9.95	19.8	9.3
$40\bar4$	4.35	5.4	
$40\bar6$	10.05	6.95	
$40\bar8$	3.5	1.0	

hkl	$|\Phi_c|$	$|\Phi_e|$	Scaled $|\Phi_e|$
$\bar201$	8.75	19.8	9.3
$\bar203$	3.35	4.4	
$\bar205$	7.2	7.65	
$\bar207$	4.85	3.1	
600	4.95	2.4	
602	2.85	1.7	
601	6.05	3.8	
603	3.1	2.95	
$\bar602$	1.35	10.1	7.3
$\bar604$	3.9	1.7	
$\bar606$	3.05	2.4	
$\bar601$	1.1	2.95	
$\bar603$	1.9	1.7	
$\bar605$		1.7	
$\bar607$		1.7	

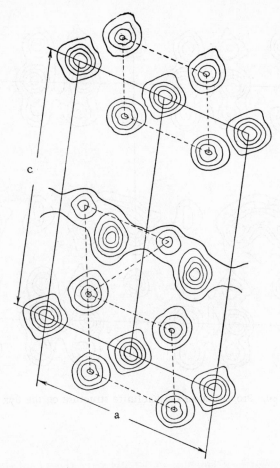

Fig. 79. Projection of the kaolinite structure on the
x0z plane.

To fix the positions of the maxima which overlapped in the projection
on the 0yz plane (O_{tetr},Si), the corresponding difference syntheses were calculated.

In calculation of the structure amplitudes corresponding to different
structure models, the origin was chosen at the most symmetrical point — the
center of an empty octahedron. Because of the lack of a symmetry plane in

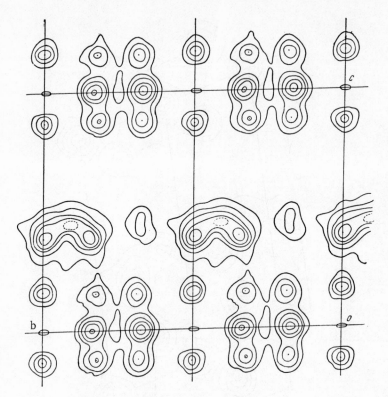

Fig. 80. Projection of the kaolinite structure on the 0yz plane.

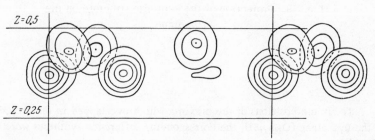

Fig. 81. Combined pattern of difference syntheses for the atoms
O$_{tetr}$ and S.

the layer, and because $\alpha \neq \pi/2$, it was necessary to use the most primitive form of the structure factor, that applicable to the group C1—P1, and carry out a summation over the 13 independent atoms given by the kaolinite formula of $Al_2Si_2O_5(OH)_4$. However, while the approximate models still contained octahedron and tetrahedron bases which could be taken as equilateral triangles lying in the same plane, with the Al and Si atoms lying on planar noncentered hexagonal motifs, the calculations could make use of the internal symmetry (C_{1V}^3 = C11m) possessed by the atoms lying in a plane parallel to xy0. In the system of coordinates of a planar network of this type, the structure factor has the form

$$A = 4\cos^2 2\pi \frac{h+k}{4} \cos 2\pi hx \cos 2\pi ky,$$

$$B = 4 \cos^2 2\pi \frac{h+k}{4} \sin 2\pi hx \cos 2\pi ky.$$

In the system of coordinates of the whole layer the atoms take on z coordinates, and the x and y coordinates undergo changes, which depend on the inclined cell:

$$\Delta x = g_1 z, \quad \Delta y = g_2 z, \qquad \qquad \text{(III,25)}$$

where

$$g_1 = -\frac{c}{a}\cos \beta, \qquad g_2 = -\frac{c}{b}\cos \alpha,$$

and also on the tetrahedral network displacements $\Delta y = -\frac{1}{3}$. These changes show up in the form of extra terms in the argument hx, so that this takes the form:

$$h(x + g_1 z) + k(g_2 z \pm n/3) + l'z \quad (n = 1 \quad \text{or} \quad 0),$$

where l' is obtained from the equation $D_{0kl}^{\pm} = lq \pm ks = l'q$, since $l' = l \pm \Delta l$, $\Delta l = ks/q$, in accordance with the fact that within its own system of coordinates each individual layer has $\alpha = \pi/2$. In this way, the number of terms in the structure factor is lowered to seven, or even five.

In the process of approximating to the true structure, explanations will be found for individual anomalies in the intensity distribution as shown by extra weakening of reflections distant from the ellipse minor axes in comparison to those lying closer, and more directly so by neighboring ones when

Table 48. Coordinates of Atoms in the Kaolinite Structure

Atom	x	y	z
O1 (H)	−0.223	0.175	−0.128
O2 (H)	−0.696	−0.003	−0.136
O3 (H)	−0.723	0.321	−0.128
O4	−0.263	0.322	0.155
O5	−0.304	0.004	0.157
O6 (H)	−0.763	0.186	0.155
O7	−0.385	−0.105	0.455
O8	−0.209	0.177	0.475
O9	0.112	−0.021	0.454
Al1	−0.500	0.171	0.002
Al2	−0.000	0.333	0.000
Si1	−0.195	0.002	0.384
Si2	−0.195	0.339	0.386

Table 49. Interatomic Distances in the Kaolinite Structure (Å)

Atoms	Å	Atoms	Å	Atoms	Å
O1 — O2	2.88	O5 — O1	2.71	O8 — O7	2.66
O1 — O3	2.88	O6 — O2	2.75	O8 — O4	2.63
O4 — O6	2.84	O8 — O5	2.66	O13 — O4	2.58
O5 — O4	2.84	O9 — O5	2.64	O14 — O4	2.65
O5 — O2	2.52	O7 — O5	2.50	O10 — O7	3.04
O4 — O1	2.42	O9 — O7	2.62	O11 — O9	2.90
O4 — O3	2.68	O8 — O9	2.58	O12 — O8	2.92
Al1 — O1	1.88	Al2 — O15	1.92	O8 — Si1	1.68
Al1 — O2	1.94	Al2 — O16	1.88	O9 — Si1	1.58
Al1 — O3	1.86	O4 — Al2	1.96	Si2 — O4	1.64
O4 — Al1	1.92	O17 — Al2	1.96	O8 — Si2	1.63
O5 — Al1	2.00	O18 — Al2	2.02	O13 — Si2	1.54
O6 — Al1	1.95	Si1 — O5	1.58	O14 — Si2	1.60
Al2 — O1	1.87	O7 — Si1	1.56		

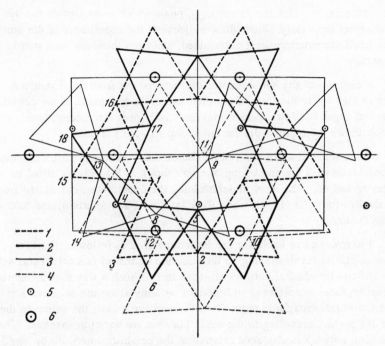

Fig. 82. Diagram of the kaolinite structure in a normal projection on the xy0 plane. (1) Lower octahedron faces; (2) upper octahedron faces; (3) tetrahedron faces; (4) lower octahedron faces of the next layer; (5) Si atoms; (6) Al atoms.

they are different for the reflections h0l and 0kl. Within each of these areas the reflections can be divided up roughly into three groups which differ in their scaling factors required to convert experimental $|\Phi_e|^2$ to calculated $|\Phi_c|^2$. Thus, if we take the scale of values for the h0l reflections with large D as our standard, then to convert the other reflections to the same scale, the $|\Phi_e|^2$ values of reflections with small D, and with D close to zero, must be multiplied by 0.51 and 0.22 respectively for the h0l reflections, or 0.71 and 0.25 respectively for the 0kl reflections, while the 0kl reflections with large D, on the other hand, must be multiplied by 1.78. The adjusted values are shown in the third column of Table 47. The probable origin of these anomalies has been discussed above (in Chapter 4).

Figures 79 and 80 show the final Fourier projections of the potential on the x0z and 0yz plane, while Fig. 81 represents a combined pattern of difference syntheses, in which the Si and O_{tetr} peaks which coincided on the 0yz plane appear separately. From these projections the coordinates of the atoms in the kaolinite structure were determined, and from these the interatomic distances.

A diagram of the kaolinite structure worked out from these results is shown in Fig. 82, in the form of a projection on the xy0 plane of the combined pattern of upper and lower octahedron faces and upper tetrahedron faces, with symbols inside these polyhedra for the Si and Al atoms present.

Numerical values of the atomic coordinates and interatomic distances, in accordance with the numbering of atoms shown in Fig. 82, are listed in Tables 48 and 49. The error in determining the coordinates, calculated from Vainshtein's formulas (1956), was 0.02 A for the Al and Si atoms, and 0.03 A for the O atoms.

The structure of kaolinite can be summarized as follows: The two-storied kaolinite layers making up the structure consist of one octahedral network and one tetrahedral network attached to it in such a way that the upper octahedron faces are oriented with corners pointing along the negative a axis, while the tetrahedron faces point along the positive a axis; the anions in the layer are packed according to the cubic law, but are not close-packed. The tetrahedral network is displaced relative to the octahedral network by $-b/3$ away from the position where their planes of symmetry would coincide, which destroys the plane of symmetry perpendicular to the b axis for the layer as a whole, as shown by the layer σ_4. From the positions which would correspond to close packing of the anions, the lower and upper octahedron faces are rotated through an angle φ of 3 and 5°, respectively, while the tetrahedron faces are turned through 20° on average ($\psi = 10°$) in such a way that their corners are displaced by different amounts. The shared edges of the octahedra are shortened, and the octahedra as a whole are somewhat compressed. The Al atoms are displaced toward the lower OH faces, and the Si atoms toward the bases of their tetrahedra, while the atoms of the polyhedron bases do not lie in a single plane, but have varying z coordinates. The closest atoms of successive layers are grouped into O−OH pairs, which vary somewhat in their lengths, however. The layers in the structure are arranged according to the formula $\sigma_4\tau + \sigma_4$ (structure I,2 in Table 6).

These structural results for kaolinite are in qualitative agreement with those derived from an X-ray study of dickite (Newnham, Brindley, 1956;

Newnham, 1961), but they differ quantitatively to some extent. These differences mainly affect the z coordinates of the O atoms and the configuration of the tetrahedron base pattern.

Calculated on the basis of the h0l and 0kl reflections (without 00l), the electron-diffraction results correspond to an R factor of 21%, while that from the X-ray investigation of dickite was 23.5%.

There is also satisfactory agreement with the results of an X-ray determination of the structure of kaolinite, carried out by Drits and Kashaev (1960).

In view of the anomalies noted above in the reflection intensities, the features described are not maintained rigorously throughout the whole structure because a strict three-dimensional periodicity is not observed in the structure. This is reflected to some extent in the projections obtained. Thus, from the residual Si maximum when y $\simeq \frac{1}{3}$, and from the height distribution of other peaks, it may be deduced that the specimen under study contains, besides its σ_4 layers, also the layers σ_2 and σ_6. From this it may be concluded that the loss of three-dimensional periodicity in the structure is largely due to variation in the structures of the layers, which differ either grossly, for example in the coupling of the networks, or minutely, in insignificant deviations of atoms from their most probable positions.

Determination of the Structure of Celadonite. The basic material for the investigation consisted of some electron-diffraction patterns of celadonite specimens prepared by Lazarenko (1956), and Malkova (1956) (who gave a detailed mineralogical description in their publications), and also some patterns prepared by V. P. Shashkina. The diffraction pattern of highest quality was obtained from Malkova's specimen 43-V. From the same specimen, the structural-chemical formula of celadonite was derived.

Celadonite gives a high-quality electron-diffraction pattern (Fig. 83), with reflections equally clear and sharp for any value of k, which indicates an extremely perfect structure. The external appearance of celadonite crystals is shown by Fig. 84. The unit cell of celadonite was found to have the constants a = 5.20, b = 9.00, c = 10.25 A, β = 100.1°.

According to the c constant, which is simultaneously a strict and a minimum repeat unit, celadonite consists of three-storied silicate layers arranged in ordered fashion with a repeat unit of one layer. Initially, syntheses of projections of the Φ^2 series were constructed on the x0z and y0z planes (Fig. 85).

Table 50. Experimental and Calculated Values of Φ for the $h0l$ and $0kl$ Reflections of Celadonite

hkl	$|\Phi_e|$	Φ_c	hkl	$|\Phi_e|$	Φ_c	hkl	$|\Phi_e|$	Φ_c	hkl	$|\Phi_e|$	Φ_c
200	38.4	35.2	$\bar{4}$02	25.9	19.2	022	13.5	16.7	063	3.8	2.8
202	23.9	24.2	$\bar{4}$06	12.4	20.0	024	4.4	−4.8	065	9.6	−10.0
204	29.3	26.0	$\bar{4}$01	21.0	−17.8	021	13.5	−16.2	067	12.3	10.2
206	11.8	−6.2	$\bar{4}$03	16.0	16.8	023	21.4	−28.7	069	2.5	2.2
201	27.6	22.8	$\bar{4}$05	10.7	13.5	025	3.9	0.2	0.6.11	3.6	7.0
203	6.7	14.8	600	5.7	7.5	040	3.5	5.0	080	1.6	−1.5
205	4.7	4.0	602	1.2	0.5	042	1.3	0.8	084	3.0	−4.5
$\bar{2}$02	7.7	−5.2	604	1.7	−0.8	044	14.2	−11.0	086	3.6	6.0
$\bar{2}$04	14.6	16.2	606	3.2	8.0	046	6.4	10.7	088	2.0	−1.5
$\bar{2}$06	22.7	28.0	603	1.5	−2.5	041	12.6	−17.2	081	3.6	−9.0
$\bar{2}$08	7.7	5.8	605	3.4	5.2	043	2.2	−4.0	083	1.3	−1.2
$\bar{2}$01	24.2	−16.8	$\bar{6}$02	11.2	12.5	045	7.8	−10.5	085	1.3	−2.8
$\bar{2}$03	13.0	12.0	$\bar{6}$04	1.2	0.2	047	9.1	−14.0	087	3.0	−8.3
$\bar{2}$05	12.4	−6.2	$\bar{6}$06	5.7	8.5	060	37.0	32.5	089	2.0	1.5
$\bar{2}$07	22.7	25.0	$\bar{6}$08	2.3	−1.5	062	5.4	5.0	0.10.0	6.5	−1.5
400	6.6	1.5	$\bar{6}$01	2.6	5.5	064	10.2	9.5	0.10.2	3.2	2.1
402	5.4	−1.8	$\bar{6}$03	6.7	3.0	066	3.4	0	0.12.0	6.8	15.0
404	11.8	20.0	$\bar{6}$07	6.1	2.5	068	11.7	11.0	0.12.2	3.6	0.8
406	4.1	5.0	$\bar{6}$09	2.5	−7.5	0.6.10	4.3	11.2	0.12.1	6.8	2.8
401	1.4	10.0	$\bar{6}$.0.10	4.7	10.0	061	17.2	11.8	0.12.3	4.2	4.2
405	4.1	8.0	020	31.6	−28.2						

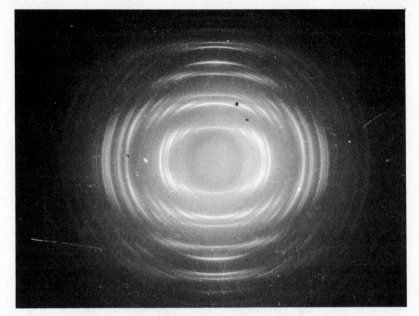

Fig. 83. Texture pattern from celadonite ($\varphi = 55°$).

These syntheses showed crudely the main features of the mineral's structure, and these were used as the basis of an approximate structure model.

In calculating the theoretical structure amplitudes the approximate formula $K_{0.8}(Mg_{0.7}Fe_{1.4})[Al_{0.4}, Si_{3.6}O_{10}](OH)_2$ was used, which neglected small quantities of other atoms, and had the $(H_3O)^*$ ions replaced by K; the structure model had a symmetry of C2/m and represented modification 1M. This model was then refined by successive processes of constructing syntheses of projections of the structure on the x0z and 0yz planes, and preparing difference syntheses for these projections and a synthesis of a section of the structure across the x0z plane, until such time as the R factor no longer improved. The calculated and experimental values of Φ used for this are given in Table 50. The final projections of the celadonite structure on the coordinate planes x0z and 0yz are shown in Figs. 86 and 87, the atomic coordinates are given in Table 51, the interatomic distances in Table 52, and the final diagram of the structure in a normal projection on the xy0 plane in Fig. 88.

Table 51. Coordinates of Atoms in the Celadonite
Structure

Atom	x	y	z
Mg, Fe	0	0.333	0
OH	0.383	0	0.114
O_1	0.366	0.328	0.109
Si, Al	0.432	0.333	0.285
O_9	—0.007	0	0.340
O_7	0.179	0.268	0.331
K	0.500	0	0.500

Fig. 84. Electron microscope photograph of celadonite
(magnification 10,000).

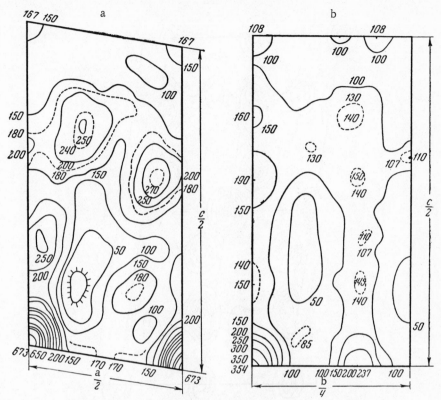

Fig. 85. Projections of syntheses of Φ^2-series for celadonite. (a) On the x0z plane; (b) on the 0yz plane.

Table 52. Interatomic Distances in the Celadonite Structure (Atoms Are Numbered as in Fig. 88)

$O_1 - O_2$	2.95	$O_7 - O_9$	2.60
$O_1 - O_3$	2.98	$O_8 - O_9$	2.64
$O_1 - O_4$	2.73	$O_1 - Mg, Fe$	2.03
$O_1 - O_5$	2.87	$O_3 - Mg, Fe$	2.06
$O_2 - O_3$	2.95	$O_2 - Si, Al$	1.78
$O_2 - O_4$	2.83	$O_7 - Si, Al$	1.71
$O_3 - O_5$	3.15	$O_9 - Si, Al$	1.61
$O_2 - O_7$	2.70	$K - O_9$	2.85
$O_2 - O_8$	2.81	$K - O_{10}$	2.96
$O_7 - O_8$	2.62 .		

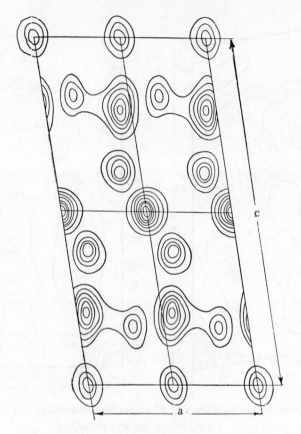

Fig. 86. Projection of the celadonite structure on
the x0z plane.

The celadonite structure shows distortion of the same type as that in kaolinite, but there are some quantitative differences. The angle of rotation of
the tetrahedra is $\varphi = 22°$ here ($\psi = 8°$), while the upper and lower octahedron
faces are turned through $\varphi \simeq 1.5°$. The cations are somewhat displaced from
their positions at the centers of the polyhedra, and the octahedra are rather
compressed in a direction perpendicular to the layers.

In should, however, be noted that these results have something of an
averaged character, since isomorphous replacements must lead to additional
deformations of the structure, although if these are randomly distributed they
may show up only as a loss of sharpness and a variation in height of the Fourier
synthesis peaks.

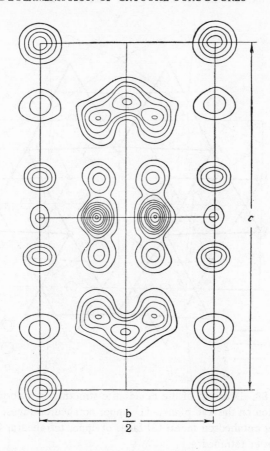

Fig. 87. Projection of the celadonite structure on
the 0yz plane.

Refining the Structure of Muscovite. In the form of an
idealized model, the structure of muscovite, $KAl_2[AlSi_3O_{10}](OH)_2$, was determined as long ago as 1930 (Jackson, West, 1930). Structural crystallographic
analyses of muscovite have appeared in Belov's works (1947, 1949, 1950,1951).
The arrangement of K cations in the interlayer spaces, and its dependence on
the configuration of the tetrahedral network, was considered by Yamzyn (1954).
Later, Gatineau and Mering (1958), using a one-dimensional projection, found
new z coordinates which showed, in particular, that the O_{oct}, Si, and Al_{tetr}
atoms assumed to lie in the same plane actually had different z coordinates.

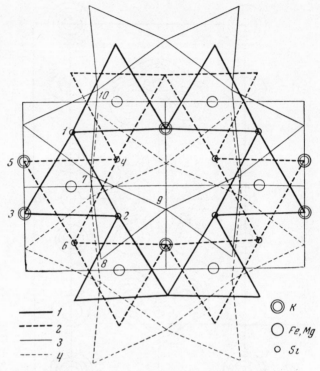

Fig. 88. Diagram of the celadonite structure in a normal projection on the xy0 plane. (1) Upper octahedron faces; (2)
lower octahedron faces; (3) faces of upper tetrahedra; (4) faces
of lower tetrahedra.

In spite of this, however, the muscovite structure still had many puzzles
left in it which could only be cleared up by a complete structure determination using Fourier analysis methods.

Such structure determinations were undertaken and carried out at approximately the same time by the electron-diffraction method (Zvyagin,
Mishchenko, 1960), and by the X-ray structural method (Radoslovich, 1960).
Electron-diffraction texture patterns from muscovite (Fig. 89) are notable for
their great clarity and contrast, weak background, and high resolution. In particular, they resolve the normally overlapping reflections $\overline{2}0l$, $1.3.l \pm 1$. They
also show the reflections $0.6.2l + 1$ and $\overline{3}.3.2l$, which are usually lacking on
layer silicate diffraction patterns.

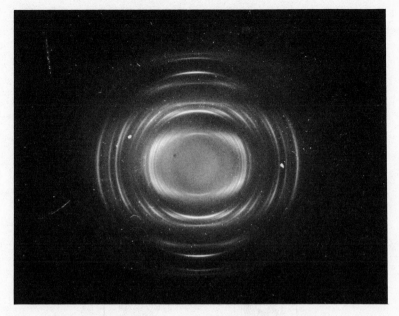

Fig. 89. Texture pattern from muscovite ($\varphi = 55°$).

From the texture patterns a unit cell was derived, with constants of $a =$ 5.18, b = 8.96, c = 20.1 A, β = 95°40'. According to Jackson and West's X-ray results (1930), a = 5.18, b = 9.02, c = 20 A, β = 95°30'. Radoslovich's latest results (1960) were a = 5.19, b = 8.996, c = 20.096 A, β = 95°11'.

From its two-layer repeat unit, the uniformity condition, and the non-polarity of its structure, the most natural structure for muscovite is one consisting of three-storied layers in which the planes of symmetry of the upper and lower (Si, Al)−O tetrahedral networks are displaced relative to the planes of symmetry of the Al−(O, OH) octahedral networks by amounts Δy equal to $1/_3$, $-1/_3$ for some layers, and $-1/_3$, $1/_3$ for other layers lying next to the first ones. In the notation we have used, these are the layers σ_2 and σ_4.

In view of the potassium cations distributed in the interlayer spaces (which means that adjacent tetrahedral networks must be coupled together in such a way as to form the appropriate hollows), the structure of muscovite and its unit cell, in particular its c-axis direction, are determined automatically and correspond to the space group $c_{2h}^6 = C2/c$.

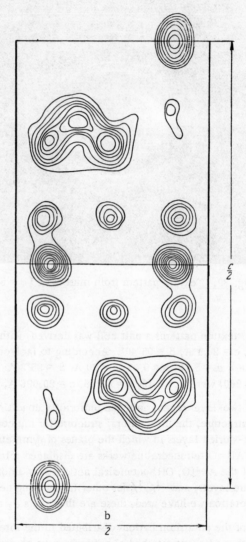

Fig. 90. Projection of the muscovite structure
on the 0yz plane.

This was the structure which resulted from the X-ray investigation of muscovite (Jackson, West, 1930), and the electron-diffraction patterns are in agreement with it.

If the origin of coordinates is chosen at one of the symmetry centers lying at the middle of an OH—OH octahedron edge, the vectors a and c will form an angle of $\beta \sim 95°$.

If the origin is chosen at one of the symmetry centers lying at the center of an empty octahedron, the vectors $a' = -a$ and $c' = c - a$ will form an angle of $\beta \sim 100°$. It is not difficult to see that the c planes of one cell are transformed into the n planes of the other, and vice versa.

In construction of the initial structure model, it was also assumed that the anions of the octahedron faces had a close-packed arrangement (s = 0), and the tetrahedron bases were rotated through an angle of $\psi = 12°$ (s = 0.158), in accordance with the equation $\cos \psi = a/2l'$, where l' is the approximate length of a tetrahedron edge. The z coordinates were taken from the X-ray data (Gatineau, Mering, 1958).

If the ditrigonal loops of adjacent layers were disposed with their centers lying exactly one above the other, then from the equation $a \cos \beta = -a/3$, an angle of $\beta = 94°55'$ would apply (see Belov, 1949). According to the experimental value of β, the K cation must be displaced by about $\Delta x = -0.013$, and its octahedron has an extra large taper.

When the experimental and calculated values of Φ are compared, it again turns out, as in the case of kaolinite, that they must be differently scaled for different values of hk$l$.

The muscovite structure was refined using projections of the structure on the x0z and 0yz planes (Fig. 90); since several of the peaks were run together, the corresponding difference syntheses were constructed.

The initial data for these syntheses are given in Table 53. The results correspond to an R factor of 23% for the h0l and 0kl reflections (without 00l and several unreliable reflections with $|\Phi_c| > |\Phi_e|$).

These syntheses resulted in an improvement of the coordinates of the atoms (Table 54), whereupon it became clear that the octahedron faces were rotated relative to one another about their normals by an angle $\varphi \sim 3°$ (s = 0.030), and the tetrahedron faces were rotated by $\varphi = 11°$ (s = 0.168). The probable errors in determining the coordinates, as calculated from Vainshtein's

Table 53. Experimental and Calculated Values of Φ for the h0l and 0kl Reflections in Muscovite

hkl	|Φe|	Φc	hkl	|Φe|	Φc	hkl	|Φe|	Φc	hkl	|Φe|	Φc	hkl	|Φe|	Φc
200	14.2	13.7	4̄02	13.8	−13.0	024	14.0	12.2	064	1.2	−1.8	0.8.10	2.0	1.7
202	7.2	7.1	4̄04	4.3	−4.3	026	2.4	−2.2	066	1.5	0.1	0.8.12	1.1	−1.6
204	3.6	−4.0	4̄06	4.3	3.4	028	1.2	1.0	068	4.5	−3.8	0.8.14	1.1	1.1
206	10.0	−7.7	4̄08	1.2	0.9	0.2.10	2.0	2.2	0.6.10	4.2	6.3	0.8.16	1.1	−0.1
208	6.5	7.8	4̄.0.10	1.3	−4.8	021	7.3	3.4	0.6.12	2.2	3.0	081	2.2	2.7
2.0.10	10.6	−12.3	4̄.0.12	8.2	9.0	023	11.9	−10.4	0.6.14	2.6	−2.1	083	2.1	−2.2
2.0.12	11.9	−14.3	4̄.0.14	5.9	7.2	025	15.0	−13.7	0.6.16	8.2	−7.5	085	0.7	0.5
2.0.14	4.5	−5.3	4̄.0.16	5.0	3.5	027	4.8	4.2	0.6.18	1.3	−0.1	087	0.7	0.1
2.0.16	2.6	2.3	600	5.8	−4.5	029	3.8	3.2	0.6.20	5.5	−4.3	089	0.5	−0.2
2.0.18	5.2	−6.0	602	1.8	1.9	040	7.2	−4.6	0.6.22	0.5	−6.0	0.8.11	3.0	−2.8
5̄02	17.0	−17.6	604	0.7	−0.3	042	—	1.8	061	—	1.0	0.8.13	4.2	3.6
5̄04	11.3	−11.8	606	1.4	−4.6	044	5.8	−5.9	063	0.7	0.2	0.8.15	1.1	−1.3
5̄06	16.8	−14.2	608	2.8	6.1	046	4.2	−4.7	065	0.6	−1.0	0.8.17	1.3	−1.0
5̄08	4.8	−3.8	6.0.10	4.1	1.3	048	1.6	−0.9	067	0.7	0.5	0.10.0	1.7	−0.5
2̄.0.10	14.6	−12.4	6.0.12	1.9	−4.7	0.4.10	1.6	−4.4	069	1.3	0.5	0.10.2	0.9	−0.7
2̄.0.12	4.8	−6.3	6̄02	5.8	−8.2	0.4.12	10.5	−7.7	0.6.11	0.6	−2.1	0.10.4	1.1	2.7
2̄.0.14	2.5	2.6	6̄04	1.5	−0.6	0.4.14	2.4	2.0	0.6.13	0.7	−0.1	0.10.6	1.2	1.3
2̄.0.16	10.1	9.5	6̄06	4.6	−3.5	041	6.5	−5.8	0.6.15	0.4	0.1	0.10.8	0.7	1.1
2̄.0.18	8.0	−5.8	6̄08	1.6	0.1	043	4.2	5.5	0.6.17	1.3	0.1	0.10.10	0.5	2.8
400	9.8	12.3	6̄.0.10	0	−0.7	045	1.1	−1.1	0.6.19	1.0	−0.3	0.10.12	2.6	3.8
402	5.3	6.7	6̄.0.12	0	0.7	047	1.1	−2.8	0.6.21	1.0	−0.3	0.10.1	1.8	0.4
404	1.2	1.5	6̄.0.14	7.1	2.0	049	1.1	0.8	080	2.4	−2.7	0.10.3	0.4	−1.8
406	2.3	5.5	6̄.0.16	0.8	0.5	0.4.11	3.4	6.2	082	—	−0.1	0.10.5	0.4	0.1
408	9.2	10.3	6̄.0.18	2.0	5.7	0.4.13	4.1	−7.2	084	0.3	−1.9	0.10.7	0.6	−0.3
4.0.10	2.9	−2.9	020	—	0.6	060	14.7	−18.2	086	0.5	−0.2	0.10.9	0.8	−0.8
4.0.12	2.9	−0.3	022	5.3	3.0	062	5.7	−5.0	088	0.5	0.9	0.10.11	1.2	−1.0
4.0.14	4.1	−2.1										0.12.0	2.2	6.8

Table 54. Coordinates of Atoms in the Muscovite Structure (Atoms Numbered as in Fig. 91)

Atom	x	y	z	Atom	x	y	z
Al_{oct}	0.250	0.090	0.000	Si_2	0.468	0.250	0.135
O_2	0.417	0.250	0.053	O_5	0.428	0.083	0.168
O_3H	—0.053	0.073	0.052	O_{5a}	0.430	0.095	0.175
O_4	0.447	—0.073	0.052	O_6	0.256	—0.190	0.162
O_{4a}	0.445	—0.073	0.048	O_7	0.756	—0.143	0.162
Si	0.468	—0.083	0.135	K	0.000	0.088	0.250
Al_{tetr}	0.470	—0.070	0.142				

formulas (1956), were 0.02 A for K, 0.03 A for Si, Al, and 0.04 A for O. Figure 91 represents a normal projection of one layer on the xy0 plane in the form of a combined pattern of tetrahedron and octahedron faces, but without the distortion caused by replacement of Si by Al.

Gatineau and Mering (1958) have established that Al atoms which replace Si atoms have somewhat greater z coordinates, and have their tetrahedra attached to O_{oct} atoms with smaller z coordinates, than those for the corresponding Si atoms. It was known previously, however, that there were four possible arrangements for these Al tetrahedra. According to the projection of the structure on the 0yz plane (Fig. 90) and the corresponding difference syntheses, the Al tetrahedra are joined only to O_4 atoms lying near the reflecting glide plane of the octahedral network which is parallel to the a axis, while the O_2 atoms lying on the mirror plane of this network parallel to the a axis are attached to Si tetrahedra only (Fig. 91).

Each set of four O_{oct} atoms consists of two atoms of each of two different kinds. In muscovite there is only one Al tetrahedron for each such set. The variability in the degree of replacement of Si by Al in layer silicates suggests that the Al atoms replace randomly among the two positions open to them. Moreover, if the Al tetrahedra were attached to any one given O atom, this would lead to infringement of the base-centering of the lattice and to the appearance of reflections with h + k = 2n + 1 on the diffraction patterns (particularly those from single crystals). However, no such reflections were observed in any of the layer silicates investigated, for example, by the microdiffraction method. From the projections obtained it follows that the randomness in the distribution of the Al_{tetr} atoms affects only two of the four positions indicated.

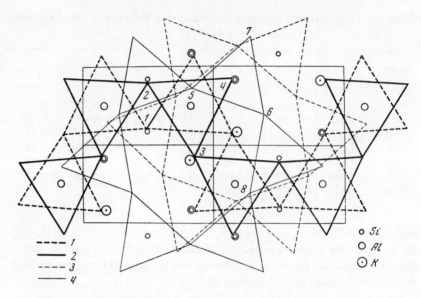

Fig. 91. Diagram of the muscovite structure in a normal projection on the xy0 plane. (1) Lower octahedron faces; (2) upper octahedron faces; (3) faces of lower tetrahedra; (4) faces of upper tetrahedra.

Table 55. Interatomic Distances in the Muscovite Structure (Atoms Numbered as in Fig. 91)

$O_2 - O_1$	2.60	$O_2 - Al$	1.92	$O_5 - O_6$	2.60	$O_6 - Si$	1.56
$O_3 - O_1$	2.86	$O_3 - Al$	1.97	$O_{5a} - O_6$	2.71	$O_6 - Al_T$	1.60
$O_2 - O_3$	2.90	$O_4 - Al$	2.00	$O_5 - O_7$	2.66	$O_{5a} - Si_2$	1.63
$O_6 - O_4$	2.71	$Si - O_4$	1.66	$O_{5a} - O_7$	2.76	$K - O_8$	2.90
$O_6 - O_{4a}$	2.76	$Al - O_{4a}$	1.88	$O_7 - O_6$	2.62	$K - O_5$	2.98
$O_5 - O_4$	2.90	$O_5 - Si_1$	1.65			$K - O_{5a}$	2.90
$O_{5a} - O_{4a}$	2.97	$O_{5a} - Al_T$	1.64				

This result is in agreement with Loevenstein's rule (1954) concerning different charges on adjacent tetrahedra.

In Table 54, the coordinates which depend on the replacement of Si by Al are given in two versions (those corresponding to Si replaced by Al are denoted by a letter "a"). The z coordinates agree essentially with those of

Gatineau and Mering (1958), but the coordinates for O_{tetr} have been corrected, those for O_{oct} differ only slightly, and the drop in z coordinates for OH has not been confirmed.

Table 55 shows the interatomic distances, with those affected by substitution of Al for Si being given in two versions also.

If we consider the results obtained, the mechanism of replacement of Si by Al becomes clear. Substitution of Al, for example in the $O_4O_5O_6O_7$ tetrahedron (Fig. 91), is expressed mainly by a shift upward and to the left of O_{5a}, with O_4 somewhat lowered and the positions of O_6 and O_7 remaining unaffected. The Al itself is displaced upward and to the left of the corresponding Si atom, while the displacement of O_5 forces the K to shift by $\Delta x = -0.013$, which is the cause of the extra taper in the K octahedron noted above and the increase in β by 1° above its theoretical value.

These deformations also bring about a loss in symmetry of the tetrahedral networks, so that, of all the relative arrangements of layers which would be possible if the deformations were absent, the only one which is still applicable is the arrangement of layers with symmetry C2/c and with a c reflecting glide plane parallel to the a axis.

In this case, the parts sticking up from the surface of one layer fit most naturally into the hollows in the surfaces of the next layer. Here, because of the nonuniform distribution of the Al tetrahedra, which differ in shape and size from the Si tetrahedra, but not because of the ditrigonal nature of the tetrahedral network, the diffraction patterns of muscovite contain the $06l$ reflections with odd l and the $\overline{33}l$ reflections with even l which were mentioned above.

The results obtained are essentially in agreement with those of Radoslovich (1960). In particular, our conclusions as to the nature of Si−Al replacement were the same. On the average, the values of φ, ψ for the rotation of octahedron and tetrahedron faces were the same. Radoslovich found the K atom positions more accurately, but the electron-diffraction results give a clearer picture of the characteristics of the Al tetrahedra. When it is considered that Radoslovich used about 900 reflections, appearing separately on single-crystal photographs, and that he used a high-speed computer, the measure of agreement obtained can be considered as evidence of the highly effective nature of the electron-diffraction approach, which with comparatively simple material (a limited number of reflections, and calculations using strips) gave the correct structural characteristics.

Table 56. Experimental and Calculated Φ Values for the h0l and 0kl Reflections of Phlogopite–Biotite

hkl	$\lvert\Phi_e\rvert$	Φ_c	hkl	$\lvert\Phi_e\rvert$	Φ_c	hkl	$\lvert\Phi_e\rvert$	Φ_c	hkl	$\lvert\Phi_e\rvert$	Φ_c	hkl	$\lvert\Phi_e\rvert$	Φ_c
200	22.5	25.4	400	2.5	−2.7	604	1.0	−2.4	022	13.2	13.0	$0.6.12$	0.8	0.3
202	20.2	19.5	402	1.8	−1.6	606	5.6	8.3	024	0.8	0.7	$06\bar{1}$	5.0	5.7
204	25.7	23.4	404	19.8	18.7	608	2.8	4.1	$02\bar{1}$	1.3	−2.7	$06\bar{3}$	1.2	−1.0
206	8.2	−7.6	406	5.3	5.0	$60\bar{2}$	10.6	11.8	$02\bar{3}$	8.8	−8.5	$06\bar{5}$	5.8	−4.6
208	9.0	9.0	$40\bar{2}$	17.7	18.3	$60\bar{6}$	5.3	6.6	040	4.0	5.8	$06\bar{7}$	4.6	4.2
$20\bar{2}$	6.8	−6.2	$40\bar{4}$	0.8	2.2	$60\bar{8}$	2.4	−3.8	042	1.8	2.1	$06\bar{9}$	1.2	−0.1
$20\bar{4}$	14.8	15.5	$40\bar{6}$	21.6	18.2	$6,0,\overline{10}$	5.0	7.4	044	5.4	−1.7	$0,6,\overline{11}$	4.4	4.3
$20\bar{6}$	15.1	17.7	$40\bar{8}$	3.8	−2.4	$6,0,\overline{12}$	4.9	6.8	046	7.8	7.8	080	2.0	3.4
$20\bar{8}$	7.6	7.5	$4,0,\overline{10}$	4.1	1.5	601	1.8	−1.8	048	2.0	−2.8	082	1.8	1.4
$2,0,\overline{10}$	4.2	8.5	401	5.3	9.0	603	1.8	−3.1	$04\bar{1}$	4.4	−5.6	084	1.4	−1.0
201	22.8	22.0	403	1.8	−4.1	605	2.4	2.7	$04\bar{3}$	3.0	2.7	086	3.7	3.9
203	5.9	5.5	405	10.2	8.8	609	3.8	5.6	$04\bar{5}$	4.0	−4.5	$08\bar{1}$	1.8	−4.8
205	5.2	7.0	407	2.8	3.1	$60\bar{1}$	3.8	3.6	$04\bar{7}$	4.5	−3.0	$0.10.0$	1.6	−1.4
207	8.6	−7.6	$40\bar{1}$	19.5	−19.4	$60\bar{3}$	6.8	6.9	$04\bar{9}$	2.8	2.9	$0.12.0$	6.6	8.0
209	2.8	−2.2	$40\bar{3}$	14.4	14.7	$60\bar{5}$	1.8	−1.6	060	15.0	14.2	$0.12.2$	2.4	−0.8
$20\bar{1}$	20.9	−17.5	$40\bar{5}$	6.0	9.2	$60\bar{7}$	3.2	−5.5	062	1.2	2.0	$0.12.4$	1.8	1.4
$20\bar{3}$	5.2	6.50	$40\bar{7}$	2.6	−3.2	$60\bar{9}$	6.0	6.55	064	3.7	4.6	$0.12.6$	1.8	1.1
$20\bar{5}$	5.0	−6.50	$40\bar{9}$	2.0	−0.8	$6,0,\overline{11}$	2.6	−1.4	066	2.8	−2.6	$0.12.\bar{5}$	2.4	−3.6
$20\bar{7}$	25.4	23.2	600	4.9	4.2	$6,0,\overline{13}$	3.3	4.8	068	6.2	6.8	$0.12.\bar{7}$	2.4	2.0
$20\bar{9}$	1.8	−5.0	602	2.0	2.6	020	6.4	−7.2	$0.6.10$	5.0	4.5			

Fig. 92. Diffraction pattern from phlogopite—biotite (trioctahedral modification 1M or 3T, $\varphi = 55°$).

Refining the Structure of Phlogopite – Biotite. The widely distributed trioctahedral mica varieties of the phlogopite—biotite type had not, until recently, been investigated by direct structural methods. In the literature, only a description of a rough theoretical model of fluor-phlogopite (Pabst, 1955) could be found. During 1961, an electron-diffraction study of this type of structure was completed (Zvyagin, Mishchenko, 1962). Subsequently, an X-ray study was made of an unusual phlogopite in which Si atoms were replaced by Fe^{3+} (Steinfink, 1962).

In electron-diffraction studies of different phlogopite and biotite specimens, the reflection intensity distributions on texture patterns did not show any individual differences for the separate varieties of these minerals, which are based on differences in chemical composition. This indicates that the isomorphous cations are randomly distributed among the permitted positions in the structure. Therefore, no matter what particular specimen is chosen to provide the diffraction pattern to be used as the experimental starting material, it is first necessary to investigate the properties which will be general for the whole of the phlogopite—biotite structure type.

The peculiar feature of electron-diffraction studies of the phlogopite—biotite structure is that the minerals belonging to this type can exist in several modifications, in particular, a monoclinic variety with a repeat unit of one layer (1M), and a trigonal modification with a repeat unit of three layers (3T), which, as we saw in general terms above (Section 4 of Chaper 4), cannot be distinguished from either their texture patterns or their X-ray powder photographs.

The fact that these diffraction patterns are the same means that no matter what modification any particular specimen actually belongs to, its electron-diffraction pattern can be used to determine the structure of both mica modifications.

It is convenient in practice to index the diffraction pattern for a monoclinic cell and determine the structure in its monoclinic version. Here the structure of an individual layer must be established, which will of course be the same as that calculated for the trigonal version. From this both the monoclinic and trigonal structures will be determined by the spatial disposition of the layers according to the symmetry and repeat unit of the individual modification.

The structure of phlogopite—biotite was investigated from the texture pattern (Fig. 92) given by specimen No. 312 from the collection of E. P. Sokolova. From photographs with multiple exposures, taken at various specimen-inclination angles, a set of 59 $h0l$ reflections and 40 $0kl$ reflections was built up (Table 56).

The distributions of the reflections on the diffraction pattern satisfied both a monoclinic cell with a = 5.28, b = 9.16, c = 10.3 A, β = 99°50', and a trigonal cell with c = 30.5 A.

The features of the atomic structure of a phlogopite—biotite layer were investigated from projections of the structure on the coordinate planes x0z (Fig. 93) and 0yz (Fig. 94) of the monoclinic cell. Here, the experimental structure amplitude moduli were assigned the signs of the amplitudes calculated from a structure model which was successively refined using the results of the preceding syntheses. In constructing the initial model, z coordinates were used which were taken from the structure of a Mg vermiculite layer (Mathieson, Walker, 1954; Mathieson, 1958), and the phlogopite formula $KMg_3[AlSi_3O_{10}](OH)_2$ was used.

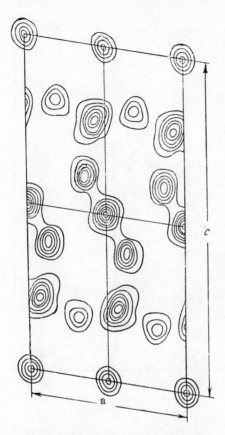

Fig. 93. Projection of the phlogopite
structure on the x0z plane.

After two cycles of refinement, the R factor, which had reached a value
of 17.2% for the h0l reflections and 20.4% for the 0kl reflections, would not
improve any further. Atomic coordinates for the phlogopite−biotite structure
corresponding to this level of refinement are given in Table 57, and interatomic distances are shown in Table 58. A diagram of the structure in a normal projection on the xy0 plane is shown in Fig. 95.

The probable error in the coordinates, determined from Vainshtein's
formulas (1956), were as follows: K, 0.02; Si, Al, Mg, 0.03; O, 0.04.

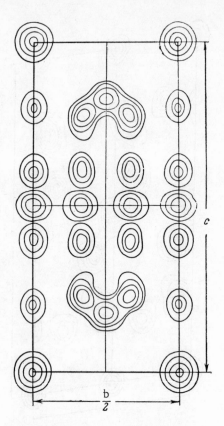

Fig. 94. Projection of the phlogopite
structure on the 0yz plane.

Table 57. Coordinates of Atoms in the Phlogopite Structure

Atom	x	y	z	Atom	x	y	z
Mg	0	0	0	Si, Al	0.423	0.333	0.271
Mg	0	0.333	0	O_{tetr}	0.176	0.265	0.325
O_{oct}	0.368	0	0.108	O_{tetr}	−0.033	0	0.315
O_{oct}	0.368	0.333	0.108	K	0.500	0	0.500

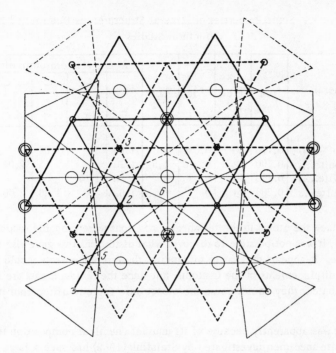

Fig. 95. Diagram of the phlogopite structure in a normal
projection on the xy0 plane (symbols used are as in Fig.88).

Table 58. Interatomic Distances in the Phlogopite Structure (Atoms Numbered
as in Fig. 95)

$O_1 - O_2$	3.05	$O_4 - O_5$	2.65	$(Si, Al) - O_2$	1.64
$O_1 - O_3$	2.80	$O_4 - O_6$	2.66	$O_4 - (Si, Al)$	1.63
$O_4 - O_2$	2.66	$O_5 - O_6$	2.65	$O_6 - (Si, Al)$	1.60
$O_6 - O_2$	2.59	$O_1 - Mg$	2.06	$K - O_4$	3.31
				$K - O_6$	3.11

According to these results, the real structure differs from the ideal model
in that the tetrahedra form a pattern with ditrigonal loops exhibiting a rotation of the tetrahedra about their body heights by $\psi \sim 5.5°$ away from the positions corresponding to strictly hexagonal loops, and the upper and lower octahedron faces do not show any rotations away from the positions corresponding
to cubic packing, although the octahedra themselves are strongly compressed.

Table 59. Some Properties of Mineral Structures, as Determined from Structural Studies

Mineral	s_{tetr}	φ_{tetr} (°)	s_{oct}	φ_{oct} (°)	Average interatomic distances				
					Si—O	Al$_T$—O	Al$_0$—O	Fe, Mg—O	K—O
Kaolinite......	0.174	20	0.038	4	1.60		1.93		
Celadonite....	0.190	22	0.016	1.5	1.70			2.04	2.92
Muscovite.....	0.168	19	0.030	3	1.60	1.67			2.93
Phlogopite....	0.208	24.5	0	0	1.62		1.96	2.06	3.11

Since the atoms in the tetrahedral positions were probably randomly distributed, it was not possible to discover what distortions this gave rise to in the structure. It appears that, as in muscovite, these replacements should cause certain displacements of the O atoms, but since the replacements are randomly distributed, they do not cause the monoclinic angle to differ from its ideal value.

It was apparently because of its unusual chemical composition that the phlogopite specimen investigated by Steinfink (1962) had such a large unit cell (a = 5.36, b = 9.29, c = 10.41 A, β = 100°0'), and large angle $\psi \simeq 12°$. The tetrahedral Fe and Si atoms present in it were randomly distributed.

As a summary of the results described, we give a table of the average interatomic distances and rotations of octahedron and tetrahedron faces found in the minerals considered (Table 59).

2. Uncertainty in the Nacrite Structure

Nacrite is a very rare layer silicate of the kaolinite group. It is found in only a few deposits in the whole world, and until recently was not known to occur in the Soviet Union. When examining the question of the occurrence of nacrite within the USSR, Nakovnik (1941) showed that the minerals accepted as nacrite in the Soviet Union were not in fact nacrite, but were dickite or kaolinite. Subsequently, some erroneous determinations of nacrite were discovered by Kovalev (1947). Because of these circumstances, for a long time it was not possible to carry out an electron-diffraction study of the mineral. Only in 1959 was a small specimen of nacrite from Freiberg obtained, which had been in the Khar'kov State University museum. The electron-diffraction

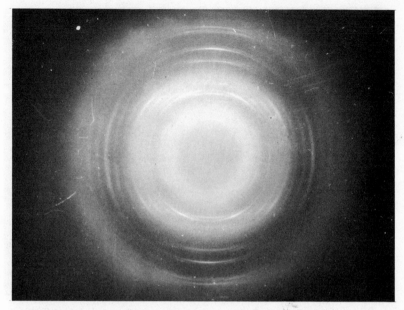

Fig. 96. Texture pattern from nacrite. Specimen from Freiberg.

patterns from this specimen [Fig. 96], while showing the features of a structure made up of two-storied kaolinite layers, at the same time differed sharply from the diffraction patterns given by other kaolinite minerals. In basic
features they did not contradict the conclusions arrived at by Hendricks (1938)
on the basis of single-crystal X-ray studies. It was suspected only that the
angle β had a greater value than the one given by Hendricks. It was not possible to decide this conclusively, however, because of the poor quality of the
diffraction pattern (Fig. 96).

At the end of 1960, in a fluorite deposit in Western Zabaikal',A. D.
Shcheglov discovered some nacrite with an extremely perfect crystal structure.
Electron-diffraction texture patterns from this (Fig. 97) showed a wealth of
clear, sharply recorded reflections of high resolution. From a purely visual
examination it was obvious that the mineral was made up of two-storied kaolinite layers (here the arrangement of the $06l$, $33l$ reflections of the fifth ellipse was particularly characteristic, these being the same for all the kaolinite
mineral varieties), and it was possible to confirm Hendrick's suggestion (1938)
that, in contrast to other kaolinite minerals, in nacrite the b axis, and not the

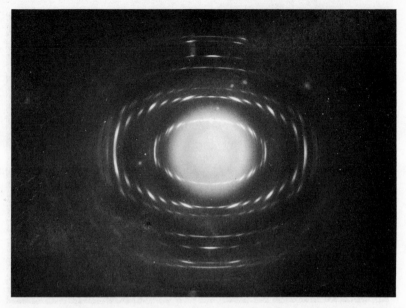

Fig. 97. Texture pattern from nacrite. Specimen from Western Zabaikal',
collection of A. D. Shcheglov.

a axis, formed the angle with the c axis which differed from $\pi/2$. Particular
proof of this was provided by the fact that in the normally indexed ellipses
with k = 0, reflections appeared on the minor axis, while in ellipses with k $\neq 0$
the reflections lay off the minor axes.

In its relative arrangement of reflections on the first ellipse $(02l,11l)$,
and in their intensity distribution, the nacrite diffraction pattern recalled the
patterns of micas of modification $2M_2$ (compare Fig. 74 and Fig. 97), which
also typically show interchange of the axes a and b. At the same time, as
demonstrated above, in texture patterns modification $2M_2$ can be confused with
modification 6H, which has hexagonal symmetry and a six-layer repeat unit.
Since nacrite diffraction patterns show weak subsidiary reflections similar to
the weak reflections which serve to distinguish 6H from $2M_2$ in the mica modifications, it is necessary to check to what extent these superficial analogies
correspond to reality.

Table 60. Comparison of Experimental and Calculated Intensities for the Three Versions of the Nacrite Structure

hkl	I calc			I_{exp}	d	hkl	Icalc			I_{exp}	d
	II, 2	II, 1	II, 3				II, 2	II, 1	II, 3		
110	17	17	17	18	4.43	$1\bar{1}5$	25	19	8	25	2.52
020	50	5	65	40	4.38	115	19	12	5	20	2.32
$1\bar{1}1$	50	95	35	40	4.34	$0\bar{2}6$	1	8	15	1	2.28
111	75	24	55	50	4.13	$1\bar{1}6$	4	4	4	2	2.20
$0\bar{2}2$	9	60	30	10	4.07	116	3	3	3	3	2.04
$1\bar{1}2$	14	14	14	23	3.94	026	9	4	2	5	1.96
112	11	11	11	18	3.63	$1\bar{1}7$	3	8	10	2	1.94
022	30	40	3	25	3.48	117	3	3	8	3	1.81
$1\bar{1}3$	30	20	57	40	3.42	$0\bar{2}8$	5	5	0	4	1.78
113	15	40	35	20	3.12	$1\bar{1}8$	2	2	2	2	1.73
$0\bar{2}4$	35	10	15	36	3.06	118	2	2	2	4	1.62
$1\bar{1}4$	7	7	7	15	2.92	028	0	3	2	1	1.57
114	6	6	6	8	2.68	$1\bar{1}9$	4	1	2	8	1.55
024	3	9	20	4	2.57						

Among the large number of regular structures which can be formed from idealized two-storied dioctahedral layers (see Fig. 11), there are six $2M_2$ structures and six pairs of 6H structures, with left- or right-handed rotations about the sixfold screw axis. In these structures, the $2M_2$ modifications have a two-layer repeat unit with $c \cos \alpha = -b/3$, while for 6H, $\alpha = \beta = \pi/2$.

If these equations are obeyed exactly, this must lead to certain reflections on the diffraction patterns coinciding, which, in particular, includes $02l$ and $\overset{\pm}{1}.\overline{1}.l+1$; $0.\overline{2}.l+1$ and $\overset{\pm}{1}1l$ for $2M_2$ (see Tables 31 and 32); and $0\overset{\pm}{2}l$, $\overset{\pm\pm}{1}1l$, $11l$ for 6H (see Tables 31 and 32). In this case, each structure may be distinguished by the $\Sigma|\Phi|^2$ values for the coincident reflections. Calculation showed that according to these criteria the nacrite experimental diffraction data agreed with three structures from modification $2M_2$ (structures II, 1,2,3 of Tables 31 and 32), but with none from modification 6H. Thus, the nacrite structure is essentially monoclinic, with a two-layer repeat unit. The corresponding unit cell, with a and b denoting the same axes as in other kaolinite minerals, has the following constants, found by the height method: $a = 5.14$, $b = 8.89$, $c = 14.6$ A, $\alpha = 100°20'$.

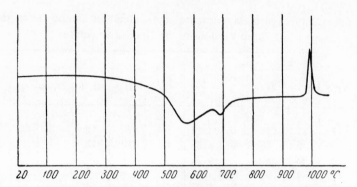

20 100 200 300 400 500 600 700 800 900 1000 °C

Fig. 98. Differential thermal analysis curve for nacrite (prepared
by V.P. Ivanova).

For a repeat unit of six layers, which changes the direction of the c axis,
we can choose a cell with c = 43.1 A, α = 91°30', which can be compared directly with the cell found by X-ray methods (Hendricks, 1938), a = 8.94, b =
5.14, c = 43.0 A, β = 90°20'. In view of the high quality of the nacrite electron-diffraction pattern, the differences observed lie outside the limits of experimental error.

The above three structures from modification $2M_2$ are distinguished by
the rotations of successive layers relative to the position shown in Fig. 11; in
one case successive rotations are of 120 and 60°, in another they are −120 and
−60°, while in the third they are 0 and 180° (counting anticlockwise rotations
positive). In all versions here the origins of layers, chosen at the centers of
empty octahedra, undergo displacements having components projected on the
ab plane of $-\frac{1}{3}$, $\frac{1}{3}$ and $\frac{1}{3}$, $\frac{1}{3}$ alternately (see Table 3). This behavior corresponds to the structures II, 1, 2, 3 (Table 6).

All these structures have the symmetry Cc, and as the c planes in them
lie parallel to the b axis, the latter is also a monoclinic axis in the basal
plane, which is the distinctive feature of modification $2M_2$. Naturally, with
such structural features present there is no need to triple the cell to give a
six-layer repeat unit, just as it is not surprising that the c axis of the tripled
cell is not strictly perpendicular to the layers in the structure.

All three versions are in agreement with the O, OH coordinates given by
Hendricks (1938) on the assumption that the nacrite structure had randomly
distributed Al atoms, and was pseudorhombohedral, with symmetry R3c.

The electron-diffraction results allow this uncertainty and ambiguity to be cleared up. Because, strictly, $c \cos \alpha \neq -b/3$, in reality the reflections indicated do not coincide completely on nacrite diffraction patterns, and because of the high degree of perfection of the specimen these reflections were separately resolved. From this it is possible to compare the three versions of the nacrite structure and determine how well each agrees with the observed intensity distribution.

From Table 60 it can be seen that the nacrite structure model closest to reality is that in which successive layers are rotated from the Fig. 11 position by -120 and $-60°$ in turn. This is the same structure as the one derived from general crystallochemical considerations by Newnham (1961), and the same as the one we obtained above by choosing the layer sequence which was compatible with the deformations in an individual layer (structure II,2 of Table 6).

It should be noted that the first ellipse on the nacrite diffraction pattern contains extra reflections, admittedly extremely weak ones, with heights of

Fig. 99. Nacrite crystals viewed with an electron microscope
(magnification 15,000).

$D = lq$, which cannot be due to a cell of modification $2M_2$, since the reflections from this must have heights of $D_{02l}^{\pm} = lq \pm 2s$ and $D_{1\bar{1}l}^{\pm} = lq \pm s$. This fact can be considered as yet another indication that in layer silicate structures any single sequence of layer positions may not be followed strictly. Normally a sequency will only predominate, being partly interrupted or displaced by another sequence. In the present case the natural assumption is that the main sequence is broken up by layers alternating at ±120°, which is characteristic, in particular, of dickite.

In spite of this irregularity, the short-range order in the structure of the nacrite specimen investigated must be high. This is in accord with the appearance of a split endothermic effect on its differential thermal analysis curve (prepared by V. P. Ivanova, Fig. 98), which may correspond to the presence of two kinds of OH groups in the kaolinite layer, positioned on the upper and lower faces of the octahedra, respectively (see Fig. 11). They are present in the ratio 3:1, and the latter, since they lie in the middle of the layer, are evidently more tightly bound in the structure. This also ties in with the shape and position of the two nacrite endotherms. With a lower degree of structural perfection, of course, there might not be such a clear distinction between the energy characteristics of these OH groups. The regular outlines of the nacrite crystals are also in accord with their structural perfection (Fig. 99).

If we were to replace the idealized layers (Fig. 11) in the structure we have obtained with real kaolinite layers, with typical atomic positions, interatomic distances, nonregular polyhedra, and distorted arrangements, we would obtain quite an accurate representation of the real structure of this mineral. This does not make an independent determination of the structure of nacrite any less valuable. On the contrary, because of the high degree of perfection of the nacrite structure, it could possibly be used to improve our picture of the kaolinite structure and its individual layers.

3. Electron-Diffraction Data on Sepiolite and Palygorskite Structures

Sepiolite and palygorskite crystals have the shape of long fibers, and this in itself must give rise to a number of distinctive features in their diffraction patterns from fine precipitates studied by transmission, with the specimens inclined at various angles to the electron beam. In particular, the arrangement of reflections in these patterns must depend on specimen properties such as the presence or absence of any preferred orientation of the fibers along their direction of elongation.

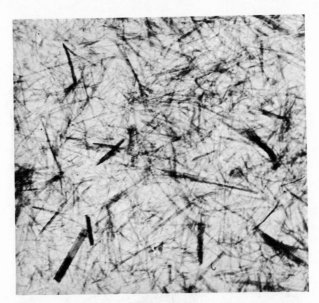

Fig. 100. Electron-microscope photograph of sepiolite
(magnification 10,000).

According to Vainshtein (1956), when preferred orientation is lacking,
the specimen has an "axial texture of the second kind," in which crystals are
aligned with certain crystallographic directions parallel to the plane of the
specimen support, but are rotated randomly about both the crystal axes and
the normals to the film.

In the other case, the specimen has an ordinary platy texture, but the
basal planes of the crystals are only very imperfectly oriented parallel to the
plane of the specimen support.

Electron-microscope studies show that sepiolite and palygorskite crystals
(Figs. 100, 101) have the shape of very narrow ribbons, rather than being
fibers of isometric cross section; therefore, like celadonite for example, they
must form platy textures.

In spite of the fact that the reflections appear as long arcs, with the
most intense arcs running together into circles, a careful examination of the
texture patterns will also show that the crystals are oriented as in a platy tex-

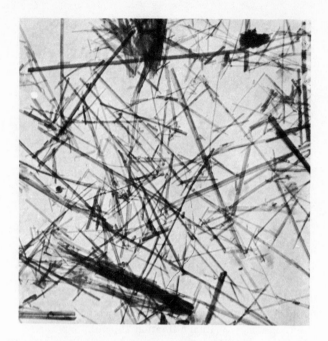

Fig. 101. Electron-microscope photograph of palygorskite
(magnification 10,000).

ture. Sepiolite crystals were found to have their bc plane* preferentially oriented parallel to the plane of the support by Brindley (1959), who studied
sepiolite by the microdiffraction method with an electron microscope. In
similar studies, we obtained patterns from crystals lying on their sides (Popov,
Zvyagin, 1959) and from crystals in the same orientation as that observed by
Brindley (Fig. 52).

Texture patterns of sepiolite and palygorskite (Figs. 102, 103) show a
number of features in common with those of the layer silicates, which is only
natural, since from X-ray results (Belov, 1958), there is a definite relationship
between the structures of these minerals.

---

*Here, c denotes the axis along the sepiolite fiber, with a unit of about 5.3 A;
but in the following, for convenience in comparing of structures and diffraction patterns of sepiolite, palygorskite, and layer silicates, it will be denoted
by a.

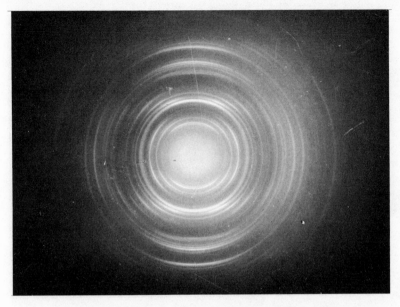

Fig. 102. Texture pattern from sepiolite ($\varphi = 55°$).

It was noted in Chapter 3 that in layer silicate texture patterns the reflections are arranged on hk ellipses which have minor axes of length

$$b_{hk} = \frac{L\lambda}{a} \sqrt{h^2 + \frac{k^2}{3}},$$

and where, because the ab face is centered, the indices hk are unmixed, i.e., both even or both odd. In accordance with the structural concepts propounded in Chapter 1, and the relationships $b_1 = 3\sqrt{3a}$ and $b_2 = 2\sqrt{3a}$ given for these minerals, sepiolite and palygorskite diffraction patterns must show sets of ellipses with respective minor axes of

$$b'_{hk} = \frac{L\lambda}{a} \sqrt{h^2 + \frac{k^2}{27}} = \frac{L\lambda}{b_1} \sqrt{27h^2 + k^2},$$

$$b''_{hk} = \frac{L\lambda}{a} \sqrt{h^2 + \frac{k^2}{12}} = \frac{L\lambda}{b_2} \sqrt{12h^2 + k^2},$$

where b_1, b_2 are the b constants of sepiolite and palygorskite, respectively, and hk may be mixed or unmixed. As a result, these sets of ellipses include the

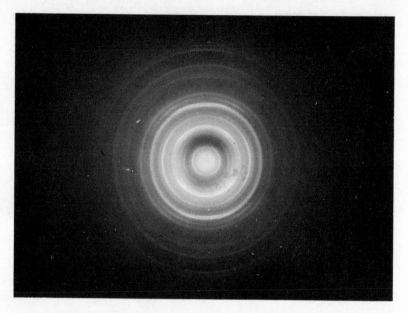

Fig. 103. Texture pattern from palygorskite ($\varphi = 55°$).

ellipses characteristic of layer silicates, but with k indices multiplied by three and two, respectively.

From these circumstances it would appear easy to determine the a and b constants, and then the spatial characteristics of the unit cells in the normal way. Here, however, certain technical difficulties first arise.

With such a wealth of ellipses on the diffraction patterns, as given by the expressions for b'_{hk} and b''_{hk}, it is necessary for analysis that the reflection arcs be as short as possible, so that they can be located reliably on the appropriate ellipses. In actual fact, because the crystals show a large range of orientations about their direction of elongation (about the a axis in Chapter 1 notation), the reflections behave in opposite fashion, being much drawn out and partially overlapping. This running together of reflections can lead to displacement along the arcs of the intensity maxima of each reflection, or to the formation of a maximum in a position which cannot agree with any unit cell which would simultaneously satisfy the positions of all the other intensity maxima. For this reason, the impression arose, and was current for a long time, that the electron-diffraction results were in complete disagreement with those of X-ray studies.

Table 61. Unit Cells and Space Groups for Sepiolite and Palygorskite

Mineral	a(c)*	b	c(a)	β	Space group
Sepiolite					
a) Results of Brauner and Preisinger (1956)	5.28	26.8	13.4	90°	D_{2h}^6 = Pncn
b) Results of Nagy and Bradley (1955).........	5.3	27.0	13.4	?	C_{2h}^3 = C2/m
c) Results of Brindley (1959).	5.25	26.96	13.50	90°	
d) Electron-diffraction results	5.24	27.2	13.4	90°	D_{2h}^6 = Pnan
Palygorskite					
a) Results of Bradley (1940) .	5.3	18.0	12.9	?	C_{2h}^3 = C2/m
b) Electron-diffraction results	5.22	18.06	12.75	95°50′	C_{2h}^4 = P2/a

*The axes given in brackets are those quoted in the literature. The space-group symbol differs accordingly in form, but not in nature.

Because of these difficulties, it was a matter of considerable importance to find sepiolite and palygorskite specimens which would be favorable for electron-diffraction investigation. Diffraction patterns of the highest quality were obtained from the following sepiolite specimens: E. N. Ushakova's specimen No. Sh-10, M. A. Rateev's specimen No.618, a specimen from Ampandrandava (Malagasy Republic), sent by Professor A. Preisinger, and sepiolite from Utah (USA), sent by Professor G. W. Brindley. The most successful palygorskite photograph was one obtained from E. P. Levando's specimen No. 3/3.

The better sepiolite diffraction patterns show a curious relationship between the shape of the reflections and their hkl indices, which might, however, have been expected on diffraction patterns from these platy textures, because of the elongated shapes of the crystals. This is shown in particular, as already noted in Chapter 4, by the reflections being more compact, the smaller their k indices.

Because of this feature it is easy to distinguish the h0l reflections on the diffraction pattern from other reflections lying on the same ellipses or closely adjacent ellipses, and use these as the basis for determining the unit cell constants.

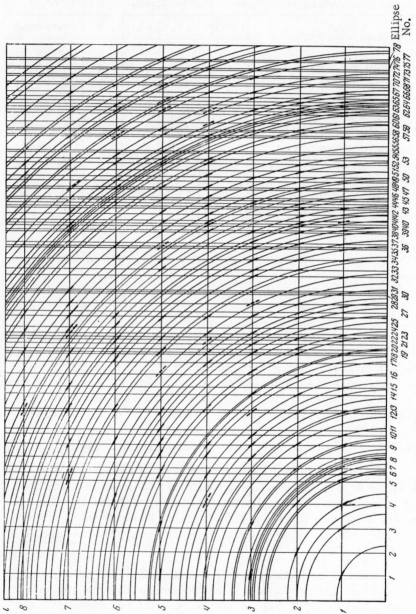

Fig. 104.　Theoretical direct texture pattern diagram for sepiolite.

In this way, using the b_{hk} minor axes noted above and the heights of the h0l reflections, unit cell constants were found for sepiolite and palygorskite, and these are given in Table 61 together with comparable X-ray results.

The electron-diffraction results for a and b are in better agreement with the relationship $b = 2\sqrt{3a}$ for palygorskite and $b = 3\sqrt{3a}$ for sepiolite, in comparison to the X-ray results. Incidentally, there is often a similar conflict of results in the case of the layer silicates.

To see how well the resulting unit cell agrees with the distribution of all the other reflections, it is most convenient, in view of the above difficulties in fixing reflection positions, to construct a theoretical direct texture pattern, which will represent a reciprocal-lattice plane section passing through the texture axis (the c* reciprocal-lattice axis), and compare this with the actual diffraction pattern. To construct such a theoretical pattern we need to draw a system of straight lines parallel to the ordinate axis and at distances from the origin which are proportional to the values of b_{hk}, the ellipse minor axes, and then draw another system of straight lines perpendicular to the first and at distances proportional to q = $L\lambda c^*$. If we then draw a system of circles with centers at the origin and radii proportional to the radii of the reflections observed on the diffraction pattern, the points where these intersect with the junctions of the rectilinear grid will show which reflections theoretically correspond to the experimental cell. When this theoretical pattern is compared with the real diffraction pattern, the theoretical reflections which actually appear on the real pattern can be picked out. The hk indices of the vertical lines and the l indices of the horizontal lines give the hk$l$ indices of the reflections directly, and from the degree of blackening of the corresponding parts of circles on the diffraction pattern, a rough visual estimate can be made of their intensities.

This operation has only been carried out in full for sepiolite. Its theoretical direct texture pattern is shown in Fig. 104, the hk indices of successive ellipses are given in Table 62, and a list of reflections, showing their values of the interplanar distance d, their hk$l$ indices, and their intensities estimated on a ten-point scale, is given in Table 63.

Of the two space groups previously proposed for sepiolite, i.e., C_{2h}^3 (Nagy, Bradley, 1955) and D_{2h}^6 (Brauner, Preisinger, 1956), only D_{2h}^6, the one found by Brauner and Preisinger, is in agreement with the indices of the experimentally observed reflections.

Table 62. Sequence of Ellipses in a Sepiolite Oblique Texture Pattern

No.	hk	$27h^2 + k^2$	No.	hk	$27h^2 + k^2$	No.	hk	$27h^2 + k^2$	No.	hk	$27h^2 + k^2$	No.	hk	$27h^2 + k^2$
0	00	0	18	19, 20	108	34	2.10	208	51	1.17	316	67	3.14	439
1	01	1	19	21	109	35	1.14	223	52	0.18, 39	324	68	0.21, 43	441
2	02	4	20	22	112	36	0.15	225	53	2.15	333	69	44	448
3	03	9	21	23	117	37	2.11	229	54	3.10	343	70	45	457
4	04	16	22	0.11	121	38	31	244	55	1.18	351	71	1.21, 3.15, 46	468
5	05	25	23	24	124	39	32	247	56	0.19	361	72	2.19	469
6	11	28	24	1.10	127	40	1.15	252	57	2.16, 3.11	364	73	47	481
7	12	31	25	25	133		2.12, 33		58	3.12	387	74	0.22	484
8	06, 13	36	26	0.12, 26	144	41	0.16	256	59	1.19	388	75	48	496
9	14	43	27	1.11	148	42	34	259	60	2.17	397	76	3.16	499
10	07	49	28	27	157	43	35	268	61	0.20	400	77	2.20	508
11	15	52	29	0.13	169	44	2.13	277	62	3.13	412	78	1.22	511
12	16	63	30	1.12	171	45	36	279	63	1.20	427	79	49	513
13	08	64	31	28	172	46	1.16	283	64	40, 2. 18	432	80	0.23	529
14	17	76	32	29	189	47	0.17	289	65	41	433	81	3.17, 4.10	532
15	09	81	33	0.14, 1.13	196	48	37	292	66	42	436			
16	18	91				49	2.14	304						
17	0.10	100				50	38	307						

Table 63. List of Reflections on a Sepiolite Diffraction Pattern (the number in brackets after each reflection is its intensity estimated on a 10-point scale; the bold numbers denote exactly coincident reflections)

d	hkl *
11.92	011 (10)
7.47	031 (2)
6.64	002 (6)
5.96	022 (6)
5.17	110 (4)
4.96	051 (2)
4.86	042, 111 (2)
4.59	121 (2)
4.52	**060, 130** (5)
4.43	013 (6)
4.28	131 (10)
3.98	033 (2)
3.95	122 (2); 141 (1)
3.74	**062, 132** (8); 071 (2)
3.51	142 (4)
3.33	004 (6), 161 (1)
3.18	133 (8)
3.04	044, 171 (2); 032, 162, 143 (4)
2.93	091 (1)
2.81	114 (1); 172 (2)
2.68	015 (6); **064, 134,** 163 (2)
2.62	144, 182, **190, 200** (4)
2.58	035 (2); **191, 201** (10)
2.55	173, 220 (4); 211 (2)
2.44	0.11.1, 240 (2); **192, 202,** 212 (8)
2.39	055, 115, 183 (2); 084, 164 (4); 1.10.1, 222 (1)
2.30	135, 242 (1)
2.26	145 (2); **0.12.0, 260** (4); **193, 203,** 213 (8)
2.19	155, 233 (1)
2.14	**0.12.2, 262,** 271 (1)
2.12	0.10.4, 165, 1.10.3, 1.11.2 (1); 046 (2)
2.06	**194, 204,** 214 (6), 1.12.1 (2), 272 (4)
2.02	234, 263 (2)
1.95	146, 185, 1.10.4 (2); 273, 291 (1)
1.88_5	2.10.0 (1)
1.88	**0.12.4, 264** (1)
1.87	086, 166, 2.10.1 (2); **195, 205,** 215 (4)
1.80_5	057, 176, 245 (2); 1.14.1.293, 2.10.2 (1)

Table 63 (continued)

d	hkl *
1.78	1.13.3, 2.11.1 (1)
1.74₅	147, 265, 2.10.3, 2.11.2 (1)
1.72	0.10.6, 311, 321 (2); **1.15.0, 2.12.0, 330** (4)
1.68₅	**0.14.4, 1.13.4**, 275, 312 (2); 0.15.3, 1.14.3, 341 (3); **206, 196, 216**(4)
1.66	167, 236, 2, 11, 3, 350(1); **1, 15, 2, 2, 12, 2, 332** (4)
1.65	0.16.2 (4)
1.62	097, 256, 313, 323, 352, 361 (1)
1.59	**0.12**.6, **266**, 187, 295, 2.11.4 (4); 0.17.1, 2.13.2, 362, 370 (2).
1.57	1.13.5, (4); 1.16.2 (2)
1.54	**197, 204**, 217, 381 (6); 2.10.5 (1), 314, 324 (2)
1.52	0.6.4, **1.15.4, 2.12.4, 334** (1)
1.51	0.15.5, 1.14.5 (1); 0.17.3 (3); **0.18.0, 390**, (8); 334 (2); 382 (6)
1.45₅	088, 168, 354 (2); 1.10.7, 257, 2.11.5 (1)
1.46	1.11.7, 267, 315, 325 (1); 1.16.4, 2.15.2 (2) 3.10.1 (3)
1.44	277 (10); 2.10.6, 345, 374, 393, 3.10.2, (1)
1.41	0.13.7, 1.12.7, 1.14.6, 287 (3); **198, 208**, 218 (6); 355 (2); 384 (1).
1.38	0.17.5, **0.18.4, 394**, 1.10.8, 248, 375 (1)
1.34₅	**0.12**.8, **268**, 1.11.8 (2); 179 (4); 2.10.7, 2.14.5, 385, 3.10.4 (1)
1.33	⁰.20.2, 1.17.5, 3.13.1 (1); 1.18.4 (1)
1.31	**2.18.0, 400** (2)
1.29₅	1.20.1, **401, 2.18.1, 411** (7)
1.29	**199, 209**, 219 (5); 317, 327, 3.10.5 (2)
1.28₅	0.21.1, 431, 3.14.1 (2)
1.28	**2.18.2, 402, 412** (4)
1.27₅	**1.15.7, 2.12.7, 337**, 1.18.5, 1.19.4, 2.14.6, 386, 3.12.4, 3, 14, 2, 432 (1)
1.25₅	0.19.5, 1.17.6, **2.16.5, 3.11.5** (1); 259 (2) **1.21.0, 3.15. 0, 460**, 442 (4)
1.25	**1.21.1, 3.15.1, 461, 2.18.3, 403, 413** (4)
1.23	0.17.7, 1.19.5, 1.20. 4, 377, 3.10.6, 3.12.5, 453 (1)
1.21	0.13.9, 1.12.9.1.18.6, 289, 2.14.7, 2.17.5, 2.19.3, 1.21.3, **3.15.3, 463** (1); **2.18.4, 404, 414, 424** (2)
1.18	0.23.1, **1.21.4, 3.15.4, 464**, 1.19.6, 3.17.1, 4.10.1, 464, 483 (2); 2.18.5, 405, 415 (4), 2.19.4, 3.12.6 (2)
1.16₅	2.13.8, 368, 3.10.7 (2)
1.16	0.20.6, 2.17.6, 445 (2); 0.22.4, 474 (1)
1.13₅	1.22.4, 2.20.4, 494 (2)
1.13	**2.18.6, 406, 416** (3)
1.12₅	3.14.6, 426, 475 (1)

Table 63 (continued)

d	hkl *
1.12	0.18.8, 3.9.8, 1.15.9, 2.12.9,339 (2); 1.19.7, 3.12.7 (1)
1.10$_5$	3, 10, 8, 456 (1)
1.08$_5$	2.16.8, 3.11.8 (1)
1.08	0.22.6, 0.23.5, 3, 17, 5, 4.10.5, 476 (1); 1.14.10 2.14.9, 389 (2); 2.18.7, 407, 417 (4)
1.05	0.16.10, 0.20, 8, 2.15.9, 2.19.7, 3.4.10, 1.21.7, 3.15.7, 467 (1)
1.03	0.19. 9,1.16. 10, 3.16.7, 487 (1); 2.18.8, 408, 418 (2)
1.00	2.17.9 (2)
0.99$_5$	3.1.11, 3.2.11 (4)
0.98^5	2.18.9, 409, 419 (1)
0.93$_5$	2.18.10, 4.0.10, 4.1.10 (1)
0.89	2.18.11, 4.0.11, 4.1.11 (1)
0.85	2.18.12, 4.0.12, 4.1.12 (1)

* The hkl indices correspond to the axis notation used by us. If the symbols for the a and c axes are interchanged to conform with the usual practice, the indices h and l should also be interchanged.

For palygorskite the diffraction pattern was only partially indexed, to include only those reflections which were amenable to fairly accurate direct measurement. Even these results, however, were sufficient to disprove the space group C_{2h}^3 found by Bradley (1940) for palygorskite, and instead agreed with the space group C_{2h}^4, which can be considered as a degenerate monoclinic form of the sepiolite space group D_{2h}^6.

4. Investigations Using the Microdiffraction Method

A characteristic property of the microdiffraction method is that the diffraction patterns produced by this method show two kinds of features, those due to the crystal structure of the object, and those due to the physical form of its crystallites. Here, features of the first kind appear in a haphazard form, depending on the orientation of the crystal and its defects, while those of the second kind supplement and correct the morphological results from electron-microscope studies.

Texture patterns provide a representation of diffraction and structural characteristics averaged over a vast number of crystals. They show the general features of the crystal structure of the object, providing a direct, experimentally averaged picture within a single diffraction pattern. In averaging the diffraction effects due to the structures of individual crystals, the general features (such as platy or needle-like shapes) are reflected by the form of the reflections, while special or individual features give the general background to the pattern.

In contrast to this, microdiffraction patterns show the local characteristics of a specimen, which could be averaged out by taking a sufficiently large number of photographs from which important conclusions may be drawn.

The first results of practical importance were obtained almost simultaneously in 1954, in Japan (Honjo, Mihama, 1954) and in France (Kulbicki, 1954). In these studies, single-crystal diffraction patterns were used to check the prevailing ideas on the structures of kaolinite, halloysite, and chrysotile, and to verify the crystallographic directions in these crystals. From the geometry of its diffraction pattern, it was concluded that halloysite had a layer repeat unit of two layers. A similar result was obtained subsequently by Popov and Zvyagin (1959). Table 32 (Chapter 4) shows that the intensity distribution observed for the $02l$ reflections was in agreement with structure I, 3 of modification $2M_1$. According to their microdiffraction patterns, elongated halloysite crystals have a discrete zonal structure, with zones having the same orientation lying on radii passing out from the axis of elongation of the particle. In all the zones the structural layers are parallel to the axis of elongation. The sequence of layers in each zone finishes up with a flat face, thus giving rise to the polygonal-prism shape of the crystals (Gritsaenko et al., 1961). From experimental studies carried out by the author and A. I. Gorshkov, it follows that in the majority of cases the direction of elongation of the particle coincides with the b axis of the structure. Rarely particles are observed in which the b axis diverges slightly away from the direction of elongation. In this case, the reflections on the diffraction pattern are split up into pairs, and here the amount of splitting gives the angle of divergence directly. Individual tubes have also been observed in which the layers do not have any general direction of elongation. This means that successive layers or groups of layers in each radial zone have a certain azimuthal rotation relative to one another.

Right from the very beginning of the development of the microdiffraction method, one very tempting aspect of its use was the mineralogical diagnosis of individual particles and crystals observed in the field of view of the electron microscope. This task was quite feasible, as long as the minerals differed significantly in structure. For the layer minerals, because the crystals were only present in one orientation, a diagnosis of this type had to be based on the values of two parameters only, a and b, and on certain qualitative features of the diffraction patterns. For this reason the scope and reliability of the method was rather limited in the diagnosis of layer silicates and clay minerals.

To establish differences of value in the diagnosis of clay mineral crystallites, a number of investigations (Suito, Uyedo, 1956; Radczewski, Balden, 1959) were devoted to calculation of the interplanar distances for the hkl reflections corresponding to the reciprocal lattice points lying in the ab plane (because the lattice was nonorthogonal, when $|h| > 1$, the index l could not be taken as equal to zero). These values were compared by the investigators with X-ray data and with results obtained from electron-diffraction texture patterns. An effort which was more fundamental to the problem of the diagnosis of layer silicates and clay minerals from single-crystal electron-diffraction patterns was that made by Brindley and De Kimpe (1961), who took a wide variety of specimens, from kaolinites to chlorites, and measured their a and b constants with great accuracy, using a sputtered Al layer as a support, which served as an internal standard.

The main virtue of the microdiffraction method is its ability to show up fine structural details inherent in individual single crystals, and to resolve various problems in the microcrystallography of minerals.

As an example, when Oberlin and Mering (1962) compared various different montmorillonite specimens, they were able to see which of the particles they observed were single crystals, and which were aggregates, and were able to establish the relative azimuthal orientation of the crystals and the value of its scatter.

A very penetrating analysis of single-crystal electron-diffraction patterns was carried out in the investigation of montmorillonite by Cowley and Goswami (1961), who found that the a and b constants increased by approximately 3% when this mineral was saturated with an organic cation. By constructing and analyzing syntheses of Φ^2 series, they showed that the pseudohexagonal character observed in three types of diffraction-pattern intensity distribution was bound up with bending in the montmorillonite crystals. A consequence of this bending was that coherent scattering occurred in diffraction by atoms which had not very different z coordinates in the silicate layer.

In studies on layer silicates, relative rotation of one part of a crystal relative to another was observed, with certain points in the reciprocal lattices of both parts of the crystal coming into coincidence. Here, peculiar diffraction effects arose due to secondary diffraction of rays formed by the first part of the crystal as they passed through the second, which gave rise to the impression that a superlattice was present. Behavior of this type was observed in the micas (Hibi, 1955; Rang, 1958; Cartraud, Zouckerman, 1960), the chlorites (Eckhardt, 1958), and in kaolinite (Zvyagin, Popov, 1959; see Fig. 43).

A theoretical possibility inherent in the microdiffraction method, that of the study of mineral intergrowths, was put into practice in an investigation of the epitaxial behavior of fireclay on kaolinite (Oberlin, Tchoubar, 1960a). Associated with the two superimposed structures there was a screening effect, which showed up directly, both in the electron microscope, in the form of a pattern of moire fringes, and also in the diffraction pattern, which contained extra irrational reflections. It was explained that the fireclay had its [100] direction oriented parallel to the [110] and [1$\bar{1}$0] directions in kaolinite. Some suggestions were incidentally confirmed concerning the transformation of kaolinite into fireclay under the action of acids.

Microdiffraction studies also showed that the completion of this mineral transformation process (Oberlin, Tchoubar, 1960b) was accompanied by changes in the intensities of the hk reflections observed; reflections with k = 3k' became stronger, and those with k ≠ 3k' became weaker. The authors also suggested that the geometry of the diffraction patterns also reveals a change from a triclinic unit cell to a monoclinic one.

By comparing diffraction patterns and electron-microscope pictures of small sectors of a specimen, Stemple and Brindley (1960) were able to establish that some crystals of talc and tremolite were intergrown so that the a axis of the former was parallel to the c axis of the latter.

To obviate certain inconsistencies which can exist between the orientations of crystals in the field of view of the electron microscope and their diffraction patterns, it is possible to follow Radczewski and Goossens' recommendation (1956), and consider the distribution of regions of the specimen which scatter in the direction of a given diffracted ray observed in the corresponding dark-field pattern. These authors used this method to establish that the direction of elongation in the kaolinite crystals they observed coincided with the b axis.

The microdiffraction method can be used to show up concealed deformations, particularly bending of crystals. An article by Ehlers (1956) shows electron-diffraction patterns obtained from mica flakes which were rotated about an axis during the exposure. They show characteristic groups of closely spaced reflections lying on arcs of hyperbolas. Similar patterns were obtained by Popov and Zvyagin (1959) from the bent edge of a kaolinite crystal (Fig.44).

Some new and extremely promising approaches to the study of layer silicate structures were opened up by the development of ultra-thin sections, from which photographs can be obtained representing different cross sections of the reciprocal lattice, including those containing the $c*$ axis. In this way, Eckhardt (1961) succeeded in obtaining point diffraction patterns from muscovite including the following axes: $a*$, $c*$, $b*$, $c*$, $[\bar{h}.0.4h]*$, $b*$, $[0hh]*$, $[hh0]*$, $[h.0.2h]*$, $b*$. For kaolinite an electron-diffraction pattern was obtained showing the axes $[h0h]*$, $b*$. These patterns are the most reliable way of identifying minerals by the microdiffraction method, and they reveal differences in their three-dimensional structures.

A very rewarding field of application for the microdiffraction method is the study of serpentine minerals. Investigations have shown (Zussman, Brindley, Comer, 1957; Popov, Zvyagin, 1959) that the electron-diffraction patterns obtained show up extremely clearly the distinguishing features of the representatives of this group; these include the clino- and ortho-chrysotiles, which have peculiar cylindrical lattices, the platy serpentines of the lizardite type, and the antigorites, with corrugated structures corresponding to different superlattice constants.

The superlattice constants of antigorite have been established both from measurements of the distances between electron-diffraction reflections, and from parallel direct electron-microscope observations of the crystal corrugations (Brindley et al., 1958, Chapman, Zussman, 1959). It was concluded here that the antigorites have three characteristic groups of values for the a superlattice constant; these cover the ranges 16-19, 35-45, and 80-110 A. This is not just of idle interest, since, according to Kunze (1959), the superlattice constant of antigorite is connected with the curvature of the corrugations in the layers and depends on certain stresses in the structure, so that it can serve as an unusual manometer and thermometer, recording the formation conditions of the corresponding rock.

According to work carried out by the present author and V. A. Shitov, the great majority of chrysotile crystal tubes have their axes along the [100] direction (the a axis), and have a monoclinic pseudocell, which from the arrangement of the confidently indexed discrete h0l reflections, principally the 20l reflections, has a c constant sometimes equal to about 7 A, and sometimes to about 14 A.

A smaller, but nonetheless appreciable, proportion of specimens, principally those from kimberlites (L. A. Popugaeva's collection), have given ortho-chrysotile patterns (Fig. 48). From the arrangement and intensities of the reflections on these, they belong to a type D structure.

In a number of cases the para form of chrysotile has been observed, and there it was especially remarkable that it did not appear alone in the tubular crystal selected for diffraction, but in combination with the clino and ortho forms of chrysotile. This means either that the particle selected consisted of two tubes lying one inside the other, with one having its tube axis along [100] (the a axis) and the other along [010] (the b axis), or that groups of layers within a single tube were rotated relative to one another by 90°. Thus, on the diffraction pattern, with a single sequence of 00l basal reflections, the other reflections appear in two versions, representing rotation patterns about both the a axis and the b axis, and the images of both these axes appear on the pattern (Fig. 49) lying perpendicular to the set of 00l reflections (the image of the c axis).

The microdiffraction method was thus a natural choice for reliably monitoring the products in the hydrothermal synthesis of chrysotile (Yang, Chi-Sun, 1961).

An interesting result was obtained from microdiffraction studies of sepiolite (Brindley, 1959; Popov, Zvyagin, 1959). The point-diffraction patterns obtained showed that a partly pseudolayered, partly pseudofibrous structure proposed for this mineral was correct, and that its crystals were ribbon-shaped. The same method was used to identify the sepiolite-type mineral loughlinite, of composition $Na_2O: 3MgO \cdot 6SiO_2 \cdot 8H_2O$. This was found to have a c constant of 5.26 A (Fahey et al., 1960).

The above brief list demonstrates the comparatively restricted scope of the microdiffraction method and the particular nature of the problems it can resolve. The scope and effectiveness of the method are increased immeasurably if it is combined with the oblique texture method.

For its own part, it can give valuable supplementary information not available from texture patterns, particularly in the study of unknown structure types.

5. Some Achievements of Electron-Diffraction Studies of Clays, Clay Minerals, and Related Layer Silicates by the Oblique Texture Method

The Distribution of Experimental Material According to Structural and Mineralogical Features of the Specimens. Over a number of years, the author and his colleagures R. A. Shakova, V. A. Shitov, and K. S. Mishchenko prepared electron-diffraction oblique texture patterns from a large number of specimens, the majority of which were sent for laboratory determination by various other people and organizations. Analysis of these diffraction patterns yielded a rich store of factual material. Derived by an independent and original method of structural analysis, this information can be used to verify, support, or supplement the concepts derived from X-ray data.

This material was published in part in the form of generalized results for minerals of particular groups and combinations of groups (Zvyagin, 1952, 1954a, 1956a, 1958a), and these results involved only that structural data which could be extracted from certain of the purely geometrical features of the texture patterns.

In subsequent studies an analysis of intensities was included. The structures of individual minerals were determined and the intensity distributions of different structure modifications were explained. Then it was demonstrated that the basic features of the intensity distributions characteristic of particular structure types and modifications were retained to some extent in texture patterns from minerals of poor quality, with not quite perfect or with imperfect structures. From this, one of the important achievements of electron diffraction came, which was that by using the results of studies of structurally perfect varieties, it was found possible to build up quite an accurate picture of the atomic structures of all layer silicates, including those for which such structural data could not be obtained directly. This meant it was possible to work over and generalize electron-diffraction results obtained earlier and not appearing in the publications mentioned, and consider also the intensities of reflections, to achieve a deeper understanding of the significance of the diffraction patterns, and obtain a better picture of the structural features and interrelationships of the specimens investigated.

The structural characteristics of the specimens were worked out from analysis of their texture patterns, using both the geometry of the reflections and their intensities. These characteristics included the type of layer making up the structure, the extent to which cations occupied the octahedral positions (dioctahedral and trioctahedral character), the order or disorder in the relative arrangement of the layers (presence or absence of polymorphic modifications), and the constants of the unit cell or pseudocell, for perfect and imperfect crystal lattices, respectively.

Depending on the similarity or dissimilarity of these characteristics, the specimens were classed into groups, and series of specimens were picked out in which the specimens were related by a gradually changing variable which reflected their probable genetic relationship and nature. The groups of related specimens and the results obtained for these are listed in Table 64 in a certain order, as shown by the corresponding subheadings.

The order in which the specimens appear in the table corresponds to a classification scheme which is derived directly from the electron-diffraction results described, and which will be examined below.

One point worth noting is that from published work on the synthesis of minerals (Yoder, Eugster, 1955; Koizumi, Roy, 1959), and also because of the lack of a continuously variable series of specimens leading from one end member to the other, it follows that the difference between the dioctahedral and the trioctahedral varieties of layer silicate minerals is more fundamental than that between minerals which have the most diverse physical and chemical properties, but which are members of the same modification or series of continuously variable specimens.

The relationships which emerge from a comparison of the different clay minerals studied are clearly illustrated by Table 64. Here we will only make a number of comments, in passing, on crystallochemical features of the minerals which cannot be clearly expressed by the table, together with a few necessary explanations of the structure of the table itself.

Two-Storied Layer Minerals. Results from electron-diffraction studies show that contrary to the usual ideas, kaolinites with perfect structures can have monoclinic as well as triclinic unit cells. Conversely, kaolinites with imperfect structures without a strict repeat unit may also possess triclinic as well as monoclinic pseudolattices.

Two points emerge from the great variety of specimens of both kinds which have been studied; these are that the shape of the unit cell and the degree of perfection of the crystal alter independently, and to some extent continuously. Thus, the intermediate position between kaolinites with triclinic cells and those with monoclinic cells is occupied by kaolinites with lattices in which the shape of the cell, or more precisely the angle α, varies between

Table 64. A Structural and Mineralogical Classification of Clay Minerals and Related Structures Investigated by Electron Diffraction

Minerals and combinations	No. of specimens	Unit cell constants (cell sides in A, angles in degrees)				Notes
		a	b	c	β	
Two-storied layer minerals						
A. Dioctahedral varieties						
Nacrite (modification $2M_2$)	2	8.89	5.14	14.6	100.5	Strict c repeat units are possessed by completely ordered structures In varieties with nonrigorous c repeat units, quantities c,β are pseudoconstants,since they are determined only by reflections with k=3k',and in particular indicate that structure is only ordered in its projection on x0z plane
Dickite (modification $2M_1$, structure I,4)	5	5.13-5.14	8.88-8.90	14.3	96.5	
Kaolinites with monoclinic cells (modification $1Tk_1$) a) varieties with strict c repeat units	10	5.13-5.14	8.89-8.90	7.2	104.5	
b) varieties with nonrigorous c repeat units	16	5.13-5.15	8.88-8.92	7.2	104.5	

Table 64 (continued)

Minerals and combinations	No. of specimens	Unit cell constants (cell sides in A, angles in degrees)					Notes
		a	b	c	α	β	
Kaolinites with triclinic cells (modification 1Tk₁) a) varieties with strict c repeat units	40	5.12-5.15	8.88-8.92	7.2-7.4	91.7	104.5	For halloysite, values of c and β refer to specimens dehydrated in vacuum
b) varieties with nonrigorous c repeat units	14	5.12-5.14	8.88-8.91	7.1-7.3	91.7	104.5	Most common values are a = 5.14, b = 8.90 A
B. Trioctahedral varieties With ordered structures Platy serpentine 1T + {2M₁}	1	5.33	9.20	7.15		90	(1T) Brackets give the
				14.5		96	(2M₁) modification to
Nepouites 1T + 2M₁	2	5.26, 5.27	9.10, 9.12	7.25, 7.12		90	(1T) which unit cell
1T + 1M	1	5.28	9.12	7.12		90	(1T) constants relate
2T + 2H (VI,I)	2	5.26	9.11	14.3, 13.7		90	(1T)
With disordered structures Platy serpentines: A	3	5.29-5.31	9.14-9.20	7.05-7.35		90	
D + B	1	5.31	9.16	14.0		90	(D)
Antigorites: A	2	~44; ~37	9.20	~7		–	
Clino-chrysotiles: B	2	5.30	9.20	14.75		93.5	
Clino-chrysotiles: B	3	5.30	9.20	7.6		106.5	One specimen contained para-chrysotile

Table 64 (continued)

Minerals and combinations	No. of specimens	Unit cell constants (cell sides in A, angles in degrees)				Notes
		a	b	c	β	
Ortho-chysotiles: D	2	5.30	9.20	14.4	90	(A)
Nepouites: A + D	1	5.26	9.10	6.9	90	(D)
A + B	1	5.28	9.12	13.7	90	(A)
Garnierites: A	2	5.26	9.12	~7	—	(A)
Three-storied layer minerals A. Dioctahedral varieties Modification 1M						
a) micas (in particular, lepidolites)	7	5.16-5.20	8.94-9.00	9.9-10.2	100.0-100.3	By celadonites and glauconites are understood dioctahedral Fe, Mg hydromicas, modification 1M, with perfect structures. Most celadonites have a = 5.20, b=9.00 A. In glauconites, a > 5.20, b > 9.00 A; in ordinary hydromicas, values of a and b are smaller, most frequently a=5.18, b= 9.96 A.
b) elongated-plate celadonites	15	5.20-5.22	9.00-9.04	10.0-10.3	100.0-100.5	
c) isometric-plate celadonites	2	5.21	9.04	10.4	100.5	
d) glauconites and hydromicas with clearly expressed strict c repeat units	29	5.17-5.23	8.94-9.08	9.9-10.3	100.0-101.0	
e) hydromicas without clearly expressed strict c repeat units	41	5.17-5.20	8.94-9.00	10.0-10.5	100.0-101.5	
f) Al montmorillonite with strict c repeat unit (beidellite)	1	5.14	8.90	10.4	100.3	

Table 64 (continued)

Minerals and combinations	No. of speci- mens	Unit cell constants (cell sides in A, angles in degrees)				Notes
		a	b	c	β	
g) montmorillonites with strict c repeat units	6	5.16-5.19	8.92-8.99	10.0-10.4	100.0-100.5	For montmorillonites, values of c and β relate to vacuum conditions. Most common values: a = 5.17, b = 8.94 A
h) montmorillonites with poorly defined strict c repeat units	7	5.17-5.19	8.94-8.97	10.0-10.2	100.0	
i) Fe montmorillonites (nontronites) with barely discernible strict c repeat units	3	5.20-5.24	9.00-9.08	~10	~100	
Modification 2M$_1$						
a) muscovites	11.	5.15-5.20	8.92-9.00	20.0-20.2	95.5	
b) sericites	20	5.16-5.22	8.94-9.04	19.7-21.0	95.0-96.0	Most often, a = 5.18, b = 8.96 A
c) hydromicas and montmorillonites with poorly defined strict c repeat units	12	5.17-5.19	8.94-8.97	19.8-20.4	95.0-96.0	
d) special varieties (K, Na hydromica of Lazarenko; allevardite)	3	5.13	8.88	19.7-19.9	95.5	
e) pyrophyllite	6	5.13-5.14	8.87-8.90	18.3-18.5(?)	95.0-96.0 (?)	

Table 64 (continued)

Minerals and combinations	No. of specimens	Unit cell constants (cell sides in A, angles in degrees)				Notes
		a	b	c	β	
Modification $3T$						Limiting values of a,b are met with rarely. Usual values are intermediate
a) hydromicas	16	5.16-5.20	8.94-9.00	29.0-30.9	90	
b) montmorillonites	9	5.16-5.20	8.94-9.04	30.0	90	
Modification $6H$ $(2M_2 + 2M_1)$						Values in brackets correct so long as specimens are not combinations of $2M_2 + 2M_1$
a) gümbelites	2	5.17,5.19	8.96	(60.0)	(90)	
Combinations of modifications						Signs $>$, $<$, $>>$, $\simeq$ in table show apparent relationship of components in diffraction patterns.
a) $3T + 2M_1$	1	5.24	9.04	29.5	90.0	($3T$)
				19.8	95.0	($2M_1$)
b) $2M_2 + 2M_1$	1	8.96	5.18	20.0	99.0	($2M_2$)
		5.18	8.96		95.5	($2M_1$)
c) $1M + 2M_1$ with $1M > 2M_1$	37	5.16-5.20	8.93-9.00	10.0-10.3	100.5-101.5	($1M$)
				19.9-20.2	95.0-96.0	($2M_1$)
d) $1M + 2M_1$ with $1M \simeq 2M_2$	8	5.17-5.19	8.94-8.98	10.0-10.3	100.0-101.5	($1M$)
				19.8-20.2	95.0-96.0	($2M_1$)

Table 64 (continued)

Minerals and combinations	No. of speci-mens	Unit cell constants (cell sides in A, angles in degrees)				Notes
		a	b	c	β	
e) 1M + 2M₁ with 2M₁ > 1M	2	5.19,5.20	8.92,9.00	20.1, 20.2	95.0,95.5	(2M₁) (1M)
f) 1M + 2M₁ with 1M >> 2M₁	1	5.18	8.98	10.2 / 10.1 / 19.8	101.0 / 100°40' / 95°30'	(1M trioct) (2M₁ dioct)
Varieties without definite layer sequences						Here the absence of a strict c repeat unit is result of disordered displacements of networks in layers along b-axis direction, by distances Δy, which are multiples of 1/3, which leads to broadening of reflections with k ≠ 3k'
a) hydromicas without strict c repeat units	5	5.17	8.95	10.0-10.2	100.0-100.5	
b) montmorillonites without strict c repeat units	10	5.16-5.19	8.94-8.99	~10	~100	
c) montmorillonites without mutually oriented layers	2	5.17-5.18	8.94-8.96	—	—	
d) nontronite	1	5.21	9.05	—	—	
B. Trioctahedral varieties Modification 1M (3T)						
a) lepidolites	6	5.19-5.24	8.97-9.08	10.0	100.25	
b) phlogopites, biotites	12	5.26-5.30	9.12-9.18	10.0-10.25	100.0-100.5	
c) vermiculites, no H₂O lost	2	5.31	9.21	14.4	96°	Nonrigorous c repeat unit

Table 64 (continued)

Minerals and combinations	No. of specimens	Unit cell constants (cell sides in A, angles in degrees)				Notes
		a	b	c	β	
d) dehydrated vermiculites	2	5.30	9.20	10.1	100.0	Values of c and β for vacuum conditions
e) montmorillonite	1	5.29	9.19	10.3	100.0	
Modification $2M_1$						
a) lepidolites	2	5.17,5.18	8.92,8.96	19.6,20.0	95.1	
b) biotite	1	5.28	9.16	20.2	94.5	
Modification $2M_2$						
a) lepidolites	3	8.92,8.96	5.15,5.17	19.7,20.5	99.8,98.5	
Varieties without definite layer sequences						
a) talcs	8	5.26-5.27	9.10-9.14	9.45-9.55	100.5-101.0	Nonrigorous c repeat unit
b) montmorillonites without strict c repeat units	2	5.25,5.28	9.11,9.15	~10	~100	
c) montmorillonites lacking three-dimensional order in their structures	3	5.23-5.27	9.08-9.18	—	—	
Chlorites						
a) Li chlorite—cookeite	1	5.17	8.93	14.3	98.75	In Li chlorite, mica-type layers are dioctahedral and brucite-type layers trioctahedral. Other chlorites wholly trioctahedral
b) chlorite showing strict c repeat unit (modification $1Tk_1$)	1	5.30	9.18	14.0	96.0	

Table 64 (continued)

Minerals and combinations	No. of specimens	Unit cell constants (cell sides in Å, angles in degrees)				Notes
		a	b	c	β	
c) chlorite with nonrigorous c repeat units	25	5.25-5.32	9.12-9.20	14.0-14.5	94.0-97.0	These chlorites diagnosed from variations in continuously distributed intensities
d) chlorites without mutual orientation of layers	4	5.28-5.31	9.12-9.18	—	—	
Mixtures of minerals Dickite + sericite	1	8.90	8.98			First constant is that of predominating mineral
Kaolinite (triclinic cell) + sericite a) mixture with K > Sr	3	8.90	8.95-8.96			For one specimen, K>> Sr
b) mixture with Sr > K	3	8.94-8.97	8.89-8.90			Signs >, <, >>, ≈ indicate "diffractional" preponderance of components in mixture
c) mixture with K ≈ Sr	1	8.90	8.94			

Table 64 (continued)

Minerals and combinations	No. of speci-mens	Unit cell constants		Notes
		b_1	b_2	
Kaolinite (triclinic cell) + hydromica 1M				
a) mixture with K > Hm	8	8.90	8.95–8.97	For one specimen, Hm >> K
b) mixture with Hm(1M) > K	5	8.94–8.98	8.90	
c) mixture with K ≃ Hm	2	8.90	8.94, 8.97	
Kaolinite (triclinic cell) + hydromica with nonrigorous c repeat unit				
a) mixture with K > Hm	3	8.88, 8.90	8.92, 9.00	
b) mixture with Hm > K	5	8.95–9.00	8.90	
Kaolinite (triclinic cell) + montmorillonite				
a) mixture with K > Mo	10	8.88–8.90	8.93–9.00	For two specimens, K >> Mo
b) mixture with K < Mo	10	8.94–9.00	8.90	For one specimen, K << Mo
c) mixture with K ≃ Mo	7	8.90	8.94–9.06	
Kaolinite (monoclinic cell) with nonrigorous c repeat unit + sericite	5	8.90	8.96	

Table 64 (continued)

Minerals and combinations	No. of speci- mens	Unit cell constants b_1	b_2	$b_{1,2}$	Notes
Kaolinite (monoclinic cell) with nonrigorous c repeat unit + hydromica					
a) probable mechanical mixtures of K + Hm ($2M_1$)	2	8.90	8.96		One specimen was Hm ($2M_1$), in three specimens K > Hm, in three, K ≃ Hm
b) probable intergrown structures (monothermites)				8.93	
Kaolinite (monoclinic cell) + montmorillonite "beidellites"					
a) probable mechanical mixtures	12			8.93	In three specimens, K > Mo, in four, Mo > K, in two, K ≃ Mo.
b) probable intergrowths — "beidellites"	21			8.93	In three specimens, K > Mo, in fourteen, Mo > K, in four, K ≃ Mo
Kaolinite + combination of 1M and $2M_1$ modifications of three-storied layer minerals	2	8.90 8.95	8.96 8.90		K + Hm Hm + K

Table 64 (continued)

Minerals and combinations	No. of specimens	Unit cell constants b_1	Unit cell constants b_2	Notes
Mixtures of three-storied layer minerals	9	8.96-9.16	8.90-8.98	Mixtures with following characteristics: $Sr \approx Hm$, $G > Mo$, $Hm \gg Mo$, $Sr \approx Hm$ (without strict c repeat unit), $Hm > Mo$, $B > Sr$, Hm (1M) $\approx Mo$, $Hm \approx Mo$
Chlorite + three-storied layer mineral				
a) C + Sr (C > Sr)	1	9.18	8.98	
b) C + Sr (C << Sr)	1	8.96	9.24	
c) C + Sr (C ≈ Sr)	3	8.96-9.00	9.12-9.24	
d) C + (1M + $2M_1$), C > ($1M+2M_1$)	2	9.18	9.00	
e) C + Hm (C > Hm)	4	9.12-9.30	8.93-9.00	
f) C + Hm (C < Hm)	3	8.93-9.00	9.15-9.17	In one case, C << Hm
g) C + Hm (C ≈ Hm)	2	8.96	9.15-9.17	
h) C + V (C > V)	1	9.20	9.20	
i) C + talc	1	9.20	9.15	

Table 64 (continued)

Minerals and combinations	No. of speci-mens	Unit cell constants			Notes
		b_1	b_2	b_3	
Three-component mixtures of layer silicates					
a) C + Hm + K	5	9.12-9.32	8.93-9.00	8.88-8.90	In different cases different components predominated, in one case the specimen was Hm ($2M_1$)
b) C + Mo + K	1	9.13	8.96	8.90	
c) Hm + Mo + K	7	8.95-9.00	8.95-9.00	8.90	The same comment applies as for (a)
Mixtures of hydromica (mont-morillonite) with palygorskite	8	8.94-8.97	17.88-17.94	—	In five specimens, Hm > P, in one, Hm ≈ P, and in one (Hm + Mo) >> P
Mixture of a layer silicate (chlorite) with a mineral of another structure type (hematite)	1	9.20	8.70	—	

certain limits. We can call these structures kaolinites with triclino-monoclinic cells.

The intermediate position between kaolinites with perfect structures and
those with imperfect structures is occupied by kaolinites which simultaneously contain elements of order and elements of disorder. We can call them kaolinites with not quite perfect structures.

Thus, from the number of possible combinations of both parameters,
each of which contains three categories, we can have a nominal total of nine
different structural forms of kaolinite. When using these parameters, it should
be borne in mind that the concept of a unit cell is strictly only applicable to
the perfect structures. If different cells in the structure are not completely
identical (through differences in the positions of atoms, or through variations
in their size or form), they should be considered as pseudocells. The concept
of the pseudocell is necessary because it serves to describe the periodicity
which does exist in the structure. In the nomenclature used for the structures
the terms "triclinic," "monoclinic," and "triclino-monoclinic" refer only to
the shape of the cell, and should not be associated with the symmetry of the
kaolinite structure. From their intensity distributions, the kaolinites, whatever
their degree of ordering, and whatever the shape of their cells, all have layer
displacements as given in Table 32, according to the formulas $\sigma_2\tau - \sigma_2$ or
$\sigma_4\tau + \sigma_4$.

Within the limits imposed by each of these forms, individual kaolinite
specimens may also differ with respect to the ordering of the structures of the
individual layers and of their relative arrangements, and also with respect to
the long-range order possessed by the structures.

This subdividing of the kaolinites is not unexpected, in spite of the fact
that X-ray studies had appeared to have provided quite a complete picture of
the structures and varieties of the kaolinites. According to the X-ray data,
the kaolinites form a continuous series, from varieties with perfect structures
and triclinic unit cells to varieties with disordered structures in which random
layer displacements lead to a unit cell which averages out as monoclinic.

This picture, however, leads to certain contradictions and inconsistencies, since a sharp demarcation line between different kaolinites according to
the shapes of their unit cells (triclinic or monoclinic) is incompatible with a
continuous range of degrees of structural perfection. It is impossible to con-
ceive how a change from kaolinite with a triclinic cell to kaolinite with a
monoclinic cell could take place within such a continuous series.

The arbitrary division into nine structural groups, based on objective data obtained from analysis of electron-diffraction patterns, thus has more justification, since it does not involve the sharp demarcation line noted above according to the shape of the kaolinite unit cells.

We should note that the diagnostic features used (cell shape and perfection of structure) are indirect and superficial in character. The shape of the cell does not determine the distribution of the atoms in it, and a qualitative estimate of the perfection of a structure cannot itself show the nature of those imperfections which occur in the kaolinite structure, and not expose the limitations which exist for all kaolinite structure varieties. It is therefore best to evaluate the above diagnostic features for kaolinite and the resulting structural forms in the light of the fine details of the kaolinite structure which have been established from detailed investigations of the kaolinite minerals (Brindley, Nakahira, 1958; Zvyagin, 1960; Drits, Kashaev, 1960; Newnham, 1961).

It was shown above (Section 4 of Chapter 4) that the kaolinite structure can exist in two enantiomorphic modifications, represented by $\sigma_2 \tau - \sigma_2 \ldots$, and $\sigma_4 \tau + \sigma_4, \ldots$, respectively. If the kaolinite layers were made up of regular tetrahedra and octahedra, successive layers would be displaced by $a/3$ in such a way that the kaolinite structure would have a monoclinic unit cell with $\beta \simeq 104°$. It will be apparent that whatever the shape of the cell, if no symmetry elements are present the symmetry of the structure will stay at triclinic.

Structural investigations of kaolinite minerals have shown that kaolinite-type layers are subject to a series of deformations. In particular, the O atoms of the tetrahedron faces lying in and outside the layer symmetry plane are displaced by different amounts along the bisectrix of the ditrigonal pattern which they form. These displacements differ in such a way, that in formation of the shortest hydrogen bonds between the O atoms and the neighboring hydroxyl groups of the next layer, the latter are displaced relative to the former not only by $a/3$ along the a axis, but also by certain amounts along the negative b-axis direction. As a result, not only β, but also $\alpha \neq \pi/2$ ($\alpha = 91°40'$).

It is perfectly obvious that strict observation of these deformations, which control the triclinic shape of the unit cell, will be directly related to the internal structural perfection of the individual layers. When loss of ordering occurs in the structure of a layer, the influence of the factors causing these deformations will be less, so that the deformations may not be maintained over the whole structure of the layer. As a result, if there are a great many defects, the layers may arrange themselves in accordance with a monoclinic cell. In the intermediate case, the triclinic angle α may not be constant over the whole structure, but may alter over a certain interval (giving a kaolinite with a triclino-monoclinic cell).

It can be assumed that in the formation of the mineral in situ, the structural perfection of its layers will depend on the duration and stability of the crystallization conditions, and on how well the thermodynamic and physicochemical conditions of the formation medium correspond with the energetic stability limits of the kaolinite structure.

The other factor determining the perfection of a kaolinite structure is the ordering in the relative arrangement of the layers.

In the ideal case, all the kaolinite layers have the same orientation. Such an arrangement is, however, unstable. According to results of a structural analysis of dickite (Newnham, 1961), in a kaolinite layer the O atoms lying on one outer surface of the layer and the OH groups on the other surface, lying in the symmetry plane of the layer and outside it, have differing z coordinates measured along the height of the layer, so that they form depressions and hollows. When these hollows and depressions come into conjunction, the arrangement of the layers in the structure is metastable. Under the influence of external forces this arrangement of layers will tend to alter, so that the hollows in the surface of one layer match up with projections on the surface of the next layer; this gives rise to the favorable layer combinations $\sigma_4\tau-\sigma_2$ or $\sigma_2\tau + \sigma_4$.

The pattern of changes in the relative arrangement of the kaolinite layers can be represented as follows:

Because of the layer nature of its structure and the weak bonding between the layers, kaolinite has a perfect cleavage. When linear displacements occur under the influence of various physical forces, through impact of mineral particles with each other, the crystals are able to split up along their cleavage planes. Displacements can also occur in which one part of a crystal becomes rotated relative to another, so that the favorable layer combinations noted above come into being in the intermediate regions between these parts. When single twinning occurs, a kaolinite layer sequence, for example $...\sigma_4\tau_+\sigma_4\tau_+\sigma_4...$, is split up into two zones separated by a "dickite" intermediate region, in the form:

$$\cdots\sigma_4\tau_+\sigma_4\cdots.\qquad \tau_+\sigma_4\tau_-\sigma_2\tau_+\sigma_2\cdots.\qquad \tau_+\sigma_2\tau_+\sigma_2\cdots$$

zone I   intermediate   zone II

or

$$\cdots\sigma_4\tau_+\sigma_4\cdots\qquad \tau_+\sigma_4\tau_-\sigma_2\tau_-\sigma_2\cdots\qquad \tau_-\sigma_2\tau_-\sigma_2\cdots$$

zone I   intermediate   zone II

In the first case, the intermediate region divides two zones of the same kaolinite modification, rotated relative to each other by 120°. In the second case, the intermediate region lies between two enantiomorphic kaolinite modifications.

For reflections with $k \neq 3k'$, both zones scatter independently and make individual contributions to the intensities of the appropriate reflections on the electron-diffraction patterns (or X-ray photographs). In the case of reflections with $k = 3k'$, both the zones and the region between them are indistinguishable, and so they all scatter coherently and make contributions to the reflection amplitudes, so that the latter will be higher in this case.

When multiple twinning occurs in the crystals, the number of independent scattering zones for reflections with $k \neq 3k'$ increases, the zones themselves become narrower, and the corresponding reflections on the diffraction patterns become weaker and connected together by an ever-increasing background, so that they may not even be separately distinguishable, but are spread out into continuous ellipses on the diffraction patterns. Because of the adjacent intermediate regions, the coherent scattering zones for reflections with $k = 3k'$ are larger than those for reflections with $k \neq 3k'$, and may give clear reflections characteristic of a triclinic or monoclinic cell (depending on the perfection of the layers) even when the reflections with $k \neq 3k'$ are no longer distinguishable. The reflections with $k = 3k'$ also weaken here, those with $k = 0$ to a lesser extent than those with $k \neq 0$.

As an accompaniment to the loss of ordering in the structure due to layer displacements, deformations can be assumed to arise within individual layers, thus weakening the forces governing the triclinic distribution of layers within a particular zone. For this reason, kaolinites with triclinic cells and imperfect structures (no reflections with $k \neq 3k'$ on the diffraction patterns) are comparatively rare.

In the limiting case of complete disorder in the displacements of layers, it is no longer possible to pick out zones which can be considered as independent lattices with a triclinic cell. The complete structure as a whole can be described in terms of a monoclinic pseudolattice. In the diffraction patterns, reflections with k ≠ 3k' are smeared out into continuous ellipses, sometimes of low intensity, reflections with k ≠ 3k' ≠ 0 are weakened or completely indiscernible, while reflections with k = 0 are retained even for maximum loss of ordering in the internal structure and relative positioning of the layers.

The formation of combinations of differently oriented zones in kaolinite crystals could also be considered to occur through dispersion of the original crystals and subsequent coupling (condensation) of the fragments into new, more favorable combinations.

The fine structural features of kaolinite crystals are indicated by the following diffraction characteristics in electron-diffraction texture patterns.

1. The comparative development of reflections with k = 3k' and with k ≠ 3k', which are grouped on different ellipses in texture patterns, and the presence or absence of reflections with k ≠ 3k'. These characteristics indicate the character of the crystal ordering, which is related to displacements of the tetrahedron and octahedron networks and the layers they form along the b axis.

2. The shape of the first-ellipse reflections $\pm 02l$, $11l$, the degree of splitting and the resolution of the reflections $0\overline{2}l$ and $02l$, the magnitude of the intermediate background; these characterize the ordering in the crystal structure and the shape of the unit cell if the structure retains any order in the displacements of its networks and layers in the b-axis direction.

3. The character of the grouping of second-ellipse reflections into fours. If the cell is triclinic, this does not affect the outer reflections of a set, $\overline{2}.0.l + 1$ and $20l$, but it does alter the inner reflections, $13l$ and $\overline{1}.3.l + 1$. As shown in Section 1 of this chapter, when $\alpha = 91°40'$ they are arranged at equal distances apart, when $\alpha = 90°$ the inner separations are considerably bigger than the outer ones, and when α is varying the inner reflections are smeared out and not separately distinguishable. This feature also characterizes the shape of the unit cell when the networks and layers are randomly displaced.

4. The relative development of the $20l$ and $13l$ reflections, which may indicate to what extent layer and network displacements differ from multiples of b/3, and may also reveal deformations not connected with these displacements.

5. The number of reflections visible in the 06l, 33l series, which indicates the degree of long-range order present in the kaolinite structure.

Varieties of two-storied layer minerals other than the kaolinites have not been analyzed in such detail. As already noted above (Chapter 1; Section 4 of Chapter 4), dickite and nacrite belong to the modifications 2M$_1$ and 2M$_2$, respectively (structures I,4 and II,2 of Table 32).

In the case of halloysite, the following general conclusions may be drawn.

1. Halloysite is a mineral of the kaolinite group, with an unusual structure which includes several different states, varying in the degree of association of their kaolinite layers with water molecules. When little or no water is present, the structure approximates to a monoclinic model with a two-layer repeat unit (structure I,3 of Table 32), which is not known for any other minerals of the group. Because the alternating layer-pair combinations are energetically equivalent (equivalent combinations in kaolinite), and also because of the influence of the water molecules, ideal ordering is not attained in practice, and the real structure is often extremely imperfect. Nonetheless, its unusualness and difference from the structures of other kaolinite minerals are demonstrated by the more perfect specimens.

2. Halloysites should not be considered as extremely poorly ordered varieties of kaolinites. They belong to a series of their own, running from well to poorly ordered forms of the structures attained in the dehydrated state; for the same degree of ordering present, kaolinites and halloysites approach completely different structure models (see point 1). When maximum disorder is present, and it is impossible to detect the appearance of features of any model, both series merge together. Thus, two branches of the kaolinite group are formed, connected together by the most highly disordered kaolinites of the "fireclay" type.

3. In nature, the halloysite structure does not crystallize in the form of platy crystals, which is the form which would be appropriate to a single lattice, but as a combination of radial zones expanding outward from a common axis (usually extremely close to the b axis), thus forming elongated, prismatic, and very frequently hollow crystallites, all faces of which are parallel to the ab plane.

4. The presence of water molecules in the structure cannot be related to any special properties of the halloysite layers, since, in other kaolinite structures, layers which are the same in principle do not contain any water.For

the same reason, when halloysite is heated the water molecules are lost ir-
reversibly. Evidently the appearance of water molecules in the halloysite
structure is tied up with specific formation conditions, which at the same time
control the features of the crystallites which have been described. It must be
supposed that all the halloysite differences, i.e., water in the structure, radial
zoned prismatic crystals, and the unusual two-layer monoclinic structure in
the dehydrated state, are not independent features, but are closely related to
each other and to the conditions of formation.

In view of the fact that each of the kaolinite group mineral varieties
mentioned has its own distinctive layer type and sequence, it is obvious that
they should never be placed in any series, as has sometimes been done, which
assumes a gradual change in structural features on going from one variety to
another.

It is noted that in both the two-storied layer minerals and in those with
layers of other types (see below), rough subdivisions have been made among
the specimens investigated according to their degree of structural perfection,
as indicated by the corresponding subheadings. However, even within a single
group of specimens, all falling under the same subheading, the qualitatively
estimated degree of structural perfection may vary considerably.

Historically, the position has been complicated by the fact that the
classification and nomenclature of trioctahedral varieties of two-storied layer
minerals, particularly the serpentines, is at present based on variations in
chemical composition, geometrical characteristics of crystal lattices, and
morphological features of the crystals.

At the same time their layer sequence behavior, as shown by their reflection intensity distributions, has not been brought out and used to any great
extent. Such a position could not but lead, and has in fact led, to the appearance of a rather superficial formulation of determinative mineral features,
and to a number of vague features and ambiguities in the terminology used.

In present practice the serpentines are divided up into three basic categories, according to the nature of their lattices and their crystal morphology.
These are the platy serpentines (rectilinear lattices, platy crystals), the antigorites (corrugated lattices with a straight-line A superlattice constant), and
the chrysotiles (cylindrical lattices, tubular crystals). Each of these three subdivisions is divided up in its turn into the individual varieties. In the platy
serpentines a distinction is drawn between the so-called one-layer and six-
layer varieties, i.e., those for which the unit cell extends over one or six

layers. The deficiencies of this purely geometrical and moreover incomplete method of defining these serpentines will be perfectly obvious. It is not specified whether the repeat unit is strict or nonrigorous. Moreover, in an article by Whittaker and Zussman (1956), one-layer serpentines are described (the authors call these lizardites) which appear to differ from one another geologically and structurally. Only two specimens gave X-ray photographs containing the 021 and 022 reflections which indicate a strict layer repeat unit (in the original work the l indices of these reflections were twice as large). The X-ray photographs of the other specimens contained only reflections with k = 3k'.

It was shown above (Chapter 4) that the four ordered serpentine structures which differ in their relative arrangements of layers, numbers of layers in the repeat unit, and spatial symmetry, all have identical hk$l$ reflections with k = 3k'. They can be distinguished by the arrangement and intensities of their reflections with k ≠ 3k'. In the diffraction patterns of serpentines with disordered structures, the reflections with k ≠ 3k' are smeared out into continuous diffuse scattering bands, which eliminates the features by which the different ordered structures could be recognized. If, therefore, disordered serpentines show the same k = 3k' reflections as ordered serpentines, they can either be considered as end-members of series of specimens of gradually decreasing structural perfection, one series for each ordered modification, or they can be considered as a separate variety. But there is no reason not to associate them with one of the ordered modifications, so long as the diffraction patterns contain any rudiments of the appropriate reflections with k ≠ 3k'. For the same reason, it would be incorrect to apply one general name to the specimens described by Whittaker and Zussman. Moreover, there is no reason to dispute the priority of these specimens with respect to any particular name. It would be more to the point to specify strictly and unambiguously what should be understood by each name.

The identification of a "six-layer serpentine" is also unsatisfactory. There exist five ordered structures, belonging to four different modifications (see Tables 9, 34), for which it is possible to select an orthogonal cell six layers in height. These differ in the symmetry and relative arrangement of their layers. An identification of these should not be limited to considering the geometrical features of the structures and the arrangement of the corresponding diffraction-pattern reflections. It is necessary to make use of the intensities of the reflections and specify the way the layers follow each other in the structure.

Gillery has shown (1959) that the manner in which platy serpentines differ from antigorites and chrysotiles may depend on the degree of isomorphous replacement of Mg cations by Al. As the number of Al cations is reduced, the difference in dimensions between the tetrahedral and octahedral networks of a two-storied layer becomes greater, and its bentness increases. The antigorites occupy an intermediate position between the serpentines with rectilinear lattices and those with cylindrical lattices. It is true that the majority of the chrysotiles have the generatrix of their cylinders parallel to the a axis, while in the antigorites the generatrix of the wave corrugations runs along the b axis. The antigorites differ among themselves with regard to their values of the superlattice constant A; these fall into three groups of values, around 15-20, 45, and 100 Å, respectively.

The chrysotiles are divided up into clino and ortho varieties, corresponding to monoclinic and orthogonal cells in their lattices. The chrysotiles with cylinder axes parallel to the b axis of their lattices are called para-chrysotiles.

Within the framework of these ideas, a number of points concerning the layer sequences remain in doubt in both antigorites and chrysotiles, involving the order—disorder relationships, symmetry and pseudosymmetry, and the extent to which the structures may vary.

In these circumstances, it is best to use the intensity distribution relationships established above (Chapter 4) to see which modifications are shown up experimentally by the diffraction patterns of platy serpentines, antigorites, chrysotiles, related minerals, etc., regardless of how their rectilinear lattices are deformed.

The diffraction characteristics given in Tables 32 and 35 were compared with electron-diffraction texture patterns and single-crystal patterns obtained by the microdiffraction method from a series of serpentine specimens and two-storied layer Ni minerals. Their structural characteristics are noted in Table 64. The high information content of the texture patterns and the unusual intensity distributions open up a real possibility of determining not only individual structure types (A, B, C, and D, see Chapter 4), but also their combinations. In Table 64 the structural components are given in the order in which they predominate. Components showing up only weakly because of their insignificant content in the specimen are shown in curly brackets. The platy serpentine specimen studied by Ushakova (1959) deserves particular attention; in this the structure IV',1 of modification 1T predominated (structure type A), and the degree of structural perfection was extremely high (see Fig. 71). This serpentine specimen showed, in a most clear and definite manner, the same structural features as those possessed by the most perfect of the lizardite specimens studied by Whittaker and Zussman (1956). Considerable interest has arisen in a specimen of platy serpentine with a superficially fibrous character, from the Central Geological Museum of the All-Union Geological Institute (P. M. Tatarinov collection, obtained on the River Laba). This is a very clear representative of a disordered structure, giving a diffraction pattern containing sharp reflections corresponding to a type-D structure, and diffuse intermediate reflections from a type-B structure (see Fig. 72).

Table 64 demonstrates the diverse structural characteristics of the two-storied layer Ni minerals. From a more detailed study of these we may expect to obtain new factual material on the crystal chemistry of these substances and on the polytypism (polymorphism) of two-storied layer minerals. It is obvious that a foundation could be laid here for a more rational and decisive classification of these as yet insufficiently studied substances.

In the specimens examined, the structure types A, B, and D were established either in pure form or in combination with one another. Structure type C is apparently unlikely to occur because the superimposition of close-lying Mg cations on the Si atoms of adjacent layers makes it energetically unfavorable.

Three-Storied Layer Minerals. When electron-diffraction patterns from the whole field of three-storied layer minerals are considered, it is seen that these minerals form several continuous series, with one set of end-members represented by the micas and mica-type minerals with perfect structures, and the opposite end-members represented by the hydromicas or montmorillonites with hardly any apparent order in the arrangement of their layers.

One very important point here is that, according to the results of Yoder and Eugster (1955), there is a difference in principle between analogous dioctahedral and trioctahedral varieties with similarly ordered structures, as is shown in particular by the lack of continuous transitions between them and by the absence of intermediate forms.

The most numerous and wide-spread of the three-storied layer silicates are those forming the series of dioctahedral monoclinic or pseudomonoclinic minerals with a one-layer repeat unit (modification 1M). This series starts off with the structurally perfect micas and celadonites, and ends up with the montmorillonites in which the variation of intensities on the texture pattern first ellipse may be barely sufficient to detect a tendency toward mutual positioning of layers. The series passes through glauconites and hydromicas with varying degrees of structural perfection.

Considered as a whole, the groups of specimens indicated in the table as showing clearly expressed and poorly expressed strict c repeat units differ considerably one from the other, but there is no sharp boundary between the groups, and the division must be purely arbitrary.

In the series under consideration, specimens 533 and 530 of D. M. Shilin deserve mention (Table 64), as these have extremely perfect celadonite-type structures but possess a crystal morphology which is unusual for celadonites; we can refer to these as "isometric platy celadonites."

A specimen of importance in the crystal chemistry of layer silicates is the Al montmorillonite which, according to the results of Weir and Greene-Kelly (1962), contains Al atoms in the tetrahedra and falls under the crystallo-chemical heading of a beidellite (received from Weir and Greene-Kelly). This montmorillonite, in agreement with its composition, has distinctively small a and b values and a three-dimensionally ordered structure. The existence of such a montmorillonite is an extremely weighty argument in favor of using the term "beidellite" to refer to this possibly extremely rare variety of montmorillonite.

Another series, also of dioctahedral varieties, has monoclinic or pseudo-monoclinic symmetry and a two-layer repeat unit (modification $2M_1$). This starts with muscovites with perfect structures, passes through sericites with ever-increasing loss of ordering in the arrangement and internal structures of layers (as shown by the intermediate diffuse background between the diffraction pattern reflections), continues through the hydromicas, and goes right through to the montmorillonites. In montmorillonites the features of the modification $2M_1$ apparently do not relate directly to the swelling part of the substance, but to fragments of the sericite structures from which the montmorillonite may have been formed. The possibility of finely dispersed sericite impurity being present also should not be ruled out, but this can usually be recognized from the qualitative differences in the character of its reflections.

Allevardite and sericite-type hydromicas (E. K. Lazarenko's specimens) lie outside the continuous series of $2M_1$ modifications, and have special chemical compositions and layer arrangements.

To complete the picture, the table also includes the pyrophyllites, although, since these do not observe a strict layer sequence, they can only approximately be assigned to modification $2M_1$. The structural varieties of the pyrophyllites are subject to correction.

The third series consists of hydromicas and montmorillonites for which the diffraction pattern intensities suggest that they correspond closely to the trigonal modification 3T, with a three-layer c repeat unit (the geometry of the reflection layout for these is the same as that for modification 1M). Some of the specimens of relevance here (Nos. 1 and 2 of V. F. Petrun') have rather high c constants, which may be due to incomplete removal of interlayer water molecules in the vacuum. It is possible that this behavior is shown by dioctahedral vermiculites. However, both this point and the very existence of such vermiculites demands a more rigorous experimental verification.

The layer sequences characteristic of the individual modifications need not be strictly maintained to any specified degree. As shown by anomalies in the scaling of experimental intensities to theoretical ones, these sequences may be randomly disordered by Belov's twinning mechanism (1949), which can apply to all layer silicate varieties.

If Belov twinning leads to the appearance of a new layer sequence, the crystallites end up as a certain combination of modifications.

Among the materials investigated, a long succession of specimens were found which showed the simultaneous appearance of several layer sequences. It should be noted that the combining modifications possess the same a, b dimensions and thus may consist of layers of identical composition, but in different orientations, forming what in a certain sense is a single structural unit.

In connection with the fact that the arrangement of reflections for the modification $2M_1$ includes within itself the particular arrangements for the modifications 1M and 3T, a combination of these modifications, in accordance with Table 39, can only be recognized from anomalies in the intensity distribution.

There is no way in which these anomalies can be related to any particular single modification as being due to a given unusual layer chemical composition. As Table 64 demonstrates, the basic diffraction properties of the individual modifications are retained over a wide range of specimen variation, which may include changes both in chemical composition and in structural perfection. This is understandable when it is remembered that both isomorphous cations and defects in layer sequence and integrity are randomly distributed in the structures. Even if we assumed the existence of some special changes in the chemical composition of the layers which would strongly affect the reflection intensities, it is not clear why this should lead to an intensity distribution which would be identical to the sum of the intensity distributions for separate modifications. Tables 43 and 44 show convincingly that there are no grounds for such a structure.

Generally speaking, similar diffraction effects may result from mechanical mixtures of minerals. Here, however, the equivalence of their a,b values as noted above should be taken into account. Moreover, certain investigations have shown (Hibi, 1955) that changes in layer sequence occur within the bounds of a single crystal, which thus consists of zones of different modifications.

The most common combination is that of the modifications 1M and $2M_1$. Their interesting feature is an increased monoclinic angle in the modification 1M (an increased angle β), which indicates a certain amount of distortion in the packing of the structure's anions. As a result of this, the $11l$ reflections of the two modifications no longer strictly coincide, which leads to the appearance of characteristic diffuse spots on the first ellipse of the diffraction pattern.

In Table 64 and in the mixtures following, the data for the different components are given in the order in which they predominate. By this, what is understood is their "diffractional" predominance, which may not correspond to the true quantitative relationship between the components of the combination or mixture, since different structure types may not show up equally strongly in the diffraction patterns.

The order in which the components are shown is not the only one, since different modifications may predominate in different specimens with the same degree of structural perfection.

In this series also, we pass from specimens with perfect structures, sometimes celadonite-type or sericite-type structures, to hydromicas with imperfect structures.

In the accepted sense, layer silicates are always, to some extent, mixed-layer structures, since both their layer sequences and the character of the layers themselves may alter.

In the extreme case where there is complete disorder in the sequence of layers, it becomes no longer possible to distinguish any modifications. This position is typical of some hydromicas and many montmorillonites. Among these varieties the ends of the series dealt with above become lost, and distinctions between the varieties also lose their significance. Within this group of specimens, however, two stages of structural imperfection may be distinguished. In one, the diffraction-pattern reflections with k = 3k' are still perceptible, indicating the presence of a pseudomonoclinic cell, and in the other the patterns show only the effects of intralayer diffraction, indicating that the structure consists of a one-dimensional sequence of randomly displaced parallel layers.

Among the montmorillonites we may distinguish varieties which, even in the absence of three-dimensional layer ordering, give extremely contrary and sharp intralayer diffraction patterns, indicating that the actual layers themselves have perfect structures. Other specimens will have a diffraction pattern with a strong background and weak reflections. Their lack of perfection is compounded from loss of ordering both in the relative arrangement of layers and within the individual layers. This loss of ordering may result in the formation of an amorphous substance.

In the trioctahedral varieties, we can note only that a sequence appears which may be characterized either by a one-layer monoclinic structure or a three-layer trigonal structure, or some combination of both, since, in texture patterns as in powder photographs, the trioctahedral modifications 1M and 3T are indistinguishable. Evidence of common diffraction properties for corresponding 1M and 3T modifications may be followed right through, starting with trioctahedral micas with perfect structures, through vermiculites, and right down to montmorillonites.

It is noted that Mg vermiculite specimens with structures which had been fully determined by Mathieson and Walker (1954) and Mathieson (1958), and which were kindly presented for electron-diffraction studies by G. F. Walker, did not show loss of interlayer water and a fall in the value of c to ~10 A. It is also noteworthy that the strict periodicity observed in the single crystal studied by Mathieson and Walker was not developed in the general mass of crystallites forming the electron-diffraction specimen.

With regard to the ambiguity between the modifications 1M and 3T, the unit cells given in the table have been nominally assigned to the monoclinic modification, but may be specified for any particular modification using relationships derived from formulas (12) and (13) of Chapter 1:

$$c_3 = 3c_1 \sin \beta, \quad \tan \beta_1 = \frac{c_3}{a}.$$

The trioctahedral modifications $2M_1$ and $2M_2$ are represented only by isolated specimens among the substances studied, and apparently are not found among the clay minerals. Table 64 also includes a specimen from P. P.

Tokmakov, which is suspected of belonging to the as yet unrepresented modification 6H. It is possible, however, that this is a $2M_2$ mica with a small amount of 1M mica impurity.

Unit cell measurements show that the values of a and b cannot themselves always decide whether a mineral is dioctahedral or trioctahedral. Trioctahedral lepidolites (specimens No. 80 of E. I. Nefedov and No. 718 of A.F. Golovachev) have comparatively small a and b constants and may be recognized only from their intensity distributions, which are characteristic of trioctahedral modifications (1M and $2M_1$). In spite of the small a and b values, the component $2M_2$ listed in the table in combination with $2M_1$ is evidently also trioctahedral.

Now that we are guided by the intensity distribution relationships for the different modifications, we can see the significance of the five types of three-storied layer mineral which were noted in an earlier review of the hydromicas (Zvyagin, 1956). These were the modifications 1M, $2M_1$, 3T, and the combination 1M + $2M_1$ with either one modification or the other predominating.

Minerals Containing Layers of Different Types. Of the minerals coming under this heading, we investigated only the chlorites. Unfortunately, we had only a limited number of specimens at our disposal, and from a study of these it was not possible to draw very general conclusions.

Almost all the specimens investigated had structures made up of layers which were randomly displaced relative to one another along the b axis, which is natural, since the brucite-type layers are insensitive to displacements relative to the mica-type layers by $\Delta y = 0, \pm^1/_3$. A perfect structure is possible in chlorites, but this may only hold for individual crystals. In a polycrystalline specimen a perfect structure may only be possessed by an insignificant proportion of the substance, as shown on the diffraction patterns by the appearance of weak reflections with $k \neq 3k'$, only distinguished with difficulty from the strong background of the corresponding ellipses (specimen No. 127 of A. N. Geisler and L. M. Myznikova; modification $1Tk_i$). On the other hand, lack of order is extremely probable in chlorite structures, and we have seen specimens which had diffraction properties similar to those of montmorillonites (specimen No. 503 of V. P. Shashkina).

In practical electron-diffraction studies, only Li chlorite—cookeite (Zvyagin, Nefedov, 1954), Mg vermiculite sent by Walker (described in Mathieson, Walker, 1954), and vermiculite specimen No. 96 of I. A. L'vova, have shown the presence of $|\sigma'|$ packets. These vermiculites did not lose water in a vacuum, and can be considered of the chlorite type. In most cases, the chlorites investigated gave diffraction patterns typical of structures with σ packets.

A specimen of jefferisite (No. 85, I. I. Ginsburg) gave a diffraction pattern according to which it occupies an intermediate position, and consists of zones of σ and | σ'| packets. In electron-diffraction patterns from specimens Zh-734 and Zh-2066 (Ya. K. Pisarchik's material), additional reflections were observed lying on the second ellipse somewhat closer to the minor axes than the $\overline{2}08$ reflection due to a structure of σ packets. According to Table 42, its appearance may be due to the presence of a structure with σ| packets in the specimen. Such structures have never been met with before. It can be expected that as chlorite studies continue, using a wider range of experimental material, chlorites of other types will be found and some idea gained of their comparative abundance.

Mixtures of Minerals. Results of electron-diffraction studies have shown that diffraction patterns will clearly and unambiguously reveal the presence of several layer silicates in the specimens, not only those belonging to different mineral types, but also, as already noted, those belonging to different polymorphic varieties of the same mineral.

The scope for identification of mixtures of layer silicates is graphically illustrated by Table 64. In view of the fact that the individual mineral types have already been considered fairly comprehensively, in the mixtures we have limited ourselves to giving the values of b, which to some extent reflect differences in chemical composition.

For brevity the following abbreviations have been used: B, biotite; V, vermiculite; G, glauconite; Hm, hydromica; D, dickite; K, kaolinite; Mo, montmorillonite; P, palygorskite; Sr, sericite; and C, chlorite.

Usually, when a diffraction pattern shows the characteristics of several structures, it is concluded that the specimen under study is a mechanical mixture, i.e., one in which the different structures relate to different crystallites, generated independently. This is a completely unjustified conception of the meaning of the term mixture, which undoubtedly has a more general sense. The term should not be restricted to one special set of circumstances, since there is no way of deciding whether the mineral components or polymorphic modifications present in a specimen were formed independently or are genetically related.

It cannot be guaranteed in advance that both these alternative cases will be diffractionally identical, and so it would be silly to give up attempts to distinguish them. On the other hand, it is natural and logical to assume the existence of certain individual structural features which will depend on the conditions of formation, and by means of which it may be possible to determine the different forms of mixtures, and at the same time form an opinion on the processes to which they must correspond in nature.

During electron-diffraction studies of a series of clay minerals which possessed the general properties ascribed to so-called beidellite (Zvyagin, 1958a), the specimens were observed to contain kaolinite and montmorillonite mineral components. At the same time, the electron microscope showed that the crystallites present were all of similar shape, usually with indistinct outlines, and from this indication alone it did not appear that the specimens could be mechanical mixtures. This view was supported by the fact that the diffraction patterns contained a number of unusual features; the reflections corresponding to kaolinite "monoclinic" varieties were of a qualitatively similar nature, and the ellipses were separated only poorly or not at all, indicating that the a and b constants of both components differed by small amounts. In respect to all these points, the diffraction patterns differed from those of either artificial or mechanical mixtures.

From this the suggestion arose that these "beidellites" were intermediate transition products of the kaolinite—montmorillonite system, so that the crystallites did not have a uniform structure, but one in which the structures of their inner parts and surfaces were different. Similarly, the impression was formed that the monothermites were analogous intermediate products, lying between kaolinite and hydromica. At the same time, it had to be recognized that the hypothesis was less likely to be correct in the latter case than in that of "beidellite." It is observed that the use of the term "beidellite" in the present case is obviously not justified (Zvyagin, Frank-Kamenetskii, 1961; Weir, Greene-Kelly, 1962). Of the mixtures of monoclinic kaolinites with imperfect structures listed in Table 64, an indication is given of those which, from the overall qualitative aspects of their diffraction patterns, could be assumed to be "beidellites" or "monothermites" in the sense indicated. It is interesting to note that the hydromica component of monothermite specimen No. 345/Ya (M. F. Vikulova) shows weak indications of the modification $2M_1$. In all other cases the mixtures are understood to be such in the most general sense.

It is particularly simple to detect mixtures containing minerals which are not layer silicates, provided they show up on the diffraction patterns, because their reflections usually have a different qualitative character and are distributed according to different rules, often in parts of the diffraction field which are free from layer silicate reflections. As an example of this, a natural mixture of chlorite and hematite was chosen (specimen No. 584/3 of V.P. Shashkina).

To investigate the scope of electron-diffraction methods in the study of clays of complex mineralogical composition, R. A. Shakhova (1962) prepared a series of binary mixtures of two kaolinite varieties and montmorillonite, nontronite, hectorite, and hydromica, and also hydromica—hectorite and montmorillonite—nontronite mixtures. By varying the concentrations of suspensions she was able to prepare electron-diffraction specimens containing predetermined proportions by weight of the mineral components. Because of the differences in the a and b constants, the presence of different minerals in the mixture led to separation of the diffraction-pattern ellipses even when one component was present only in a minor amount (5-10%), this showing up particularly clearly for the fifth ellipse (060 reflections). It was apparent at first glance that the diffraction patterns of artificial mixtures of kaolinite and montmorillonite differed from those of the "beidellites" mentioned above. When comparing the diffraction patterns corresponding to different relative proportions of minerals in the mixtures, Shakhova made a note of groups of reflections lying close to the ellipse minor axes which related to different minerals and which had roughly equal intensities. These reflections could thus be used to characterize the corresponding relative mineral proportions in mixtures. Their intensities can also be used to give a rough estimate of the amounts of different minerals present in natural mixtures.

A Classification Scheme for Clay Minerals and Related Layer Silicates.   When the clay minerals and related specimens investigated by electron-diffraction methods were tabulated to give Table 64, this led directly to a definite classification scheme. In its construction, a number of general considerations were taken into account.

A classification of this type must fulfil its purpose by correctly showing the relationships existing between minerals and by serving as a guide to their identification.

This can only be achieved if the classification is built upon a crystallochemical, structural foundation, thus excluding any ambiguity in mineral identification. The classification criteria here may be hierarchical in nature, passing from more general and fundamental features and major differences down to more specific and less fundamental properties and minor distinguishing features.

These points apply to the following hierarchy, to some extent based also on the distribution of the specimens studied by electron diffraction according to their structural and mineralogical features:

(a) type of structure (layer, pseudolayer);

(b) type of silicate layer (two-storied kaolinite type; three-storied mica-type; chlorite packet);

(c) extent of octahedron filling, tied up with the chemical composition of layers (dioctahedral, trioctahedral);

(d) polymorphic (polytypal) modification, defining the relative positions of layers;

(e) extent of layer interaction, tied up with the distribution of isomorphous cations (above all those in tetrahedral networks) and the manner in which the interlayer spaces are filled; and

(f) minor differences in the chemical composition of layers and in the spaces between them, not already dealt with in (e); and degree of structural perfection.

The first criterion separates the clay minerals into two very clearly distinguished sections; those with structures made up of individual layers in the form of definite combinations of two-dimensional tetrahedral and octahedral networks, and those without clearly expressed layers. Although Si—O tetrahedral networks also appear in the latter type of structure, because these have complete bands of inverted tetrahedra alternating on either side of the basal planes, and because there are intermediate octahedral bands joining the tetrahedral networks up into a single three-dimensional unit, these are not layer structures, and they also cannot be considered as ribbon structures. They may validly be described as pseudolayer or layer-ribbon structures.

The next criterion divides up the layer minerals according to the way in which the tetrahedral and octahedral networks are joined up into layers. This criterion should obviously be assigned secondary importance, and this usually does not lead to objections.

Opinions differ only with regard to the chlorite layers. However, the scientific definition of a layer, which implies that the atoms within a layer are bound more strongly with each other than with atoms of other layers, means that we must take it that the "brucite" network of octahedra is an independent layer, forming an integral part of the repeating packet, rather than an unimportant unit lying between the true layers; this is especially true when the structural uniformity noted above for these brucite layers is allowed for. Thus, the chlorites may be considered to consist of two types of layer (three-storied and one-storied), forming a single type of packet.

After the type of layer has been fixed, the next most general and crystallochemically significant factor is the chemical constitution of the layer, involving its cationic makeup, which largely comes down to how the octahedra are filled and to what extent; this is usually indicated by saying either that the layers are dioctahedral or that they are trioctahedral. According to Belov (1961), the nature of the octahedral structure elements is the chief controlling factor with silicates in general and layer silicates in particular. In contrast to these, the tetrahedral arrangements are more flexible, and may be adapted to the particular structure features fixed by the octahedra.

The extent to which the octahedra are filled affects the goodness of matching between tetrahedral and octahedral networks, the energy state of the layers, and the polymorphism rules in the structures, and brings about sharp distinctions between the physical and chemical properties of the different minerals (Kiffer, 1957; Bates, 1959).

Thus, structures and minerals which differ in the way their octahedral networks are assembled correspond to different conditions of formation and of thermodynamic stability. The change from one type of network to another requires much more effort and involves a much greater structural rearrangement than is the case with different tetrahedral networks. What is essentially involved here, at least where the mica-type minerals are concerned, is the complete breakdown of one structure and the crystallization of another from scratch. A reflection of this is that, as far as can be judged, minerals which fall in different classes according to the above criteria, such as muscovites and phlogopites, do not have a continuous transitional series between them, the gap between them being natural and based on principle. The gap cannot even be filled by the artificial synthesis of minerals (Yoder, Eugster, 1955), although this can give a number of varieties which are unstable in nature (Koizumi, Ray, 1959). On the other hand, minerals which fall in the same class according to these criteria, such as the Al hydromicas and Al montmorillonites, will be found to have a number of relationships, forming continuous transitional series and mutual products, the existence of which is the most important factor in evaluating the closeness of relationships between minerals (Zvyagin, Frank-Kamenetskii, 1961).

The important fourth criterion in the above list, the polymorphic modification, is usually underestimated as a characteristic of clay minerals with regard to both their structure and classification. This was apparently because of the difficulties involved in establishing the modification. These difficulties have now been largely obviated. It is, of course, quite obvious that a given layer sequence is a very constant feature of a given mineral. To bring about a change to another relative layer arrangement, complete layers must be rotated and displaced relative to one another in a certain way, which is difficult to imagine for a structure made up of layers with coupling centers forming a two-dimensional network. It is not by chance that mineral varieties with a given type of layer sequence can be traced over a wide range of changes in structural perfection, corresponding to widely diverse external forces and changes in the surrounding medium, and the gaps between minerals of different modifications are of the utmost significance.

The mineral combinations dealt with above, if they are not just mixtures, as might be assumed, either arise simultaneously under special conditions and continue to coexist, or are the products of an extended and slow transformation process.

Thus, varieties which are similar according to the above criterion can be considered as falling under more closely related classification subheadings than varieties which are not similar, and this is also supported by the features noted above in the structural and mineralogical distribution of the three-storied layer minerals.

Within the limits circumscribed by each of the criteria already stated, minerals may be found with different degrees of layer interaction, and consequently with both stable and unstable structures. This property thus links varieties which are more closely related than those in the previous category, and occupies the fifth place in the series of classification criteria.

The sixth criteria, minor differences in chemical composition and in degree of structural perfection, will in some cases take the classification to its ultimate point, and define the individual minerals, and sometimes lead to alternative possibilities, which could apply to the same mineral and just relate to different specimens.

It should be noted that these classification criteria are not completely independent. The compositions and structures of layers, their degrees of layer interaction, and the polymorphic modifications formed are usually closely interrelated.

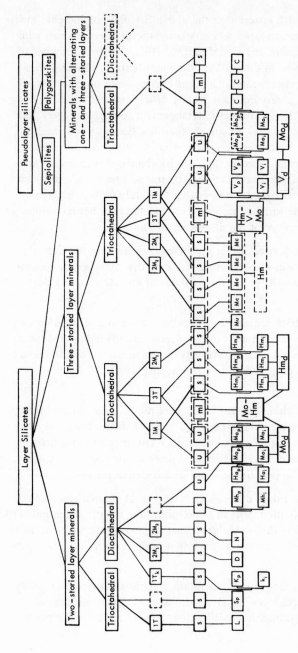

Fig. 105. Classification scheme for clay minerals and related layer silicates. Symbols have the following meanings: s, stable; u, unstable; ml, mixed layer; d, disordered; L, lizardite; Sp, serpentines; K, kaolinites; D, dickite; N, nacrite; Mh, metahalloysites; Ha, halloysites; Mo, montmorillonites; Hm, hydromicas; Mu, muscovites; Mc, micas; V, vermiculites; C, chlorites; 1M,3T,2M₁,etc., polymorphic modifications.

Because of this interrelation, there is a certain amount of freedom in the way a classification can be constructed, and different classifications may all have a rational basis. Frank-Kamenetskii, in particular (1960), considered layer structure and layer interaction to be equal-rank classification criteria in his classification.

The distinctive feature of the classification determined by the series of criteria examined here is that one of the most important links in the series depends on the polymorphic modification, which is revealed by the exploitation of electron-diffraction methods in determining the layer sequences present in clay minerals.

To a large extent, this link eliminates the disagreement between the historically devised mineral groups and the groups which the classification would lead to if criterion (d) was not present between (c) and (e). So, although modification $2M_1$ can be followed right through to the montmorillonites, it is very likely that in this final stage it appears by virtue of vestiges of hydromica present. In this case, the main mass of material will lack a strict c repeat unit, and such specimens will merge into other montmorillonite varieties via the montmorillonites with imperfect structures in the classification.

Thus, the disagreement noted above expresses itself in the classification only by the fact that the dioctahedral and trioctahedral varieties of the micas, hydromicas, and montmorillonites are separated into independent classes, which is as natural and rational as the separation between kaolinites and serpentines.

At the same time, the classification clearly distinguishes between the mica—hydromicas of modification $2M_1$ on the one hand, and the hydromica—montmorillonites of modification 1M on the other.

The classification scheme for clay minerals and related layer silicates, which are best considered together to obtain a unified and complete picture, is given in Fig. 105.

This is only an outline scheme, and does not pretend to cover exhaustively all the relevant minerals. Its basic purpose is to show the relationships between minerals which have been revealed by electron-diffraction experiments. The parts of the scheme which have been insufficiently clearly defined by the available experimental evidence appear dotted. The scheme also includes mixed-layer products which occupy an intermediate position between minerals with uniform structures.

Fig. 106. Texture pattern from boehmite (φ = 55°).

6. Possible Fields of Application of the Electron-Diffraction Method in the Study of Minerals Which Are Not Layer Silicates

It has already been pointed out that layer silicates, and clay minerals in particular, are very well suited to electron-diffraction investigation. The advantages offered by electron diffraction in the study of other minerals are usually coupled with experimental difficulties. Nonetheless, in the cases where these difficulties have been overcome, the use of electron diffraction has justified itself by the worth of the new factual material obtained with its aid. In order to appreciate the scope and possibilities of the electron-diffraction approach in the study of such minerals, a short review of investigations already carried out to date will be valuable.

A very promising field of application of the oblique texture method is in the study of the oxides and hydroxides of Fe, Al, and other metals. The author has succeeded in obtaining good electron-diffraction patterns from boehmite (Fig. 106) and hematite. By analyzing these, the unit cell constants of the minerals were found. For boehmite, a = 12.2, b = 2.86, c = 3.70 Å. For hematite, a = 5.03, b = 8.70, c = 13.7 Å.

From studies on zircons from Ceylon and Oklahoma, Christ, Dwornik, and Tishler (1954) came to the conclusion that electron-diffraction methods were especially favorable for investigation of metamict minerals; from poly-crystalline specimens of these they were able to obtain quite clear diffraction patterns, demonstrating the crystalline nature of specimens which to X-rays appeared amorphous. This conclusion is also supported by the work of Habel (1958), who obtained an electron-diffraction pattern from a film preparation of powdered ampangabéite from the Malagasy republic, whereas the X-ray method gave the pattern of an amorphous substance. It should, however, be recognized that an explanation of the effectiveness of electron-diffraction methods in the study of metamict minerals awaits further basic investigations.

Studies by the microdiffraction method are notable for the great diversity and range of their specimens. These studies have mostly been concerned with oxides and hydroxides of metals. Mackey (1960a,b) used the diffraction of electrons from micro single crystals with great success in the study of transformations, mostly on heating, of different phases of the β, δ-FeOOH type and of so-called green rust, into lepidocrocite or spinel. Using single-crystal diffraction patterns, he analyzed the mechanisms of topotactic, epitactic, and reconstructive transformations in this richly varied group of minerals, and established the controlling role played by the oxygen frameworks of the structures during these transformations.

In another investigation (Oosterhout, 1960), the microdiffraction method was used to find the crystallographic directions in needle-shaped crystals from the α- and γ-FeOOH and γ-Fe_2O_3 system, which is essential for effective use of the magnetic properties of these substances in magnetic recording devices.

During studies of an intermediate stage in the goethite—hematite transformation (McConnell, Lima-de-Faria, 1961), the appearance of satellite reflections was observed in the microdiffraction patterns. When separated off with the aid of an aperture diaphragm they led to a banded electron-microscope picture showing dislocations and nonuniformities related to the nature of the heavy Fe-atom distribution.

A number of studies have been devoted to aluminum oxides and hydroxides. Thus, diffraction patterns from boehmite crystals in different orientations were used to derive the reciprocal lattice structure of this mineral (Bosmans, Michel, 1959). Yamaguchi and Sakamoto (1960) were able to prove that two varieties of bayerite were identical, showing that their differences only extended to the morphology of their crystals. Brindley (1961) obtained type patterns for minerals of the gibbsite—khi-alumina—kappa-alumina —corundum II series, measured their unit cell constants, and found the geometrical relationship between the lattices of the $\varkappa_1$- and $\varkappa_2$-Al_2O_3 modifications. Wyart, Oberlin, and Tchoubar (1963), when making transmission studies of the hydrothermal alteration products of albite by separating them from the surface with carbon replicas, established that boehmite was formed, described its structural features at different crystallization stages, and determined the orientation of the crystals.

Gard (1960) used microdiffraction methods to investigate different hydrated Ca silicates, i.e., β-$CaSiO_3$, foshagite, xonotlite, and tobermorite, which differ even in the case of disordered structures. The authors explained that the different modifications, in particular those with A and F type lattices, depend on the coordination of the Ca relative to O, OH. This work showed that the hydrated Ca silicates also were amenable to electron-diffraction study.

High-quality electron-diffraction patterns have been obtained for MoS_2 by the microdiffraction method (Stabenow, 1959a). In this case, some parts of the crystals were found to be rotated relative to other parts, forming a discrete set of relative orientations in analogous fashion to those noted above for layer silicates. Subsequently the same author (Stabenow, 1959b) used the relationship between the moire patterns and the electron-diffraction patterns to analyze the scheme of layer superimposition in nondefect MoS_2. In another investigation (Gillet, 1960), a relationship was established between the Moiré fringes caused by the relative rotations of different parts of the crystals and the effects of secondary diffraction in crystals of molybdenite and gold. The author confirmed the above view that the Moiré fringes gave a magnified picture of the lattice planes and the breakdowns in sequence order shown by these.

Microdiffraction and Moiré patterns have also been used to analyze defects in the layer packing in ZnS containing alternating wurtzite and zinc blende zones. These defects were observed in both the (001) planes and in the planes lying at 60° to the (2$\overline{1}$0) face, i.e., (110) and (1$\overline{2}$0) (Chadderton et al., 1963).

The microdiffraction method has also been applied to the study of nepheline. For this mineral, McConnell (1962) observed satellites close to the main reflections, showing that the presently accepted nepheline structure is only valid as an averaged structure. The different forms of these satellites, which sometimes are degenerated into continuous bands, indicate a wide range of structural varieties of this mineral and also show the loss of ordering present.

An unusual electron-diffraction application has been described in the works of Dwornik, Ross, and Christ (Dwornik, Ross, 1954; Ross, Christ, 1958; Ross, 1959). In carrying out studies sponsored by the Raw Material Division of the U. S. Atomic Energy Commission, they obtained a series of electron-diffraction patterns intended to serve as standards for use with minerals and deposits connected with uranium occurrences. The materials studied included sooty pitchblendes, carnotite, tyuyamunite, vanadates, ultrafine fractions separated from lignites, specimens from the oxidized zone, leached-zone material from phosphate deposits, schröckingerite, torbernite, and others.

Later, after the acceptance of the microdiffraction method, unit cell results derived during the program were published for colemanite, $KClO_3$, and a series of vanadium minerals: hewettite, corvusite, navajoite, fervanite, steigerite, etc.

Microdiffraction patterns were used in combination with polycrystalline specimen photographs in the study of ranquilite, a calcium uranyl silicate of composition $1.5CaO \cdot 2UO_3 \cdot 5SiO_2 \cdot 12H_2O$ (Jimenez, Benjagen, 1960). The single-crystal pattern gave the a and b constants and helped in the indexing of the polycrystalline specimen photograph. The final results included the unit cell constants a = 17.64, b = 14.28, c = 18.48 Å, corresponding to the low-temperature phase of this mineral, which indicated that weakly bound H_2O molecules were retained under the electron-diffraction conditions used.

With electron-diffraction apparatus it is possible to observe and record a diffraction pattern with a very brief exposure (about 1 sec). Because of this, the method is favorable for the study of rapidly occurring structural transformations. Descriptions have been published of apparatus especially constructed for this purpose (Trillat, Takahashi, 1953). At present only one investigation can be quoted (Selma, Croissant, 1960) in which this valuable approach has been successfully put into practice. Electron-diffraction patterns were obtained from cassiterite, SnO_2, and calcium stannate, $CaSnO_3$, corresponding to different single-crystal orientations. It was found in most cases that the SnO_2 had its (101) and (110) planes oriented perpendicular to the electron beam, and the $CaSnO_3$ had its (010) and (100) planes perpendicular. The unit cell constants were determined and the three-dimensional symmetry of the structure found. The electron beam was then focused on the specimens, and under the influence of the intense irradiation the specimens vaporized, with successive condensation of the phases $Sn + SnO_2$, $SnO_2 + SnO$, SnO. From the Ca stannate a textured specimen was first obtained, and then a CaO polycrystalline specimen.

From the above review of applications of electron diffraction in the study of minerals, and from a consideration of its theoretical possibilities, some conclusions may be drawn as to the prospects and likely applications of the method in the area under consideration.

1. There is every reason to suppose that electron diffraction may develop into a valuable and effective method of investigating a wide range of finely dispersed rocks and minerals. The method will need to gain approval and undergo appraisal in specific fields of application.

2. The possible uses of electron diffraction in the reflection approach must be pursued, both for reflection from powdered films and from the surfaces of massive rock and mineral specimens. It is possibly in this area that electron diffraction may find application outside the finely dispersed mineral field.

3. Development must occur in the ultrathin section method, which will make the electron-diffraction method more universal in application and lead to a significant increase in the diffractional information obtainable from minerals with its aid.

4. The greatest efforts must be made to exploit the possibilities of diffraction of ultrahigh-energy electrons in all types of investigation, using polycrystalline, textured, and single-crystal specimens, in transmission and reflection studies, and employing electron-diffraction cameras and electron microscopes.

LITERATURE CITED

Akhundov, Yu. A., Mamedov, Kh. S., and Belov, N. V. The crystal structure of brandisite, Dokl. Akad. Nauk SSSR, 137(1) (1961).

Alekseev, A.G., Alekseev, V.A., Boyandina, L.G., Bul'ba, I.A., Vertsner, V.N., Tulyi, A.M., Zelenetskaya, E.V., and Tregubov, M.I. An electron-diffraction apparatus with direct registration of the intensities of scattered electrons, Summary of Papers Presented at the Second Conference on Electron Diffraction, Izd. MGU (1962).

Amelinckx and Dekeyser, W. Le polytypisme des mineraux micaces et argileux, Compte rend. rech. IRSYA, No. 14 (1955).

Bagdyk'yants, G. O. The industrial model of the EM-4 electron-diffraction apparatus, Izv. Akad. Nauk SSSR, Ser. Fiz. 17(2):255 (1953).

Bates, T. Morphology and crystal chemistry of 1:1 layer lattice silicates, Am. Mineralogist 44(1-2) (1959).

Belov, N. V. Structures of Ionic Crystals and Metallic Phases, Izd. Akad. Nauk SSSR, Moscow (1947).

Belov, N. V. Essays on structural mineralogy, Min. Sb. L'vov. Geol. Obshch., No. 3:29 (1949); No. 4:21 (1950); No. 5:18 (1951a); No. 12 (1958).

Belov, N. V. Structural Crystallography, Izd. Akad. Nauk SSSR, Moscow (1951b).

Belov, N. V. Nomographic methods of calculation in X-ray structural analysis, Tr. Inst. Kristallogr., 9:277 (1954).

Belov, N. V. Crystal Chemistry of Large-Cation Silicates, Consultants Bureau, New York (1964).

Bosmans, H., and Michel, P. Étude de cristaux de bemite par microscopie et diffraction electronique, Compt. Rend. 249:16 (1959).

Boyandina, L. G. Types of electron-diffraction apparatus produced by the Sumy electron microscope factory, and prospects for factory production of electron-diffraction apparatus, Summary of Papers Presented at the Second Conference on Electron Diffraction, Izd. MGU (1962).

Bragg, W. L. The Structure of Silicates, Fundamental Ideas in Geochemistry, III [Russian translation], ONTI (1937).

Bradley, W. E. The structural scheme of attapulgite, Am. Mineralogist 25(6) (1940).

Bragg, W. L., and West, G. The structure of certain silicates, Proc. Roy. Soc. 114A : 450 (1927).

Brauner, K., and Preisinger, A. Struktur und Entstehung des Sepioliths, Tschermarks Mineral. Petrog. Mitt. 6(1-2) (1956).

Brindley, G. W. (ed.). X-ray Identification and Crystal Structures of Clay Minerals [Russian translation], IL, Moscow (1955).

Brindley, G. W. X-ray and electron diffraction data for sepiolite, Am. Mineralogist 44(5-6) (1959).

Brindley, G. W. The kaolin minerals, in: X-ray Identification and Crystal Structures of the Clay Minerals, Brown, G. (ed.), London (1961), Chapter 2.

Brindley, G. W. The reaction series gibbsite → chi alumina → kappa alumina → corundum II, Am. Mineralogist 46(9-10) (1961).

Brindley, G. W., Comer, J. J., Ueda, R., and Zussman, J. Electron-optical observation with crystals of antigorite, Acta Cryst. 11(2) (1958).

Brindley, G. W., and De Kimpe, C. Identification of clay minerals by single crystal electron diffraction, Am. Mineralogist 46(9-10) (1961).

Brindley, G. W. and Nakahira, M. Further consideration of the crystal structure of kaolinite, Mineral. Mag. 31(240) (1958).

Brindley, G. W., Oughton, B. M., and Robinson, K. Polymorphism of the chlorites, Acta Cryst. 3(6) (1950).

Brown, B. E., and Bailey, S. W. Chlorite polytypism I. Regular and semi-random one-layer structures, Am. Mineralogist 47(7-8) (1962).

Brown, G. (ed.). X-ray Identification and Crystals Structures of Clay Minerals, London (1961).

Cartraud, R., and Zouckerman, R. Glissements rotationels et pseudostructure du mica muscovite, J. Phys. Radium 21(1) (1960).

Chadderton, L. T., Fitzgerald, A. G., and Yoffe, A. D. Stacking faults in zinc sulfide, Phil. Mag. 8(85) (1963).

Chapman, J. A., and Zussman, J. Further electron optical observations on crystals of antigorite. Acta Cryst. 12(7) (1959).

Christ, C. L., Dwornik, F. G., and Tishler, M. S. Crystalline regions in metamict minerals, Science, 119(3094) (1954).

Cowley, J. M. Structure analysis of single crystals by electron diffraction, II. Disordered boric acid structure, Acta Cryst., 6(6): 522 (1953).

Cowley, J. M. On order—disorder structures, Acta Cryst. 10(2) (1957).

Cowley, J. M. Diffraction intensities from bent crystals, Acta Cryst., 14(9) (1961).

Cowley, J. M., and Goswami, A. Electron diffraction patterns from mont-
    morillonite, Acta Cryst. 14(10) (1961).

Cowley, J. M., and Rees, A. L. G. Design of a high-resolution electron dif-
    fraction camera, J. Sci. Instr. 30(20) (1953).

Cowley, J. M., and Rees, A. L. G. Fourier methods in structure analysis by
    electron diffraction, Rept. Progr. Phys., Vol. 28 (1958).

Drits, V. A. The nature of defects in the structures of kaolinite minerals,
    Application of X-rays to the study of materials (Summary of Reports),
    Izd. Akad. Nauk SSSR, Moscow (1961), p. 142.

Drits, V. A., and Kashaev, A. A. An X-ray study of a kaolinite single crystal,
    Kristallografiya, 5(2) (1960).

Dwornik, E., and Ross, M. Application of the electron microscope to mineral-
    ogic studies, Am. Mineralogist, 40(3) (1954).

Eckhardt, F.-J. Elektron-optische Untersuchungen an Einkristallen aus tonigen
    Sedimenten, Neues Jarhb. Mineral., Monatsh. 1 (1958).

Eckhardt, F.-J. Über die Anwendung von Ultramikrotomschnitten bei der
    elektronenoptischen Untersuchung von Tonen, Fortschr. Mineral. 38(2)
    (1961).

Ehlers, H. Zur Feinstruktur von Elektroneninterferenzen, Z. Naturforsch.
    11a(5) (1956).

Ewald, P. P. Das "reziproke Gitter" in der Strukturtheorie, Z. Krist. 56(2) (1921).

Fahey, J. J., Ross, M., and Axelrod, J. Loughlinite, a new hydrous sodium
    magnesium silicate, Am. Mineralogist 45(3-4) (1960).

Fok, V. A., and Kolpinskii, V. A. Diffraction of waves from a curved lattice,
    Zh. Eksperim. Fiz. 10(2): 211 (1940).

Frank-Kamenetskii, V. A. X-ray methods of studying clays, in collection:
    Study and Uses of Clays, Izd. L'vov Univ. (1958), p. 713.

Frank-Kamenetskii, V. A. A crystallochemical classification of simple and
    interstratified clay minerals, Clay Minerals Bull. 4(24) (1960).

Gard, J. A. Electron diffraction studies of stacking modifications in the
    calcium silicates, Acta Cryst. 13(12) (1960).

Gatineau, L., and Mering, J. Precisions sur la structure de la muscovite,
    Compt. Rend. 246(6) (1958).

Gillery, F. H. The X-ray study of synthetic Mg—Al serpentines and chlorites,
    Am. Mineralogist 44(1) (1959).

Gillet, M. Étude des defauts cristallins par les moires sur des cristaux d'or et
    de molibdenite, Bull. Soc. Franc. Mineral. Crist. 83(10-12) (1960).

Grim, R. E. Clay Mineralogy, McGraw-Hill, New York (1953).

Grim, R. E. Some applications of clay mineralogy, Am. Mineralogist 45(3-4)
    (1960).

Gritsaenko, G. S., Rudnitskaya, E. S., and Gorshkov, A. I. Electron
    Microscopy of Minerals, Izd. Akad. Nauk SSSR, Moscow (1961).

Habel, V. Elektronenbeugung auf metamikten Ampangabeite aus Madagaskar,
    Naturwissenschaften 45(9) (1958).

Hendricks, S. B. The crystal structure of nacrite and the polymorphism of the
    kaolin minerals, Z. Krist. 100(6) (1938).

Hendricks, S. B., and Jefferson, M. E. Polymorphism of the micas with op-
    tical measurements, Am. Mineralogist 24(12) (1938).

Hibi, T. Electron diffraction patterns of mica of various thicknesses, Nature
    175(4454) (1955).

Honjo, G., Kitamura, N., and Mihama, K. A study of clay minerals by means
    of single-crystal electron diffraction diagrams: the structure of tubular
    kaolin, Clay Minerals Bull. 2(12) (1954).

Honjo, G., and Mihama, K. A study of clay minerals by electron diffraction
    diagrams due to individual crystallites, Acta Cryst. 7(6-7) (1954).

International Tables for X-Ray Crystallography, Kynoch Press, Birmingham,
    England (1952).

Jackson, W. W., and West, J. The crystal structure of muscovite, $KAl_2 \cdot$
    $(AlSi_3O_{10})(OH_2)$, Z. Krist. 76(3) (1930); 85(1-2) (1933).

Jagodzinski, H., and Künze, G. Die Röllchenstruktur des Chrisotils, Neues
    Jahrb. Mineral. A, No. 4/5: 95 (1954); No. 6: 113 (1954).

Jimenez, M., and Benjagen, M. R. Ranquilite, a calcium uranyl–silicate,
    Am. Mineralogist, 45(9-10) (1960).

Kiffer, C. Interaction de la structure et de la texture sur les proprietes des
    mineraux philliteux, Bull. Soc. Franc. Ceram., No. 37 (1957).

Kitaigorodskii, A. I. X-ray Structure Analysis, Gos. Izd. Tekh. Teoret. Lit.,
    Moscow-Leningrad (1950).

Kitaigorodskii, A. I. X-ray Structure Analysis of Finely Crystalline and
    Amorphous Bodies, Gos. Izd. Tekh. Teoret. Lit, Moscow-Leningrad
    (1952).

Kitaigorodskii, A. I. Theory of Structure Analysis, Izd. Akad. Nauk SSSR,
    Moscow (1957).

Koizumi, M., and Roy, R. Synthetic montmorillonoids with variable exchange
    capacity, Am. Mineralogist 44(7-8) (1959).

Kovalev, G. A. X-ray determination of dickite, a modification of a kaolinite
    group mineral from the Crimea, Zap. Vses. Mineralog. Obshchestva,
    76(4) (1947).

Kulbicki, G. Diagrammes de diffraction electronique de monocristaux de
    kaolinite et d'halloysite et observations sur la structure de ces
    mineraux, Compt. Rend., 238(25) (1954).

Kunze, G. Zur Bildung diskreter Überstrukturvariationen der antigoritschen
    Serpentine, Fortschr. Mineral., 37(4) (1959).

Lazarenko, E. K. Problems in the nomenclature and classification of
    glauconite; Problems in the Mineralogy of Sedimentary Structures,
    Books 3 and 4, Izd. L'vov Univ. (1956), p. 345.

Levkin, N. P., and Kushnit, Yu. M. A new model 100-kV universal elec-
    tron-diffraction camera with armored inlet, Izv. Akad. Nauk SSSR,
    Ser. Fiz., 23(4):531 (1959).

Lobachev, A. N., and Vainshtein, B. K. An electron-diffraction study of urea,
    Kristallografiya, 6(3) (1961).

Loevenstein, W. The distribution of aluminum in the tetrahedra of silicates
    and aluminates, Am. Mineralogist, 39(1-2) (1954).

Mackey, A. L. Structural transformations in the iron oxide—hydroxide system,
    Acta Cryst., 13(12) (1960a).

Mackey, A. L. Some aspects of the topochemistry of the iron oxides and
    hydroxides, Proceedings of the Fourth International Symposium on
    Reactivity of Solids, Amsterdam (1960b).

Malkova, K. M. Celadonite from Pobuzh'e, Min. Sb. L'vov Geol. Obshch.,
    Vol. 10 (1956).

Mathieson, A. McL., Mg-vermiculite: a refinement and reexamination of
    the 14, 36 A phase, Am. Mineralogist, 43(3-4) (1958).

Mathieson, A. McL., and Walker, G. F. Crystal structure of magnesium—
    vermiculite, Am. Mineralogist, 39(3-4) (1954).

McConnell, J. D. C. Electron diffraction study of subsidiary maxima of scat-
    tered intensity in nepheline, Mineral. Mag., 33(257 (1962).

McConnell, J. D. C., and Lima-de-Faria, J. Electron-optical and electron-
    diffraction study of a disordered structural state in the transformation
    geothite—hematite, Mineral. Mag., 32(254) (1961).

McMurchy, R. C. Structure of chlorites, Z. Krist., 88:420 (1934).

Mikheev, V. I. Effect of isomorphous replacement in micas on the character
    of their powder photographs, Min. Sb. L'vov Geol. Obshch., Vol. 8
    (1954).

Mitra, R. P., and Rao, M. V. R. V. Basal reflection of electron waves by
    oriented aggregates of clay minerals, Naturwissenschaften, 42(5) (1955).

Mott, N. Proc. Roy. Soc., 127:658 (1930).

Muller, G. Zur Kenntnis dioktaedrischer Vierschichte-Phylosilikate (Sudoit-
    Reihe der Sudoit-Chlorit-Gruppe), Proceedings of the International
    Clay Conference, Stockholm, 1963. Pergamon Press, New York (1963).

Nagy, B., and Bradley, W. F. The structural scheme of sepiolite, Am.
    Mineralogist, 40(9-10) (1955).

Nakovnik, N. I. Nacrite and other kaolinite minerals of the USSR, Zap. Vses. Mineralog. Obshchestva, 70(1) (1941).

Newnham, R. E. A refinement of the dickite structure and some remarks on polymorphism in kaolin minerals, Mineral. Mag., 32(252) (1961).

Newnham, R. E., and Brindley, G. W. The crystal structure of dickite, Acta Cryst., 9(9) (1956).

Oberlin, A., and Mering, J. Observation en microscopie et microdiffraction electronique sur la montmorillonite Na, J. Microscopie, 1(2) (1962).

Oberlin, A., and Tchoubar, K. Étude en microscopie et microdiffraction electronique de l'alteration des cristaux de la kaolinite par une solution acide, Compt. Rend., 250(4) (1960).

Oberlin, A., and Tchoubar, K. Étude en microscopie et microdiffraction electronique des epitaxies du fireclay sur la kaolinite, Compt. Rend., 250(5) (1960b).

Oosterhout, G. W. Morphology of synthetic submicroscopic crystals of α- and γ-FeOOH and of γ-Fe_2O_3 prepared from FeOOH, Acta Cryst., 13(11)(1960).

Pabst, A. Redescription of the single-layer structure of the micas, Am. Mineralogist, 40(11-12) (1955).

Pauling, L. The crystal structure of topaz, Proc. Natl. Acad. Sci. US., 14:1036 (1928).

Pauling, L. The principles determining the structure of complex ionic crystals, J. Am. Chem. Soc., 51:1010 (1929).

Pauling, L. The structure of micas and related minerals, Proc. Natl. Acad. Sci. US, 16:123 (1930).

Pauling, L. The structure of the chlorites, Proc. Natl. Acad. Sci. US, 16:578 (1930b).

Pauling, L. The Nature of the Chemical Bond, Cornell, New York (1960).

Pines, B. Ya., and Bublik, A. I. A high-temperature electron-diffraction camera, Zh. Tekh. Fiz., 24(6):1139 (1954).

Pinsker, Z. G. Diffraction of Electrons, Izd. Akad. Nauk SSSR, Moscow (1949).

Pinsker, Z. G. Electron-diffraction determination of the unit cell and space group of kaolinite, Dokl. Akad. Nauk SSSR, 73(1) (1950).

Pinsker, Z. G. Electron-diffraction and electron-microscope studies of clay minerals, Tr. Biogeokhim. Lab., Vol. 10 (1954).

Pinsker, Z. G. Modern electron-diffraction apparatus, Pribory i Tekhn. Eksperim., No. 1:3 (1959).

Pinsker, Z. G. Electron-diffraction structure analysis and the investigation of semiconducting materials, Advances in Electronics and Electron Physics, Academic Press, Inc., New York (1959).

Pinsker, Z. G. Disordered structures and the ordering process, Kristallografiya, 5(4) (1960).

Pinsker, Z. G., Lapidus, E. L., and Tatarinova, L. I. An electron-diffraction study of the structure of kaolinite, Zh. Fiz. Khim., 22(9) (1948).

Popov, N. M. An electron microscope—electron diffraction camera with an accelerating voltage of 400 kV, Izv. Akad. Nauk SSSR, Ser. Fiz., 23(4) (1959).

Popov, N. M., and Zvyagin, B. B. Use of a 400-kV electron-diffraction apparatus in the study of single crystals, Kristallografiya, 3(6) (1958).

Popov, N. M., and Zvyagin, B. B. Study of minerals by the microdiffraction method in a 400-kV electron microscope—electron diffraction camera.

Porai-Koshits, M. A. A Practical Course of X-Ray Structural Analysis, Vol. 2, Izd. MGU (1960).

Radczewski, O. E., and Balden, H. J. Röntgenographische und elektronen-optische Untersuchungen an der schletaer Erde. Interferenzbilder hoher Auflösung von einzelnen Kristallen, Fortscher. Mineral., 37(1) (1959).

Radczewski, O. E., and Goosens, H. Die kristallographische Orientierung von Mineralen auf Grund ihren Elektronenbeugung, Optik, 13(7) (1956).

Radoslovich, E. W. Structural control of polymorphism in micas, Nature, 183(4656) (1958).

Radoslovich, E. W. The structure of muscovite, $KAl_2(Si_3Al)O_{10}(OH)_2$, Acta Cryst., 13(11) (1960).

Radoslovich, E. W. Surface symmetry and cell dimensions of layer lattice silicates, Nature, 191 (4783) (1961).

Radoslovich, E. W. Some relations between composition, cell dimensions, and structure of layer silicates, Proceedings of the International Clay Conference, Stockholm, 1963. Pergamon Press, New York (1963).

Rang, O. Überstruktur in Elektronenbeugungsdiagrammen eines Glimmer-kristalls, Z. Phys., 152(2) (1958).

Riecke, W. D. Über die Genauigkeit der Übereinstimmung von ausgewählten und beugenden Bereich bei der Feinbereichs—Elektronenbeugung im Le Pooleschen Strahlengang, Optik, 18(6) (1961).

Ross, M. Mineralogical applications of electron diffraction, II. Studies of some vanadium minerals of the Colorado plateau, Am. Mineralogist, 44(3-4) (1959).

Ross, M., and Christ, C. L. Mineralogical applications of electron diffraction I. Theory and techniques, Am. Mineralogist, 43(11) (1958).

Rumsh, M. A. A simplified electron-diffraction camera, Zh. Teor. Fiz., 25(14) (1955).

Selma, P., and Croissant, O. Étude par diffraction electronique de cristaux de cassiterite et de stannate de calcium et de leur dissociation sour l'effet du bombardement electronique, Compt. Rend., 251(4) (1960).

Shakhova, R. A. An electron-diffraction study of artificial mixtures of clay minerals, Zap. Vses. Mineralog. Obshchestva, 9(5) (1962).

Shubnikov, A. V. Symmetry and Antisymmetry of Finite Figures, Izd. Akad. Nauk SSSR, Moscow (1961).

Smith, I. V., and Yoder, H. S. Experimental and theoretical studies of the mica polymorphs, Mineral. Mag., 31(234) (1956).

Stabenow, J. Elektroneninterferenzen an übereinanderliegenden Kristall-schichten I. Orientierte Verwachsung von Mikrokristallen aus Molib-dänsulfid, Z. Krist, 112(1) (1959a); II. Zusammenhang zwischen Mehr-fachbeugung und Kristall—Moire, Z. Phys., 156:503 (1959b).

Steinfink, H. The crystal structure of chlorite. I. A monoclinic polymorph. II. A triclinic polymorph, Acta Cryst., 11(3) (1958).

Steinfink, H. Crystal structure of a trioctahedral mica: phlogopite; Am. Mineralogist, 47(7-8) (1962).

Stemple, I. S., and Brindley, G. W. A structural study of talc and talc—tremolite relations, J. Am. Ceram. Soc., 43(1) (1960).

Steadmen, R. The structure of the trioctahedral kaolin-type silicates, Acta Cryst., 17(7) (1964).

Suito, E., and Ueda, N. A study of the clay minerals from Kurata mine by the electron microdiffraction method, Proc. Japan. Acad., 33(3) (1956).

Takeuchi, Y., and Sadanaga, R. The crystal structure of xanthophyllite, Acta Cryst., 12(11) (1959).

Trillat, J. J., and Takahashi, N. Diffractographe electronique enregistreur permettant l'etude des transformations chimique ou physiques, Compt. Rend., 235(8) (1953).

Ushakova, E. N. Some hydrous silicates from Zaval'e, Central Pobuzh'e, Min. Sb. L'vov. Geol. Obshch., No. 13 (1959).

Vainshtein, B. K. Structure Analysis by Electron Diffraction, Izd. Akad. Nauk SSSR, Moscow (1956).

Vainshtein, B. K. The intensities of electron-diffraction pattern reflections (general case), Kristallografiya, 2(3): 340 (1957).

Vainshtein, B. K. The antisymmetry of Fourier transformation of figures with a singular point, Kristallografiya, 5(3) (1960).

Vainshtein, B. K., and Zvyagin, B. B. Representation in reciprocal space of crystal lattice symmetry, Kristallografiya, 8(2):147 (1963).

Vainshtein, B. K., and Pinsker, Z. G. Electron-diffraction determination of the structure of barium chloride monohydrate, $BaCl_2 \cdot H_2O$, Fiz. Khim., 23(9) (1949).

Vainshtein, B. K., and Pinsker, Z. G., The EG horizontal electron-diffraction apparatus, Kristallografiya, 3(3): 358 (1958).

Vikulova, M. F. (ed.). Systematic Manual on Petrographic and Mineralogical Study of Clays, Gosgeoltekhizdat (1957).

Waser, J. Fourier transforms and scattering intensities of tubular objects, Acta Cryst., 8(3) (1954).

Weir, A. W., and Greene-Kelly, R. Beidellite, Am. Mineralogist, 47(1-2) (1962).

Wyart, J., Oberlin, A., and Tchoubar, K. Étude en microscopie et microdiffraction electronique de la boemite formee lors de l'alteration de l'albite, Compt. Rend., 256(3) (1963).

Whittaker, E. J. W. The diffraction of X-rays by a cylindrical lattice, Acta Cryst., 7(12) (1954); 8(5) (1955).

Whittaker, E. J. W. Fine structure within the diffraction maxima from chrysotile, Acta Cryst., 16(16) (1963).

Whittaker, E. J. W., and Zussman, J. The characterization of serpentine minerals by X-ray diffraction, Mining Mag., 31(233) (1956).

Yamaguchi, G., and Sakamoto, K. The identity of bayerite-a and bayerite-b. Canad. J. Chem., 38(8) (1960).

Yamzin, I. I. The structure of the network of oxygen—silicon tetrahedra in micas, Tr. Inst. Kristallografiya, Vol. 9 (1954).

Yamzin, I. I., and Pinsker, Z. G. Atomic scattering of electrons, Dokl. Akad. Nauk SSSR, 65: 645 (1949); Tr. Inst. Kristallografii, 5: 69 (1949).

Yang, J., and Chi-Sun. The growth of synthetic chrysotile fiber, Am. Mineralogist, 46 (5-6) (1961).

Yoder, H. S., and Eugster, H. P. Synthetic and natural muscovites, Geochim. Cosmochim. Acta, 8(5/6) (1955).

Zhdanov, G. S. Numerical symbols for close-packing of spheres and their use in close-packed sphere theory, Dokl. Akad. Nauk SSSR, 48(1) (1945).

Zussman, J., Brindley, G. W., and Comer, J. J. Electron-diffraction studies of serpentine minerals, Am. Mineralogist, 42(3-4) (1957).

Zvyagin, B. B. Electron-diffraction study of montmorillonite group minerals, Dokl. Akad. Nauk SSSR, 86(1) (1952).

Zvyagin, B. B. Electron-diffraction study of kaolinite group minerals, Dokl. Akad. Nauk SSSR, 96(4) (1954).

Zvyagin, B. B. Some features of layer silicate diffraction patterns, Dokl. Akad. Nauk SSSR, 97(2) (1954).

Zvyagin, B. B. Some diffraction properties of clay minerals as shown by electron-diffraction oblique texture patterns, Tr. Inst. Kristallogr., No. 11 (1955).

Zvyagin, B. B. Electron-diffraction study of hydromicas, Kristallografiya, 1(2): (1956).

Zvyagin, B. B. Determination of clay minerals by the electron-diffraction method, Problems in the Mineralogy of Sedimentary Structures, Books 3 and 4, Izd. L'vov Univ. (1956), p. 691.

Zvyagin, B. B. Electron-diffraction analysis, in: Systematic Handbook on Petrographic-Mineralogical Study of Clays, Gosgeoltekhizdat, Moscow (1957), Chapter 7.

Zvyagin, B. B. Principles of mineral diagnosis by electron diffraction, Methods of Investigating Sedimentary Rocks, Vol. 1, Gosgeoltekhizdat, Moscow (1957).

Zvyagin, B. B. Electron-diffraction determination of the structure of celadonite, Kristallografiya, 2(3) (1957).

Zvyagin, B. B. Electron-diffraction study of beidellite and monothermite, in collection: Study and Uses of Clays, Izd. L'vov Univ. (1958), p. 102.

Zvyagin, B. B. Possibilities and achievements of electron diffraction in the investigation of clays and clay minerals, in collection: Study and Uses of Clays, Izd. L'vov Univ. (1958), p. 769.

Zvyagin, B. B. New possibilities in the structural study of clay minerals by electron diffraction, Materials on the Geology, Mineralogy, and Use of Clays in the USSR (Proceedings of the International Conference on Clays, Brussels, 1958), Izd. Akad. Nauk SSSR, Moscow (1958).

Zvyagin, B. B. The contribution of electron diffraction to the crystal chemistry of clay minerals (Fedorov Session on Crystallography, Leningrad, May 21-27, 1959), Summary of Papers, Izd. Akad. Nauk SSSR, Moscow (1959), p. 80.

Zvyagin, B. B. The structures of clay minerals in the light of electron-diffraction data, Reports to the Assembly of the International Commission on the Study of Clays, Izd. Akad. Nauk SSSR, Moscow (1960), p. 5.

Zvyagin, B. B. Electron-diffraction determination of the structure of kaolinite, Kristallografiya, 5(1) (1960); Dokl. Akad. Nauk SSSR, 130(5) (1960).

Zvyagin, B. B. Results and prospects in the study of finely dispersed minerals by electron diffraction, Proceedings of the Eighth Conference of Geological Organization Laboratory Workers, No. 6, Nauchn.-Tekhn. Gorn. Obshch., Moscow (1961), p. 6.

Zvyagin, B. B. The classification of clay minerals and related layer silicates, Materials on the Classification of Clay Minerals (Information Bulletin of the Commission on the Study of Clays), Moscow (1961).

Zvyagin, B. B. The theory of mica polymorphism, Kristallografiya, 6(5) (1961).

Zvyagin, B. B. The theory of polymorphism in minerals made up of two-storied (kaolinite-type) layers, Kristallografiya, 7(1) (1962).

Zvyagin, B. B. The effect of elongated crystal form on the distribution of reflection intensities in platy-texture electron-diffraction patterns, Kristallografiya, 7(6) (1962).

Zvyagin, B. B. The theory of chlorite polymorphism, Kristallografiya, 8(1): 32 (1963).

Zvyagin, B. B., Lapidus, E. L., and Petrov, V. P. The nature of askanite clays and their parent rocks, Dokl. Akad. Nauk SSSR, 68(2) (1949).

Zvyagin, B. B., and Mishchenko, K. S. Electron-diffraction refinement of the structure of muscovite, Kristallografiya, 5(4) (1960).

Zvyagin, B. B., and Mischenko, K. S. Electron-diffraction data on the phlogopite—biotite structure, Kristallografiya, 7(4) (1962).

Zvyagin, B. B., Mishchenko, K. S., and Shitov, V. A. Electron-diffraction data on the sepiolite and palygorskite structures, Kristallografiya, 8(2) (1963).

Zvyagin, B. B., and Nefedov, E. I. Cookeite, Dokl. Akad. Nauk SSSR, 95(6) (1954).

Zvyagin, B. B., and Pinsker, Z. G. An electron-diffraction study of the structure of montmorillonite, Dokl. Akad. Nauk SSSR, 68(1) (1949).

Zvyagin, B. B., and Pinsker, Z. G. Electron-diffraction determination of the unit cells of pyrophyllite and talc and the structural relationship between these minerals and montmorillonite, Dokl. Akad. Nauk SSSR, 68(3) (1949).

Zvyagin, B. B., and Popov, N. M. Some results and possibilities in the investigation of minerals by the microdiffraction method using a 400-kV electron microscope—electron diffraction camera, in collection: Materials on the Mineralogy of Mineral Deposits, Mineralogical Series No. 26, Gostoptekhizdat, Leningrad (1959), p. 61.

Zvyagin, B. B., and Frank-Kamenetskii, V. A. Apropos of beidellite, Min. Sb. L'vov Geol. Obshch., No. 14 (1959).

Zvyagin, B. B., and Frank-Kamenetskii, V. A. The principles of construction, evaluation, and significance of various clay mineral classifications, Zap. Vses. Mineralog. Obshchestva, 90(6) (1961).

Zvyagin, B. B., and Shakhova, R. A. Electron-diffraction reflection studies
    of powdered celadonite specimens, Kristallografiya, 2(1) (1957).
Zvyagin, B. B., Shakhova, R. A., and Shitov, V. A. Some rules for the classi-
    fication of clay substances according to structural and mineralogical
    criteria dependent on electron-diffraction results, in collection: Ma-
    terials on Paleogeography and Lithology, Otdel Nauchno-Tekh. Inf.
    VSEGEI, Vol. 72 (1962).
Zvyagin, B. B., and Shcheglov, A. D. Nacrite from a fluorite deposit in
    Western Zabaikal' and its structural features according to electron-
    diffraction data, Dokl. Akad. Nauk SSSR, 142(1) (1962).

INDEX

357